AF770587

Transboundary Pollution

NUS CENTRE FOR INTERNATIONAL LAW

The NUS Centre for International Law series aims to provide critical analysis on issues of international law relevant to the region, including ocean law and policy, ASEAN law and policy, trade and investment law and international dispute resolution. The series will examine international law issues from a legal, political and policy perspective, and will serve as a valuable reference for international and regional organizations, government officials, scholars and practitioners. More information about Centre for International Law can be found at www.cil.nus.edu.sg.

Titles in the series include:

Piracy and International Maritime Crimes in ASEAN
Prospects for Cooperation
Edited by Robert Beckman and J. Ashley Roach (2012)

Beyond Territorial Disputes in the South China Sea
Legal Frameworks for the Joint Development of Hydrocarbon Resources
Edited by Robert Beckman, Ian Townsend-Gault, Clive Schofield, Tara Davenport and Leonardo Bernard (2013)

The South China Sea Disputes and Law of the Sea
Edited by S. Jayakumar, Tommy Koh and Robert Beckman (2014)

Transboundary Pollution
Evolving Issues of International Law and Policy
Edited by S. Jayakumar, Tommy Koh, Robert Beckman and Hao Duy Phan (2015)

Transboundary Pollution

Evolving Issues of International Law and Policy

Edited by

S. Jayakumar

Professor, Chairman of the International Advisory Panel of the Centre for International Law (CIL), National University of Singapore (NUS)

Tommy Koh

Professor, Chairman of the Governing Board of the Centre for International Law (CIL), National University of Singapore (NUS)

Robert Beckman

Associate Professor, Director of the Centre for International Law (CIL), National University of Singapore (NUS)

Hao Duy Phan

Senior Research Fellow, Centre for International Law (CIL), National University of Singapore (NUS)

NUS CENTRE FOR INTERNATIONAL LAW

EE Edward Elgar PUBLISHING

Cheltenham, UK • Northampton, MA, USA

© The Editors and Contributors Severally 2015

All rights reserved. No part of this publication may be reproduced, stored in a retrieval system or transmitted in any form or by any means, electronic, mechanical or photocopying, recording, or otherwise without the prior permission of the publisher.

Published by
Edward Elgar Publishing Limited
The Lypiatts
15 Lansdown Road
Cheltenham
Glos GL50 2JA
UK

Edward Elgar Publishing, Inc.
William Pratt House
9 Dewey Court
Northampton
Massachusetts 01060
USA

A catalogue record for this book
is available from the British Library

Library of Congress Control Number: 2014959496

This book is available electronically in the **Elgar**online
Law subject collection
DOI 10.4337/9781784715793

FSC www.fsc.org MIX Paper from responsible sources FSC® C013056

ISBN 978 1 78471 578 6 (cased)
ISBN 978 1 78471 579 3 (eBook)

Typeset by Columns Design XML Ltd, Reading
Printed and bound in Great Britain by T.J. International Ltd, Padstow

Contents

Editors

S. Jayakumar was Singapore's former Deputy Prime Minister and Senior Minister. Professor Jayakumar has held various portfolios such as Minister for Home Affairs, Minister for Foreign Affairs, Minister for Labour, and the longest Ministership as Minister for Law, some 26 years. He served as Singapore's Permanent Representative to the United Nations, was the High Commissioner to Canada from 1971–1974 and was a member of Singapore's delegation to the Third Law of the Sea Convention. In addition, he oversaw the *Pedra Branca* case before the International Court of Justice as well as the *Land Reclamation* case before the International Tribunal for the Law of the Sea. Professor Jayakumar is the author of several books, the most recent being *Pedra Branca – The Road to the World Court* and *Diplomacy – A Singapore Experience*, and has written more than 30 articles on topics of constitutional law, international law and legal education. He served as Dean of the National University of Singapore (NUS) Faculty of Law. He now serves as the Chairman of the Law Faculty's Advisory Council and Chairman of the International Advisory Panel of the Centre for International Law (CIL). Professor Jayakumar is also the Co-Chairman of the International Advisory Panel on Transboundary Pollution appointed by the Government of Singapore in 2014.

Tommy Koh is Ambassador-At-Large at the Singapore Ministry of Foreign Affairs, Chairman of CIL and Rector of Tembusu College at NUS. He is the Co-Chairman of the China-Singapore Forum, the India-Singapore Strategic Dialogue, the Japan-Singapore Symposium and the International Advisory Panel on Transboundary Pollution. Professor Koh was the President of the Third United Nations Conference on the Law of the Sea and chaired the Preparatory Committee for and the Main Committee at the Earth Summit. He was Singapore's Permanent Representative to the United Nations for 13 years, Ambassador to the United States for six years, and served as the United Nations Secretary General's Special Envoy to Russia, Estonia, Latvia and Lithuania. He was also

Singapore's Chief Negotiator for the US-Singapore Free Trade Agreement and has chaired two dispute panels for the World Trade Organization. Professor Koh acted as Singapore's Agent in the case concerning *Pedra Branca* before the International Court of Justice and in the *Land Reclamation* case before the International Tribunal for the Law of the Sea. He is a renowned expert on international law and in particular on law of the sea. Professor Koh was recently named the recipient of the 2014 Great Negotiator Award, an award given out by an inter-university consortium comprising Harvard University, the Massachusetts Institute of Technology, Tufts University and the Future of Diplomacy Project at Harvard Kennedy School.

Robert Beckman is the Director of CIL. He is also an Associate Professor at the NUS Faculty of Law, where he has taught for more than 36 years. He is a member of the Governing Board and a lecturer at the Rhodes Academy of Oceans Law and Policy. Professor Beckman is an expert on issues of law of the sea. He served for several years as a regional resource person in workshops on Managing Potential Conflicts in the South China Sea. As a member of the National Executive Committee of the Council for Security Cooperation in the Asia Pacific (CSCAP) Singapore, he has represented Singapore in various CSCAP meetings on maritime security and has worked on legal issues relating to the Straits of Malacca and Singapore. He is an Adjunct Senior Fellow in the Maritime Security Programme at the S. Rajaratnam School of International Studies, Nanyang Technological University. He has written extensively on issues of law of the sea, including a recent publication *Beyond Territorial Disputes in the South China Sea: Legal Frameworks for the Joint Development of Hydrocarbon Resources* and contributed to the Agora on the South China Sea that appeared in the January 2013 issue of the *American Journal of International Law*.

Hao Duy Phan is a Senior Research Fellow at CIL. He received a BA from the Institute for International Relations of Vietnam, an LLM summa cum laude from the University of Notre Dame Law School, and an SJD from the American University Washington College of Law. He is the author of *A Selective Approach to Establishing a Human Rights Mechanism in Southeast Asia: The Case for a Southeast Asian Court of Human Rights* (Martinus Nijhoff, 2012) and several articles on various issues of international law. Prior to joining CIL, he worked as Assistant Director-General of the Department of International Law and Treaties, Ministry of Foreign Affairs of Vietnam, and a visiting fellow at the East-West Center in Washington DC and the Institute of Southeast Asian Studies (Singapore).

Contributors

Alan Boyle is Professor of Public International Law at the University of Edinburgh School of Law. He teaches international law, international environmental law and law of the sea. His publications include, among others, *International Law and the Environment* (with Patricia Birnie and Catherine Redgwell) (Oxford University Press, 2009) and *The Making of International Law* (with Christine Chinkin) (Oxford University Press, 2007). He practises international law from Essex Court Chambers, London, in recent or current cases including *Pulp Mills on the River Uruguay* (International Court of Justice), *Aerial Spraying* (International Court of Justice), *Bay of Bengal Maritime Boundaries* (International Tribunal for the Law of the Sea and Permanent Court of Arbitration), *Japanese Whaling* (International Court of Justice), *Chagos Islands* (Permanent Court of Arbitration) and *South China Sea Case* (Permanent Court of Arbitration).

Hans Christian Bugge is Professor Emeritus of Environmental Law at the Department of Public and International Law, University of Oslo. He holds a doctorate in regional planning from the Université de Paris II and is Juris Doctor from the University of Oslo on a thesis on responsibility for pollution in Norwegian law. In 1998 he was appointed Professor of Environmental Law. Prior to becoming an academic lawyer in 1991, he held various senior positions including Director-General of Pollution Control Authority and Director-General of the Department for Urban and Regional Planning, Norway's Ministry of the Environment (1972–1982); Secretary General of Redd Barna running humanitarian and development projects in Asia, Africa and Central America (1982–1991); and State Secretary (Deputy Minister) for Development Cooperation and personal advisor to Prime Minister Gro Harlem Brundtland in her work as Chair of the World Commission on Environment and Development (1986–1987). He has been Vice Chair of the Board of the National Research Council of Norway, Dean for Research at the Faculty of Law and Chair of the Faculty's Interdisciplinary Research Group in Natural Resources Law. He has lectured and written extensively on national and international environmental law and sustainable development.

Günther Handl is the Eberhard P. Deutsch Professor of Public International Law at Tulane University Law School. He holds law degrees from the University of Graz (Dr. iur.), Cambridge (LLB) and Yale (SJD). He has been a Fellow at the Max Planck Institute for Comparative Public Law and International Law in Heidelberg, Germany and has taught at several other law schools in the United States, Canada, Japan and Europe. He is the founder and former editor-in-chief of the *Yearbook of International Environmental Law*, and author of many books and articles in the field of public international law, international environmental law and law of the sea. Professor Handl is the recipient of a number of awards, including the Prix Elisabeth Haub 1997. He has served in an advisory capacity to a number of governments, as a consultant to various international organisations, including the United Nations Environment Programme, the Organisation for Economic Co-operation and Development, the Asian Development Bank and the World Bank, as well as a lecturer for the United Nations Codification Division, Office of Legal Affairs and The Hague Academy of International Law.

Lan Hua is Associate Professor of Faculty of International Law, China University of Political Science and Law (CUPL). She holds law degrees from Wuhan University Law School (LLM and JSD) and Nanchang University Law School (LLB). She teaches international law, EU case study, international organisations, legal English, and Chinese and WTO law. She has served for many years as the faculty advisor and coach of the CUPL teams in the Philip C Jessup International Law Moot Court Competition. Her current research interests include international environmental law, European law and WTO law. She has published numerous articles and book chapters and been involved in many research projects on international environmental law.

Youna Lyons is a Senior Research Fellow at CIL. Her primary research areas are offshore oil and gas in Southeast Asia, including decommissioning and the prospects for rigs-to-reefs, the use of satellite imagery to inform marine environmental management through habitat mapping and maritime boundary disputes on remote features, the protection of marine sensitive environments in the seas of Southeast Asia and the underlying legal and institutional governance framework. Prior to this, she created and led the private international law practice of the Litigation and Arbitration Group at Clifford Chance (Paris office), one of the world's leading law firms, for more than 12 years. Passionate about the ocean, she was trained in marine ecology and policy at the College of Ocean and Fisheries of the University of Washington where she obtained a Master's degree in Marine Affairs. Her dual expertise in law and marine sciences uniquely position her to integrate these fields for marine policy-making purposes. In addition, she focuses on developing industry-relevant research and on involving the industry in marine policy proposals.

Stephen C McCaffrey is Distinguished Professor of Law at the University of the Pacific, McGeorge School of Law. He served two terms on the International Law Commission, chaired that body for one of its sessions and was the Commission's special rapporteur on international watercourses. He has served as Counsel to States in cases before the International Court of Justice and the Permanent Court of Arbitration.

Jacqueline Peel is Professor at the Melbourne Law School. She holds a Bachelor of Science and a Bachelor of Laws (Hon I) from the University of Queensland, a Master of Laws from New York University (NYU) where she was a Fulbright Scholar, and a PhD in law from the University of Melbourne. Her teaching and research interests lie in the areas of environmental law, risk regulation and the role of science in international law and climate change law. She has published numerous articles and several books on these topics, including the third edition of *Principles of International Environmental Law* (with Philippe Sands), *Australian Climate Law in Global Context* (with Alexander Zahar and Lee Godden), *Environmental Law: Scientific, Policy and Regulatory Dimensions* (with Lee Godden), *Science and Risk Regulation in International Law* and *The Precautionary Principle in Practice*. Professor Peel was an expert member of the International Law Association's Committee on the Legal Principles relating to Climate Change, acting as Rapporteur of that Committee during the latter half of 2012. She has also been the recipient of several awards and fellowships, including a Hauser Research Scholarship and Emile Noel Fellowship at NYU, a Research Associateship from the United States Studies Centre, and two Australian Research Council grants on climate change law-related topics. During 2013–2015, Professor Peel was a Visiting Scholar at the Woods Institute for the Environment at Stanford University.

Catherine Redgwell is Chichele Professor of Public International Law at the University of Oxford and a Fellow of All Souls College. Other affiliations include membership of the Greenwich Forum, the Council of the British Branch of the International Law Association, the International Bar Association's Academic Advisory Group of the Section on Energy, Environment, Natural Resources and Infrastructure Law, and the International Bar Association's Climate Change Justice and Human Rights

Task Force. Formerly the general editor of the *International and Comparative Law Quarterly,* she is currently joint general editor of the *British Yearbook of International Law* and co-editor of *Oxford Monographs in International Law*. Professor Redgwell has published extensively in public international law, particularly in the areas of international environmental law, international energy law, law of the sea and treaty law. Her recent publications include the co-authored third edition of *International Law and the Environment* (Oxford University Press, 2009), *Lyster's International Wildlife Law* (Cambridge University Press, 2010) and *Energy Law in Europe* (Oxford University Press, 2014). She is currently co-investigator in a two-year (2012–2014) research project on Climate Geoengineering Governance funded by the Economic and Social Research Council and Arts and Humanities Research Council.

Nicholas A Robinson is the University Professor for the Environment at Pace University, the Gilbert and Sarah Kerlin Distinguished Professor of Environmental Law Emeritus at Pace University School of Law and the Professor Adjunct at the Yale University School of Forestry & Environmental Studies. He is also the Co-Director of the Center for Environmental Legal Studies. Professor Robinson has developed environmental law since 1969 when he was named to the Legal Advisory Committee of the President's Council on Environmental Quality. He has practised environmental law in law firms for municipalities and as general counsel of the New York State Department of Environmental Conservation. He drafted New York's wetlands and wild bird laws and was the first chairman of both the statutory Freshwater Wetlands Appeals Board and Greenway Heritage Conservancy for the Hudson River Valley. He has served as legal advisor and chairman of the Commission on Environmental Law of the International Union for the Conservation of Nature and Natural Resources, engaged in drafting treaties and counselled different countries on the preparation of their environmental laws. He founded Pace's environmental law programmes, edited the proceedings of the 1992 United Nations Earth Summit and is author of several books and articles. He teaches a number of environmental law courses. He served as James D. Hopkins Professor of Law during the 1991–1993 academic years.

Laode M Syarif is a Senior Lecturer at Hasanuddin University Faculty of Law. He also serves as a guest lecturer at the University of South Pacific, Vanuatu and a principal trainer for the Supreme Court of Indonesia in environmental law since 2007. Syarif is a Senior Advisor on Justice and Environmental Governance at the Partnership for Governance

Reform in Indonesia, where he develops and manages several environmental projects on sustainable forest management, climate change and environmental law enforcement. He also develops and manages several anti-corruption and good governance programmes with various national agencies and non-governmental organisations. As a senior advisor, he works closely with donor communities, such as US Agency for International Development, Australian Agency for International Development, World Bank, UN Development Programme (UNDP), UN Office on Drugs and Crime, Danida and the Norwegian Embassy, among others. He has published widely on atmospheric pollution, forestry, fisheries, biodiversity and REDD+. He is also one of the drafters of Indonesian Law No 32/2009 on Environmental Management and Protection. He has a Bachelor of Laws (SH) from Hasanuddin University, a Master of Laws (LLM) from Queensland University of Technology, Brisbane and his PhD from Sydney Law School at the University of Sydney.

Alan Khee-Jin Tan is Professor of Law at NUS. An alumnus of the NUS and Yale Law School, he researches and teaches aviation law, maritime law and environmental law. His book on marine environmental law and the shipping industry, *Vessel-Source Marine Pollution: The Law and Politics of International Regulation*, was published by Cambridge University Press. Professor Tan has also written on Indonesia's State responsibility for the transboundary impact of forest fires and the problems of decentralisation and natural resource management in Indonesia. These writings have appeared in journals such as the *International and Comparative Law Quarterly* and New York University's *Environmental Law Journal*. In 2007, he was co-editor of a landmark multi-disciplinary work, *Biodiversity and Human Livelihoods in Protected Areas: Case Studies from the Malay Archipelago*, published by Cambridge University Press. He has also advised the UNDP and the government of Vietnam on legal issues pertaining to the environment and natural resource management.

Simon Tay is a public intellectual as well as private advisor to major corporations and policy-makers. He is concurrently Chairman of the Singapore Institute of International Affairs, the country's oldest think tank and founding member of the ASEAN network of think tanks. He is also an Associate Professor, teaching international law at NUS. His commentaries feature regularly in newspapers such as TODAY, Jakarta Post and the China Daily. He also appears on television with the BBC, CNN, Bloomberg, CNBC and Channel News Asia. Professor Tay is also a Senior Consultant at WongPartnership, a leading Asian law firm. He serves on the Global Advisory Boards for Toyota Motor Corporation and

MUFG. Furthermore, he sits as a board member for two major Singaporean corporations, Hyflux Ltd and the Far East Organization. From 1992 to 2008, he served in public positions for Singapore, including as Chairman of the National Environment Agency, reporting to the Minister as an independent Member of Parliament and coordinating the Singapore Volunteers Overseas. He has taught as a visiting professor at Yale University, the Fletcher School and Harvard Law School. Professor Tay is a prize-winning author of stories and poems and his novel *City of Small Blessings* was awarded the Singapore Literature Prize in 2010. He graduated in law from NUS (1986), where he was president of the student union for three terms, and from Harvard Law School (1993), where he won the Laylin Prize for the best thesis in international law.

Acknowledgements

The editors are grateful to all the authors for their contributions to this book. We would like to record our special thanks to Kristine Dalaker Kraabel, Research Associate at the Centre for International Law (CIL), National University of Singapore, for her excellent work and invaluable support with the editing, proofreading and preparation of the manuscripts. Her dedication and professionalism played an important role in helping us maintain the high standards for this project. We thank Monique Page, former Research Associate at CIL, for her support during the initial stage of the project. We wish to thank Heidi Tan and Stephanie Teh as well for their assistance in checking the references. Finally, we would like to express our gratitude to Edward Elgar Publishing for their cooperation and support on this project.

Introduction

S. Jayakumar, Tommy Koh, Robert Beckman and Hao Duy Phan

Pollution does not recognise national borders. It may originate in one country and damage the environment in another. An oil spill in one State may affect the marine environment of its neighbours. Smoke haze from agricultural practices or other economic activities in one country may pollute the air in other countries. The pollution of an international river in an upstream country may reduce the water quality in downstream countries. The leaking of radioactive materials from a nuclear plant in one country may raise serious environmental concerns across the region.

As the international community has become more cognizant of the social and economic costs of transboundary pollution, it has sought to establish international legal principles to address the issue. The *Trail Smelter* arbitral tribunal in 1941 stated that 'no State has the right to use or permit the use of its territory in such a manner as to cause injury by fumes in or to the territory of another'.[1] Principle 21 of the 1972 Stockholm Declaration of the United Nations Conference on the Human Environment declared that States have 'the responsibility to ensure that activities within their jurisdiction or control do not cause damage to the environment of other States'.[2] The International Court of Justice confirmed that the obligation of States to ensure that their activities respect the environment of other States has become 'part of the corpus of international law relating to the environment'.[3] At issue is how this well-settled principle has been applied and what mechanisms have been

---

1 *Trail Smelter Case (United States of America, Canada)* (1938 and 1941) 3 RIAA 1905 (*Trail Smelter*) at 1965.

2 Stockholm Declaration on the Human Environment, in 'Report of the United Nations Conference on the Human Environment', UN Doc A/CONF.48/14, at 2 and Corr.1 (1972), 11 ILM 1416 (1972 Stockholm Declaration), Principle 21.

3 *Legality of the Threat or Use of Nuclear Weapons*, Advisory Opinion, [1996] ICJ Rep 226 at 241, para 29.

developed to deal with specific types of transboundary pollution, including pollution of the marine environment and shared water resources, nuclear pollution and air pollution.

This book developed out of an international conference hosted by the Centre for International Law (CIL) at the National University of Singapore (NUS) in February 2014. The Conference brought together world leading scholars and experts on environmental law to engage in discourse and discussion on various aspects of transboundary pollution.

The book provides a comprehensive overview of the international legal principles governing transboundary pollution and examines the practical applications of the State responsibility doctrine in the context of transboundary pollution. It analyses the international legal framework that has developed to address all major types of transboundary pollution. The book also includes case studies of Asia and Southeast Asia to demonstrate how international law and policy governing transboundary pollution have evolved and have been applied in practice within a regional context.

The book consists of 13 chapters, which are organised into three parts. Part I examines 'General Principles and State Responsibility', Part II focuses on 'Problems and Prospects' and Part III addresses 'Cooperative Mechanisms for Addressing Transboundary Pollution'.

In Part I on 'General Principles and State Responsibility', Chapter 1 by Catherine Redgwell charts the development of major customary law principles of international environmental law, providing a comprehensive overview of the general principles, policies and practices that govern transboundary pollution. It also reviews the lawmaking activities of the last decades that have resulted in a number of pollution-specific treaties, especially with regard to marine pollution and nuclear pollution. The chapter highlights the tension between the competing concepts of permanent sovereignty over natural resources and the common concern of transboundary environmental damage, but concludes that States have the obligation under international law to exercise due diligence to regulate and monitor activities within their territory, jurisdiction and control that may cause significant harm to other States.

Chapter 2 by Simon Tay looks at the challenges of applying general principles governing transboundary pollution in a global economy where countries are deeply interconnected both environmentally and economically. The chapter advances the argument that transboundary pollution should be seen not only through the lens of environmental harms that are traceable from one country to another but also through links of economic interdependence. It calls for global law approaches that include both State and non-State actors and argues that these emergent and innovative approaches should be applied to tackle the regional challenge of smoke haze from forest fires in Southeast Asia.

As Chapters 1 and 2 indicate, the legal requirements regarding State responsibility for transboundary pollution are well settled in international law. However, the limited number of cases on transboundary pollution that have been decided by international courts and tribunals leaves open many questions concerning the practical application of the State responsibility doctrine. Chapter 3 by Jacqueline Peel seeks to analyse the elements of a State responsibility claim for transboundary pollution, focusing on those elements that represent the most complex issues in practice. These include questions concerning how to attribute polluting actions of private entities to States, the level and content of a State's obligation to act with 'due diligence' in preventing transboundary harm, and the methodologies to prove the causation of injury through transboundary pollution. The chapter draws on the example of transboundary air pollution to illustrate how these questions are applied in practice to the issue of how the emission of greenhouse gases contributes to climate change.

Part II examines in detail various specific transboundary pollution problems facing the international community today. Chapter 4 by Stephen C McCaffrey examines the problem of transboundary pollution of shared freshwater resources. His chapter discusses two international treaties that regulate the problem of transboundary water pollution. It also looks at major cases before international courts and tribunals and argues that these judicial decisions hold valuable lessons for governments regarding the use, protection and management of international watercourses. His careful examination highlights the emphasis that international law places on preventing pollution of international watercourses and protecting the aquatic ecosystem.

Chapter 5 by Lan Hua examines the use of Environment Impact Assessments (EIAs) in international treaties and case law as a means to prevent transboundary pollution of freshwater resources. In particular, the chapter comments on pollution control and environmental cooperation with respect to the Lancang-Mekong River and makes recommendations as to how EIAs can be utilised to protect shared freshwater resources.

Chapter 6 by Robert Beckman focuses on the problem of transboundary pollution of the marine environment. It offers a detailed analysis of the relevant substantive provisions and the dispute settlement mechanisms under the 1982 United Nations Convention on the Law of the Sea (UNCLOS). As the chapter forcefully demonstrates, UNCLOS treats transboundary marine pollution differently from other problems of transboundary pollution in three important respects. First, the principles and provisions governing the protection of the marine environment are more specific than those in other areas. Second, the rules in UNCLOS on the responsibility of States for internationally wrongful acts give injured States the right to seek remedies to prevent transboundary pollution of the marine environment. Third, the dispute settlement regime in UNCLOS allows States to institute compulsory proceedings against other States and hold them accountable when they fail to meet these obligations.

Chapter 7 by Youna Lyons provides a case study of a particular instance of transboundary marine pollution, the 2009 Montara offshore oil spill in a field in the Timor Sea. In this chapter, Lyons investigates the transboundary nature of the oil spill and explores the ability and effectiveness of the administrative and legal mechanisms that were in place to deal with the event. The chapter draws attention to the gap that exists in the international regulatory regime for managing offshore oil and gas activities. It also offers a number of comments on the appropriate levels of oil spill preparedness, the response processes, and the funding mechanisms for clean-up activities and the payment of compensation.

Chapter 8 by Günther Handl examines the role of international law in establishing the normative foundation for transboundary nuclear pollution prevention, especially in the post-Fukushima period. Handl reviews transboundary nuclear risk management from the perspective of customary law and from the viewpoint of the international nuclear safety system. He discusses the complex legal issue raised by the 'border siting' of nuclear facilities whose emergency planning zones extend into neighbouring States' territories. Finally, the chapter examines the challenges arising from the maritime transportation of radioactive materials through foreign States' territorial seas. It suggests that the 'notification without authorisation' approach may be an acceptable compromise if appropriate safeguards are put in place. The chapter also argues for a more effective system of international safety review and recourse to third party dispute settlement in the event the concerned parties fail to reach a solution.

In Chapter 9, Alan Boyle addresses the international legal regime pertaining to transboundary air pollution and uses recent case law to illustrate the practical applications of the regime. The chapter suggests that interstate litigation can contribute effectively to addressing transboundary air pollution. It also examines transboundary air pollution within a broader regional context through the mechanism of evolutionary regimes of multilateral regulation exemplified by the 1979 Geneva Convention on Long-Range Transboundary Air Pollution and the 1991

United States-Canada Air Quality Agreement. The chapter raises interesting points of discussion in light of the increasing levels of emissions and fossil fuel consumption in China and India.

The book concludes with Part III on 'Cooperative Mechanisms for Addressing Transboundary Pollution'. Chapter 10 by Hans Christian Bugge discusses the principle and duty of cooperation in international environmental law. The chapter presents some key points on the principle of cooperation, including its legal basis, content and application in practice. It also offers two examples of regional cooperation in Europe on the basis of framework conventions – the 1979 Convention on Long-Range Transboundary Air Pollution and the 1992 Convention for the Protection of the Marine Environment in the North-East Atlantic. The chapter concludes with a note on the lessons that may be learned from the operation of these European conventions by other regions in crafting cooperative mechanisms to respond to transboundary pollution.

Since 1991, haze has become a serious problem in Southeast Asia, with the main source coming from forest fires in Indonesia that have resulted from the agricultural practice of using fire to clear land. In response to this problem, the Association of Southeast Asian Nations (ASEAN) has adopted a number of soft law and hard law instruments, including in particular the 2002 ASEAN Agreement on Transboundary Haze Pollution (2002 ASEAN Haze Agreement). Chapter 11 by Laode M Syarif examines the various instruments that ASEAN has adopted and cooperative mechanisms that it has established to deal with regional transboundary haze pollution. In particular, Syarif's chapter draws attention to the 2002 ASEAN Haze Agreement and various substantive and procedural provisions therein that may affect its effectiveness. The chapter also includes suggestions on the steps that countries in the region can take towards remedying the issue of transboundary air pollution through legal and political measures.

Chapter 12 by Alan Khee-Jin Tan further highlights the systemic weaknesses of the 2002 ASEAN Haze Agreement and numerous political and socio-economic factors within Indonesia that may constrain the effectiveness of the Agreement. His chapter then evaluates the promises and pitfalls of cooperative mechanisms that have been proposed to deal with the haze problem, including the proposal for Indonesia to share its palm oil plantation concession maps, a measure that would supposedly enable the perpetrators of the forest fires to be identified and prosecuted. The chapter argues that until the 2002 ASEAN Haze Agreement is fully implemented and concrete cooperation mechanisms are fully contemplated, regional States have limited options outside acting under existing bilateral cooperative mechanisms or taking unilateral extraterritorial action against those entities and individuals involved in causing the forest fires.

Against this background, the last chapter by Nicholas A Robinson concludes the book by offering a comprehensive and integrated set of solutions to address the root causes of transboundary pollution in general and haze in particular. Assertions of State responsibility and political negotiations alone are not likely to end the air pollution because they address the haze only as a symptom and do not reach underlying causes. Even though Indonesia has ratified the 2002 ASEAN Haze Agreement, there would still be a need to use additional and complementary legal tools to avert new fires. Numerous alternatives to the setting of forest fires exist and could be pursued in mutually reinforcing ways. Indonesia, for example, could designate more of Sumatra and Kalimantan as world heritage areas and utilise cooperative mechanisms under the United Nations Educational, Scientific and Cultural Organization (UNESCO) World Heritage Convention and the ASEAN Heritage Program for Protected Area Management to protect its natural peat forests. It could also work with other countries to develop a special programme under the Ramsar Convention on Wetlands of International Importance to identify and protect its coastal mangroves and inland peatland wetlands. The Convention on International Trade in Endangered Species of Wild Fauna and Flora, the Convention on Biological Diversity, the International Union for Conservation of Nature (IUCN) Commission on Protected Areas and the IUCN World Parks Congress and regional mechanisms at ASEAN also offer avenues for cooperation. Efforts could be made to strengthen the capacity of regional States to undertake EIAs. Other remedies include the Soils Charter adopted by the United Nations Food and Agricultural Organization (FAO) to end burning as a technique of forest clearing, the approach employed by the World Health Organization (WHO) to curb urban air pollution and the use of trade agreements to control the production of palm oil. Actions could also be taken under the 'Reducing Emissions from Deforestation and Forest Degradation, Plus Conservation' (REDD+) to preserve primary forests and peatlands. As it is not likely that these measures will be employed in Indonesia in the near future in a way that can effectively prevent the forest fires, donor States, regional States, their businesses, international organisations and NGOs could work with Indonesia to establish REDD+ investments in Kalimantan and Sumatra to push back against the commercial investors who seek to destroy forests and peatlands for palm oil plantations. Regionally, States can encourage their companies to support financially environmental projects in Sumatra. The chapter concludes that States have the capacity and authority – and the duty – to cooperate more holistically, to make it a common cause to protect and sustain the Earth, its people and its fragile ecosystems.

PART I

Transboundary pollution: general principles and state responsibility

1. Transboundary pollution: principles, policy and practice

Catherine Redgwell

I. SETTING THE CONTEXT: A GENERAL OBLIGATION TO PROTECT AND PRESERVE THE ENVIRONMENT?

In the past half century or so, there has been significant growth in the body of general and particularised rules governing State conduct in respect of the environment. Despite this growth, it remains the case that there is not yet any general customary or treaty law obligation on States to protect and preserve the environment per se. To be sure, there is a relevant negative obligation found in customary international law and a central focus of this chapter: the so-called 'no harm principle' or the obligation imposed on States not to allow their territory to be used in such a manner so as to cause significant harm to the territory of other States, or to the global commons. This principle may be traced back to the seminal *Trail Smelter* case[1] and, had it proceeded to the merits, would have been revisited by the International Court of Justice (ICJ) in the *Aerial Spraying* case[2] in the light of late 20th and early 21st century concerns. Positive obligations, however, remain largely sectoral in focus,

---

[1] *Trail Smelter Case (United States of America, Canada)* (1938 and 1941) 3 RIAA 1905 (*Trail Smelter*). See also Stockholm Declaration on the Human Environment, in 'Report of the United Nations Conference on the Human Environment', UN Doc A/CONF.48/14, at 2 and Corr 1 (1972) (1972 Stockholm Declaration), Principle 21; Declaration of the UN Conference on Environment and Development, Rio Declaration on Environment and Development, in 'Report of the United Nations Conference on Environment and Development', UN Doc A/CONF.151/26 (Vol I), 12 August 1992, Annex I, UN Doc A/CONF.151/26/Rev.1 (1992 Rio Declaration), Principle 21.

[2] *Case concerning Aerial Herbicide Spraying (Ecuador v Colombia)* ICJ (*Aerial Spraying*). For settlement of the dispute shortly before oral hearings were due to commence, see http://www.icj-cij.org/docket/index.php?p1=3&code=

the most outstanding examples being the obligation in Article 192 of the 1982 UN Convention on the Law of the Sea (UNCLOS) to protect and preserve the marine environment[3] and the obligation in Article 2 of the 1991 Protocol on Environmental Protection to the 1959 Antarctic Treaty to comprehensively protect the Antarctic environment and dependent and related ecosystems.[4] But as for a general obligation to protect and preserve the environment wherever situated, one looks in vain.

This is partly explained by the piecemeal development of the subject. Early international regulation of environmental activities dealt with conservation of common property resources subject to over-exploitation and customary international law first developed to restrain State actions causing transboundary harm, primarily economic harm, in the territory of another State. International environmental regulation of State behaviour penetrating within the State has been less rapid to develop, particularly since it encounters the twin, yet related, obstacles of State sovereignty and permanent sovereignty over natural resources. As will be discussed further below, the issue of transboundary harm arising from polluting activities within and beyond the State remains a key preoccupation of international environmental law.

In this lawmaking activity of the last decades, there has been an evolution in lawmaking focus from environmental regulation incidental to the primary focus, such as economic regulation of a resource, to a holistic approach to environmental protection within and beyond State borders. These stages are not necessarily sequential, however, with evidence in the present of international rules reflecting each of these stages of evolution. This is particularly evident in the tension between permanent sovereignty over natural resources and the common concern of humankind contained in the Convention on Biological Diversity,[5] and in the continuing resistance to robust international regulation of forest and desertification issues because of the perceived threat to State sovereignty. Moreover, what is absent from the corpus of international

---

ecol&case=138 (accessed 5 January 2015). The author was counsel for Colombia, but the views expressed herein are her own and not those of any government which she has represented.

[3] United Nations Convention on the Law of the Sea, adopted 10 December 1982, 1833 UNTS 3 (entered into force 16 November 1994) (UNCLOS).

[4] Protocol on Environmental Protection to the Antarctic Treaty, adopted 4 October 1991, 30 ILM 1445 (entered into force 14 January 1998).

[5] Convention on Biological Diversity, adopted 5 June 1992, 1760 UNTS 79 (entered into force 29 December 1993) (1992 Convention on Biological Diversity).

rules thus developed is a comprehensive codification of the basic rules and principles applicable to international regulation of the environment[6] analogous to the 1948 Universal Declaration of Human Rights or to the 'constitution for the oceans', UNCLOS.[7] The 1972 Stockholm Declaration and the 1992 Rio Declaration were political attempts to articulate such principles, though only some of the principles contained in them are of a sufficiently norm-creating character, and enjoy the widespread and consistent State practice and buttressing *opinio juris*, for recognition as customary international law. Given the wide range of environmental issues confronting humankind, many of which require a detailed, standard-setting regulatory response, it is not surprising that the vast bulk of international environmental law is found in tailor-made treaty arrangements. This is particularly so with respect to regulation of polluting activities likely to cause transboundary harm.[8] The 2002 ASEAN Agreement on Transboundary Haze Pollution is one such example.[9]

II. SOME RELEVANT PRINCIPLES OF CUSTOMARY INTERNATIONAL LAW

A. The 'No Harm' Principle

Nonetheless, State practice has given rise to a number of customary law principles,[10] buttressed by the process of treaty and customary law interaction. Of these the most significant is the so-called 'no harm' principle, pursuant to which States have a duty to prevent, reduce, and control pollution and significant transboundary environmental harm arising from activities within their territory, jurisdiction or control. It has

---

[6] One such attempt is the International Union for Conservation of Nature's (IUCN) *Draft International Covenant on Environment and Development*, accessed 5 January 2015 at http://www.uncsd2012.org/content/documents/IUCN%20Intl%20Covenant%20on%20Env%20and%20Dev%20EPLP-031-rev3.pdf.

[7] Tommy Koh, 'A Constitution for the Oceans', Remarks of the President of the Third United Nations Conference on the Law of the Sea (UNCLOS III) accessed 5 January 2015 at http://www.un.org/depts/los/convention_agreements/texts/koh_english.pdf.

[8] This is discussed further in several sectoral contexts in Part III below.

[9] ASEAN Agreement on Transboundary Haze Pollution, signed 10 June 2002 (entered into force 25 November 2003) accessed 5 January 2015 at http://agreement.asean.org/media/download/20140119170340.pdf.

[10] See generally, Nicolas De Sadeleer, *Environmental Principles: From Political Slogans to Legal Rules* (Oxford: Oxford University Press, 2002).

been enunciated in soft law declarations,[11] endorsed inter alia by the General Assembly,[12] the International Law Commission (ILC)[13] and in various multilateral environmental agreements,[14] and in judicial decisions.[15] Thus, for example, in the *Pulp Mills* case, which involved the siting of a pulp mill on a shared watercourse, the River Uruguay, the ICJ observed that '[a] State is … obliged to use all the means at its disposal in order to avoid activities which take place in its territory, or in any area under its jurisdiction, causing significant damage to the environment of another State'.[16]

This obligation not to cause *significant* harm has achieved widespread recognition, particularly (though not exclusively) in the contexts of shared resources and of hazardous activities.[17] The threshold of significant harm or damage was confirmed by the ICJ in the *Pulp Mills* case in

[11] See e.g., Stockholm Declaration, *supra* note 1, Principle 21; 1992 Rio Declaration, *supra* note 1, Principle 2.

[12] 1974 Charter of Economic Rights and Duties of States, GA Res 3281, UNGAOR, 29th Sess, Supp No 31, UN Doc A/9631 (1974) 50, Art 30.

[13] 'International Liability for Injurious Consequences Arising out of Acts Not Prohibited by International Law (Prevention of Transboundary Harm from Hazardous Activities)' in 'Report of the International Law Commission to the General Assembly covering the work of its fifty-third session, with commentaries, 2001' (UN Doc A/56/10), Ch V, in *Yearbook of the International Law Commission 2001*, vol II, Part Two (UN, 2001) (Draft Articles on Prevention).

[14] See e.g., United Nations Framework Convention on Climate Change, adopted 5 May 1992, 1771 UNTS 107 (entered into force 21 March 1994) (UNFCCC), Preamble; 1992 Convention on Biological Diversity, *supra* note 5, Art 3.

[15] See e.g., *Legality of the Threat or Use of Nuclear Weapons*, Advisory Opinion, [1996] ICJ Rep 226 (*Legality of Nuclear Weapons*) at 241–2, para 29.

[16] *Case concerning Pulp Mills on the River Uruguay (Argentina v Uruguay)*, [2010] ICJ Rep 14 (*Pulp Mills*) at para 101. See also the Draft Articles on Prevention, which accurately reflect the current state of international law in this regard, *supra* note 13 at 152. And, as noted by the ICJ in *Legality of Nuclear Weapons*, States have the general obligation to ensure that activities within their jurisdiction and control respect the environment of other States. See *ibid.*

[17] It is acknowledged that there are disagreements between States over the identification of shared natural resources, and in particular on what, if any, rights and responsibilities flow in relation to such resources. International watercourses, migratory species and mountain chains, are some common illustrations. See e.g., 1997 Convention on Watercourses, *infra* note 21, Art 7; 'Shared Natural Resources' in 'Report of the International Law Commission to the General Assembly covering the work of its sixtieth session' (UN Doc A/63/10), Ch IV, in *Yearbook of the International Law Commission 2008*, vol II, Part Two (UN,

the passage cited above, which explicitly refers to avoiding activities '*causing significant damage* to the environment of another State'.[18]

Whatever similar qualifier is used – serious, significant, substantial or appreciable – it is generally agreed that States are enjoined from causing transboundary harm that is so qualified. On this point, the ILC Draft Articles on Prevention of Transboundary Harm for Hazardous Activities (Draft Articles on Prevention) accurately reflect the current state of international law in likewise referring to the threshold of 'significant'.[19] Assistance in determining whether this threshold of significant harm has been crossed may be derived from international undertakings by States, for example, to publicise national pollution release and transfer data[20] or certain types of activities which involve radiological, toxic or highly dangerous substances may *a priori* be deemed significantly harmful.[21] Many international instruments refer to the need to take measures 'to prevent, reduce and control' pollution[22] and the significant adverse transboundary environmental impact from proposed activities.[23] Some examples of such treaty provisions are considered in Part III below. Thus, far from imposing an absolute prohibition on any transboundary harm as

---

2008); *The Law of Transboundary Aquifers,* GA Res 63/124, UNGAOR, 63rd Sess, UN Doc A/RES/63/124 (2008), Art 6.

[18] *Pulp Mills*, *supra* note 16 at para 101 (emphasis added).

[19] See e.g., Draft Articles on Prevention, *supra* note 13 at 152. The threshold criteria for their application are that the hazardous activity in question has 'a high probability of causing significant transboundary harm' or 'a low probability of causing disastrous transboundary harm'. *Ibid,* Art 2(a).

[20] See e.g., Protocol on Pollutant Release and Transfer Registers to the 1998 Aarhus Convention, an UNECE treaty now open to general participation, adopted 21 May 2003 (entered into force 29 October 2009) accessed 5 January 2015 at http://www.unece.org/fileadmin/DAM/env/documents/2003/pp/ch_XXVII_13_ap.pdf (2003 Protocol on Pollutant Release).

[21] See e.g., invasive alien species under the 1992 Convention on Biological Diversity, *supra* note 5, and Convention on the Law of the Non-navigational Uses of International Watercourses, adopted 21 May 1997, 36 ILM 700 (entered into force 17 August 2014) (1997 Convention on Watercourses), Art 21; and restricted pesticides and chemicals under the Rotterdam Convention on the Prior Informed Consent Procedure for Certain Hazardous Chemicals and Pesticides in International Trade, adopted 10 September 1998, 2244 UNTS 337 (entered into force 24 February 2004) (1998 Rotterdam Convention).

[22] See e.g., UNCLOS, *supra* note 3, Art 194(1); the 1997 Convention on Watercourses, *ibid.*

[23] See e.g., Convention on Environmental Impact Assessment in a Transboundary Context, adopted 25 February 1991, 30 ILM 1461 (1991) (entered into force 10 September 1997) (1991 Espoo Convention), Art 2(1).

the 'no harm' shorthand appears to suggest, the obligation to prevent significant transboundary harm is widely viewed as imposing an obligation of due diligence.[24] The existence of breach of the 'no significant harm obligation' thus depends on establishing the failure to exercise 'reasonable due diligence', the assessment of which is flexible and context-specific. 'Reasonable due diligence' means that a State is obliged to apply the restraints on transboundary injurious activities that it may reasonably be expected to adopt. This is where the flexibility in the concept operates – what is reasonable under the circumstances will vary from case to case. The standard of due diligence against which the polluting State's conduct should be examined is that which is generally considered to be appropriate and proportional to the degree of risk of transboundary harm in the particular instance. Thus, for example, where the probability of harm is low, and the adverse consequences arising from any future event minimal, the obligation is not one of 'absolute prevention'[25] but rather to minimise the risk of significant transboundary harm.

B. Obligations to Consult and Notify Transboundary Harm and to Conduct Environmental Impact Assessment

State practice further supports the customary law obligation to consult and to notify of potential transboundary harm, particularly where there are shared resources or hazardous activities being carried out, and the requirement to conduct a prior transboundary environmental impact assessment (EIA). In the *Pulp Mills* case, the ICJ found the requirement to conduct a transboundary EIA to be a distinct and freestanding obligation in international law where significant transboundary harm is threatened.[26] Although the specific content of such an EIA is left to the State's discretion, international law *requires* that an EIA is conducted *and* that it bears a relation to the 'nature and magnitude of the proposed development and its likely adverse impact on the environment'.[27] The requirement of an EIA is thus not an empty shell to be filled (or not)

---

[24] 'The obligation of the State of origin to take preventive or minimization measures is one of due diligence': Draft Articles on Prevention, *supra* note 13, Commentary to Art 3, at 154, para (7).

[25] Ecuador argued in *Aerial Spraying* that the appropriate standard of care in the circumstances of that case was one that required the elimination of *all* risk of transboundary pollution: 'Ecuador Memorial' (28 April 2009), [2013] ICJ Pleadings (vol I) at para 8.27.

[26] *Pulp Mills*, *supra* note 16 at para 204.

[27] *Ibid* at para 205.

willy-nilly by the State proposing the potentially harmful activity. As the ILC has noted, '[t]he assessment should include the effects of the activity not only on persons and property, but also on the environment of other States'.[28]

The trigger for any requirement to conduct a transboundary EIA is 'significant adverse transboundary impact' from proposed activities. This is the requirement found, for example, in Article 2(1) of the Convention on Environmental Impact Assessment in a Transboundary Context (1991 Espoo Convention). Though clearly not directly applicable to non-parties, as the ICJ observed in the *Pulp Mills* case, the Convention may be used for illustrative purposes.[29] The threshold is a 'high degree of probability' of such adverse impact. As the 1991 Espoo Convention explains, EIA is 'a national procedure for evaluating the likely impact of a proposed activity on the environment'.[30] This is a purposive definition – the procedure must 'evaluate the likely impact of a proposed activity' – and arguably does not depend on national designation of such procedures as 'EIA' per se. Indeed, impact assessment is a ubiquitous feature of national law, under a number of guises and names. The key criterion is the existence of a national procedure for the *ex ante* assessment of the environmental impact of a proposed activity and, once undertaken, regular assessment. As stated in the *Pulp Mills* case, the scope and content of such assessment are for national law to determine in the absence of general international rules or specific treaty provision, subject to the requirement to exercise due diligence in conducting such an assessment.[31]

There is thus a clear link between EIA and the exercise of due diligence. Indeed, there the ICJ found an EIA to be a necessary element of due diligence, which is interpreted so as to entail the 'adoption of appropriate rules and measures' as well as 'a certain level of vigilance in their enforcement and the exercise of administrative control applicable to public and private operators, such as the monitoring of activities undertaken by such operators'.[32] It will be recalled that a bilateral treaty existed between the parties, the 1961 Treaty between Argentina and Uruguay for the delimitation of the river boundary between them and

---

28 Draft Articles on Prevention, *supra* note 13, Commentary to Art 7.

29 *Pulp Mills*, *supra* note 16 at paras 205, 210.

30 1991 Espoo Convention, *supra* note 23, Art 1(vi).

31 *Pulp Mills*, *supra* note 16 at para 205.

32 *Ibid* at para 197.

establishing a regime for the use of the river, a shared resource.[33] This regime was fleshed out in a 1975 Statute 'in order to establish the joint machinery necessary for the optimum and rational utilization of the River Uruguay, in strict observance of the rights and obligations arising from treaties and other international agreements in force for each of the parties'.[34] It is in the context of this international cooperation establishing an institutional architecture for joint management of the River Uruguay (the Administrative Commission of the River Uruguay) that the ICJ made its observations regarding notification and EIA. And it was in the context of the 1961 Treaty, and in particular the construction of its 'protect and preserve' provision[35] that the ICJ observed that 'it may now be considered a requirement under general international law to undertake an environmental impact assessment where there is a risk that the proposed industrial activity may have a significant adverse impact in a transboundary context, in particular, on a shared resource'.[36] In essence, the ICJ was applying the due diligence requirement in the flexible and context-specific manner suggested above. The requirement to conduct transboundary EIA here is thus linked to significant adverse impact in a transboundary context upon a shared resource from a (large-scale) industrial project.

As for the trigger for the application of any general procedural obligation of prior notification and consultation in good faith, once again it is that the activities in question may have 'significant adverse transboundary environmental effects' or the existence of an accident or emergency.[37] Under the only international treaty to address transboundary EIA, the 1991 Espoo Convention, where the trigger for assessment is (i) listing in Appendix I and (ii) likely to cause significant adverse transboundary impact, the additional procedural requirements of the Convention of prior notification and consultation presuppose the existence of such impact from the nature, scale and location of the activity to be undertaken. However, even where there is a duty to notify and to consult, such a duty does not entail the obligation to agree nor provide

---

33 Treaty of Montevideo, signed 7 April 1961, 635 UNTS 98 (entered into force 19 January 1966).

34 Statute of the River Uruguay, signed 26 February 1975, 1295 UNTS 340 (entered into force 18 September 1976).

35 Treaty of Montevideo, *supra* note 33, Art 41.

36 *Pulp Mills*, *supra* note 16 at para 204.

37 See e.g., Convention on the Early Notification of a Nuclear Accident, adopted 26 September 1986, 1457 UNTS 133 (entered into force 17 October 1986).

neighbouring States with a veto over the conduct of the activity contended to have adverse transboundary effects.

C. Prevention and Precaution

While the principle of preventive action may be said to be well-established, more controversial is the customary law status of the precautionary principle or approach. Arguments range from lack of normative content to the absence of a uniform understanding of the meaning of the principle, and widely varying consequences of its application depending on the specific context. Whilst it may lack legally binding force as customary international law, its impact may nonetheless be considerable when further concretised in a treaty text (for example, the precautionary principle under the 1995 Fish Stocks Agreement)[38] or used as a 'general guideline' or aid to judicial interpretation of treaty obligations between the parties (for example, as was the case with the concept of sustainable development and the bilateral agreement between Hungary and Slovakia in the *Gabčíkovo-Nagymaros* case).[39]

In a wide-ranging assessment of the environmental consequences of deep seabed mining activities, the Seabed Disputes Chamber of the International Tribunal for the Law of the Sea (ITLOS) in an Advisory Opinion affirmed the obligation of sponsoring States to apply a precautionary approach, relying inter alia on provisions of the Nodules and Sulphides Regulations.[40] It was prepared to go further, however, in noting that 'the precautionary approach is also an integral part of the general obligation of due diligence of sponsoring States, which is applicable even outside the scope of the Regulations'.[41]

Apart from this Advisory Opinion, instances of international judicial recognition of the principle are muted and few. For example, there was

---

[38] Agreement for the Implementation of the Provisions of the United Nations Convention on the Law of the Sea of 10 December 1982 relating to the Conservation and Management of Straddling Fish Stocks and Highly Migratory Fish Stocks, adopted 4 August 1995, 2167 UNTS 3 (entered into force 11 December 2001) (1995 Fish Stocks Agreement), Art 5(c).

[39] See Vaughan Lowe, 'Sustainable Development and Unsustainable Arguments' in Alan E Boyle and David Freestone (eds), *International Law and Sustainable Development: Past Achievements and Future Challenges* (Oxford: Oxford University Press, 1999).

[40] *Responsibilities and Obligations of States Sponsoring Persons and Entities with respect to Activities in the Area,* Advisory Opinion, [2011] ITLOS Rep 10 (*Advisory Opinion on Seabed Activities*) at paras 121–122, 125–127.

[41] *Ibid* at para 131.

no mention of it by the ICJ in the *Gabčíkovo-Nagymaros* case nor was it generally recognised as a principle of customary international law in the *Pulp Mills* case where the Court considered that 'a precautionary approach *may* be relevant *in the interpretation and application of the Statute*' between the parties.[42] This statement is correctly regarded as 'fall[ing] well short of any confirmation as to the requirement of precaution in customary law'.[43] In the *Beef Hormones* case, the WTO Appellate Body found the legal status of the precautionary approach to be uncertain in general international law,[44] and eight years later in the *EC-Biotech* case, a WTO Panel still found its status 'unsettled'.[45]

As a matter of general principle, Principle 15 of the non-binding Rio Declaration is phrased in very general terms and lacks the normative character of a rule of law.[46] Leading authorities have rightly expressed scepticism regarding the potential for the customary law status of the precautionary approach,[47] not least because '[t]here is no clear and uniform understanding of the precautionary principle among States and other members of the international community'.[48] Its application has chiefly been in consequence of treaty or other provisions which set forth the content and manner of its application binding only upon parties to that instrument and in that particular context.[49] Such treaty obligations between the parties requiring the application of the precautionary approach provide the necessary concrete guidance in doing so.

Even if applicable, the precautionary approach does not reverse the burden of proof, either as a matter of general principle, as the ICJ observed in the *Pulp Mills* case[50] or in treaties adopting a precautionary

---

42 *Pulp Mills*, *supra* note 16 at para 164 (emphasis added).

43 Philippe Sands and Jacqueline Peel, *Principles of International Environmental Law*, 3rd edn (Cambridge: Cambridge University Press, 2012) at 224.

44 *European Communities – Measures Concerning Meat and Meat Products* (1998), WTO Doc WT/DS26/AB/R; WT/DS48/AB/R (Appellate Body Report) at paras 120–125.

45 *European Communities – Measures Affecting the Approval and Marketing of Biotech Products* (2006), WTO Doc WT/DS291/R (Panel Report) at para 7.89.

46 1992 Rio Declaration, *supra* note 1.

47 Patricia Birnie, Alan Boyle and Catherine Redgwell, *International Law & The Environment*, 3rd edn (Oxford: Oxford University Press, 2009) at 160–1.

48 Sands and Peel, *supra* note 43 at 222.

49 See e.g., 1995 Fish Stocks Agreement, *supra* note 38; 1996 Protocol to the 1972 Convention on the Prevention of Marine Pollution by Dumping of Wastes and Other Matter, adopted 7 November 1996, 2006 ATS 11 (entered into force 24 March 2006) (1996 Protocol to the 1972 London Convention).

50 *Pulp Mills*, *supra* note 16 at para 164.

approach,[51] nor does it remove the evidentiary burden on a State to prove harm where such is alleged in a transboundary context (for example, Ecuador would have been required in the *Aerial Spraying* case to prove that Colombia's aerial herbicide programme to destroy coca cultivation within Colombian territory had caused harm in Ecuador). Its effect is limited to reducing the evidentiary standard required to prove that an activity poses a risk of harm. As ITLOS noted in its *Advisory Opinion on Seabed Activities*, and in what must currently rank as the high water mark of international judicial recognition of the precautionary approach, '[the due diligence] obligation applies in situations where scientific evidence concerning the scope and potential negative impact of the activity in question is insufficient but where there are plausible indications of potential risks'.[52] Thus, said the Tribunal, '[a] sponsoring State would not meet its obligation of due diligence if it disregarded those risks [as] [s]uch disregard would amount to a failure to comply with the precautionary approach'.[53]

III. SELECTED POLLUTION-SPECIFIC TREATY CONTEXTS[54]

It was noted above that assistance in determining whether the threshold of significant harm has been crossed may be derived from international undertakings by States, for example, to publicise national pollution release and transfer data;[55] or certain types of activities involving radiological, toxic, or highly dangerous substances may *a priori* be deemed significantly harmful.[56] There is of course extensive regulation of pollution by treaty, which serves the purpose, inter alia, of setting forth concrete standards against which State compliance may be assessed. Multilateral regulatory regimes take the place of general principles and the uncertainty, cost and time of international litigation. Marine and

---

[51] See e.g., Stockholm Convention on Persistent Organic Pollutants, adopted 22 May 2001 [2004] ATS, 40 ILM 532 (2001) (entered into force 17 May 2004) (2001 Stockholm Convention); 1995 Fish Stocks Agreement, *supra* note 38.

[52] *Advisory Opinion on Seabed Activities*, *supra* note 40 at para 131.

[53] *Ibid.*

[54] This part draws in part on Catherine Redgwell, 'International Environmental Law' in Malcolm D Evans (ed), *International Law*, 4th edn (Oxford: Oxford University Press, 2014).

[55] See e.g., 2003 Protocol on Pollutant Release, *supra* note 20.

[56] See e.g., *supra* note 21.

nuclear pollution standards and liability are amongst the most developed while air pollution, and protection of the atmosphere more generally, is less well regulated by treaty, and, in the absence of a general global instrument, is more apt to be subject to the general principles noted above. Each will be considered briefly in turn.[57]

A. Marine Pollution

The protection of the marine environment was one of the key issues at the 1972 Stockholm Conference, and is clearly reflected in the flurry of lawmaking in this area which occurred in the early 1970s in particular. The negotiation of UNCLOS, which commenced in 1973, likewise had an influence upon, and was influenced by, these developments.[58] Thus, Part XII of the resulting UNCLOS with its 46 Articles devoted to the marine environment implicitly acknowledges existing marine environment treaties in the areas of dumping at sea[59] and vessel-source pollution in particular.

Pollution of the oceans, and concerns about their limited absorption capacity, formed a key thrust of these 1970s lawmaking activities: both the global 1972 Convention on the Prevention of Marine Pollution by Dumping of Wastes and Other Matter (1972 London Dumping Convention)[60] and the regional 1972 Convention for the Prevention of Marine Pollution by Dumping from Ships and Aircraft[61] (1972 Oslo Dumping Convention) date from this period, as does the 1973 International

---

[57] For further details, see Robert Beckman, Günther Handl and Alan Boyle's chapters in this book (Chapters 6, 8 and 9 respectively).

[58] See UNCLOS III, 1973–1982, accessed 5 January 2015 at http://legal.un.org/diplomaticconferences/lawofthesea-1982/lawofthesea-1982.html.

[59] For analysis of the relationship between UNCLOS and both prior and subsequent treaties regulating ocean dumping, see Catherine Redgwell, 'From Permission to Prohibition: the 1982 LOSC and Protection of the Marine Environment' in David Freestone, Richard Barnes and David Ong (eds), *The Law of the Sea: Progress and Prospects* (Oxford: Oxford University Press, 2006); and, more generally, Catherine Redgwell, 'Mind the Gap in the GAIRS: The Role of Other Instruments in LOSC Regime Implementation in the Offshore Energy Sector' (2014) 29(4) *International Journal of Marine and Coastal Law* 600.

[60] Convention on the Prevention of Marine Pollution by Dumping of Wastes and Other Matter, adopted 29 December 1972, 1046 UNTS 120 (entered into force 30 August 1975) (1972 London Dumping Convention).

[61] Convention for the Prevention of Marine Pollution by Dumping from Ships and Aircraft, adopted 15 February 1972, 932 UNTS 5 (entered into force 7 April 1974).

Convention on the Prevention of Pollution from Ships, as modified by the Protocol of 1978 (MARPOL 73/78).[62] The United Nations Environment Programme (UNEP), established after the Stockholm Conference, undertook the establishment of regional seas programmes for which the 1976 Barcelona Convention for the Protection of the Mediterranean Sea against Pollution,[63] and subsequent protocols, formed the prototype for many other regional seas areas (only the Baltic, the North-east Atlantic/North Sea, Antarctic and Arctic oceans are regulated by regional instruments outside UNEP's programme). The Protocols to the Barcelona Convention range in subject matter from pollution caused by dumping, land-based sources, and seabed activities, to cooperation in emergencies, specially protected areas, and transboundary movement of hazardous waste.[64] The Convention was substantially revised in 1995 following the Rio Conference, the impact of which is observable in many other areas of regulatory activity where such 'second generation agreements' have been adopted.

Dumping is one area of regulatory activity where this progression is particularly marked. Initially, regional and global dumping conventions adopted the regulatory approach of listing prohibited, dangerous, and other substances and relying on nationally-implemented licencing schemes for their enforcement. Coastal States exercised the jurisdiction afforded them under international law to do so: territoriality and nationality of the vessel. There was no presumptive general ban operating in respect of all dumping at sea, with the regulatory approach essentially one of 'permitted unless prohibited'. Wastes were divided into three categories: Annex I, the 'black list', contained a list of substances prohibited from dumping; Annex II, the 'grey list', those substances the dumping of which required a prior special permit; for all other substances, Annex III (the 'white list') required a prior general permit.[65] However, with the replacement of the 1972 Oslo Convention and 1974

---

[62] Protocol relating to the 1973 International Convention on the Prevention of Pollution from Ships, adopted 17 February 1978, 1340 UNTS 61 (entered into force 2 October 1983) (MARPOL 73/78).

[63] Barcelona Convention for the Protection of the Mediterranean Sea against Pollution, adopted 16 February 1976, 1102 UNTS 27 (entered into force 12 February 1978) (1976 Barcelona Convention).

[64] For the list of the protocols to the 1976 Barcelona Convention, see http://www.unepmap.org/index.php?module=content2&catid=001001001 (accessed 5 January 2015).

[65] 1972 London Dumping Convention, *supra* note 60.

Paris Convention for the Prevention of Marine Pollution from Land-Based Sources by the 1992 Convention on the Protection of the Marine Environment of the North-east Atlantic (OSPAR)[66] and the negotiation of the 1996 Protocol to the 1972 London Convention,[67] this philosophy has been replaced by a 'prohibited unless permitted' approach. Dumping is not permitted unless it falls within one of the permitted exceptions to a general ban on dumping.[68] The list is quite restrictive, including bulky matter such as dredged material and sewage sludge and (at least for the moment) offshore installations.[69] The precautionary approach is very much in evidence here, with a significant shift in the burden of proof to the polluter to demonstrate that dumping at sea will not have significant harmful consequences for the marine environment. There has also been a noticeable shift away from an approach based on pollution prevention and control, the focus of UNCLOS, to a more integrated approach to marine protection under these instruments.[70]

Vessel-source pollution, though a relatively minor contributor to marine pollution, has nonetheless been subject to extensive international regulation following highly publicised oil spills such as the *Torrey Canyon* in 1967.[71] MARPOL 73/78 adopts a framework approach in that six accompanying annexes embody the technical details of regulating discharges of the harmful substances: oil (Annex I), noxious liquids in bulk (Annex II), harmful packaged substances (Annex III), sewage (Annex IV), garbage (Annex V) and air pollution (Annex VI).[72] This last is the only *global* instrument addressed to air pollution, albeit only from

---

[66] Convention on the Protection of the Marine Environment of the North-east Atlantic, adopted 22 September 1992, 2354 UNTS 67 (entered into force 25 March 1998) (OSPAR).

[67] 1996 Protocol to the 1972 London Convention, *supra* note 49.

[68] *Ibid*, Art 4.

[69] Recent (October 2013) amendments were adopted addressed to marine geo-engineering, introducing a new Article 6bis and Annexes 4 and 5.

[70] For general discussion, see Yoshifumi Tanaka, 'Zonal and Integrated Management Approaches to Ocean Governance: Reflections on a Dual Approach in International Law of the Sea' (2004) 19(4) *International Journal of Marine and Coastal Law* 483.

[71] For further information about the Torrey Canyon oil spill, see, e.g., Patrick Barkham, 'Oil Spills: Legacy of the Torrey Canyon', *The Guardian* (24 June 2010) accessed 5 January 2015 at http://www.theguardian.com/environment/2010/jun/24/torrey-canyon-oil-spill-deepwater-bp.

[72] For the list of annexes to MARPOL 73/78, see http://www.imo.org/About/Conventions/ListOfConventions/Pages/International-Convention-for-the-Prevention-of-Pollution-from-Ships-(MARPOL).aspx (accessed 5 January 2015).

ships. Any State wishing to become a party to MARPOL 73/78 must also adopt Annexes I and II as a minimum.[73] Annex I reflects a particular preoccupation with vessel-source oil pollution, focusing on limiting discharges of oil as part of routine tanker operations. It sets forth oil discharge and tanker design criteria[74] and requires coastal States to provide adequate reception facilities for oily residues. Some maritime areas have been designated as 'special areas' under MARPOL 73/78 in which no discharge is permitted, including areas such as the North-west Atlantic Ocean and Mediterranean Sea areas, and the Antarctic Southern Ocean.[75]

A key regulatory device under MARPOL 73/78 is the use of standardised International Oil Pollution Prevention Certificates, the issuance of which is linked to regular surveying and inspection of vessels.[76] This is supplemented by a requirement for tankers and other ships to carry an Oil Record Book itemising all operations involving oil.[77] In addition to the requirement that this record may be inspected by any other party to MARPOL 73/78, it is these documents which in certain circumstances coastal and port States are entitled to inspect under UNCLOS, whether or not a party to MARPOL 73/78.[78] Coastal States' enforcement powers are also enhanced under UNCLOS, including the power to investigate, inspect, and, in limited circumstances, to arrest vessels navigating in the Exclusive Economic Zone (EEZ) when a violation of applicable international rules and standards for the prevention, reduction and control of pollution from vessels, including MARPOL 73/78, has occurred which threatens or causes environmental damage.[79] Apart from prompting the regulation of marine pollution through routine operational discharges at sea, the oil tanker catastrophes of the 1960s and 1970s in particular led to the development of separate liability conventions, the 1969 International

---

[73] MARPOL 73/78, *supra* note 62, Art 14.

[74] Thus, constituting the 'generally accepted international rules or standards', which coastal States may apply to third party shipping traversing the territorial sea fully in accordance with UNCLOS, *supra* note 3, Art 21(2). See further Redgwell, 'Mind the Gap in the GAIRS: The Role of Other Instruments in LOSC Regime Implementation in the Offshore Energy Sector', *supra* note 59.

[75] See International Maritime Organization, Special Areas under MARPOL, accessed 5 January 2015 at http://www.imo.org/OurWork/Environment/Pollution Prevention/SpecialAreasUnderMARPOL/Pages/Default.aspx.

[76] MARPOL 73/78, *supra* note 62, Annex I, Regs 5–8.

[77] *Ibid*, Annex I, Reg 20.

[78] UNCLOS, *supra* note 3, Art 218.

[79] *Ibid*, Art 220, especially paras (3), (5), and (6).

Convention on Civil Liability for Oil Pollution Damage (1969 CLC)[80] and the 1971 International Convention on the Establishment of an International Fund for Compensation for Oil Pollution Damage (1971 Fund Convention).[81] These follow a pattern found in the nuclear civil liability conventions, which is to limit and channel liability.[82] Risk is assumed by, and liability is channelled through, the shipowner, which is strictly liable for oil pollution damage as defined under the conventions. A monetary ceiling on liability is fixed, enabling the shipowner to obtain insurance cover in the market through Protection & Indemnity Clubs. If this ceiling is exceeded under the 1969 CLC, then the 1971 Fund Convention may provide a further source of compensation funds for claimants. The fund is based on levies from oil-importing countries, thus spreading the risk between the shipowner and the risk-creating States. The advantage for claimants in States party to the CLC/Fund regime is the ability to obtain compensation swiftly without recourse to the courts in their own or another State. Both the 1969 CLC and the 1971 Fund Convention were updated by protocols in 1992, with an increase in the compensation limits (further increased in 2000 and 2003), an extension of geographic coverage (to incidents in the EEZ) and the inclusion of pollution prevention costs in the recoverable heads of damage.[83] In 2001, a further convention was concluded on pollution damage from bunker fuel from ships, the International Convention on Civil Liability for Bunker Oil Pollution Damage.[84] Modelled on the 1969 CLC, it entered into force in 2008.

Further responses to catastrophic oil spills are reflected in the 1969 International Convention Relating to Intervention on the High Seas in Cases of Oil Pollution Casualties, which permits the coastal State to

---

[80] International Convention on Civil Liability for Oil Pollution Damage, adopted 29 November 1969, 973 UNTS 3 (entered into force 19 June 1975) (1969 CLC).

[81] International Convention on the Establishment of an International Fund for Compensation for Oil Pollution Damage, adopted 18 December 1971, 11 ILM 284 (entered into force 16 October 1978).

[82] See Part III.B of this chapter.

[83] Protocol to Amend the International Convention on Civil Liability for Oil Pollution Damage, adopted 27 November 1992, 1956 UNTS 255 (entered into force 30 May 1996).

[84] International Convention on Civil Liability for Bunker Oil Pollution Damage, 23 March 2001, 2009 ATS 14 (entered into force 21 November 2008).

intervene beyond its territorial sea[85] and the 1990 International Convention on Oil Pollution Preparedness, Response and Cooperation which requires, inter alia, the preparation of emergency response plans for oil spill incidents.[86] These were international legislative responses to the 1967 *Torrey Canyon* and 1989 *Exxon Valdez* oil spill emergencies respectively. In recognition of the fact that the threat to the marine environment from vessel-source pollution extends beyond petroleum, in 1973 the International Convention relating to Intervention on the High Seas in Cases of Oil Pollution Casualties was extended by protocol to apply to hazardous substances other than oil[87] and, in 2000, the OPRC was extended by Protocol[88] to cover hazardous and noxious substances.[89]

Despite this extensive legislative activity, significant gaps remain in marine environmental protection. In addition to the problem of declining fish stocks, the two most notable gaps are the absence of robust global regulation of the largest source of marine pollution, land-based pollution, and international regulation of the marine environment in areas beyond national jurisdiction. The former is the subject of weak provisions in UNCLOS and addressed in protocols to four UNEP regional seas agreements, and in the 1992 Convention on the Protection of the Marine Environment of the Baltic Sea Area,[90] but there is no present prospect of further regulation, not least because regulating this source of marine pollution most heavily intrudes on a wide range of human activities under the sovereignty of States.

---

[85] Convention Relating to Intervention on the High Seas in Cases of Oil Pollution Casualties, adopted 29 November 1969, 970 UNTS 211 (entered into force 6 May 1975).

[86] International Convention on Oil Pollution Preparedness, Response and Cooperation, adopted 30 November 1990, 1891 UNTS 51 (entered into force 13 May 1995).

[87] Protocol Relating to Intervention on the High Seas in Cases of Pollution by Substances Other Than Oil, adopted 2 November 1973, 1313 UNTS 4 (entered into force 30 March 1983).

[88] Protocol on Preparedness, Response and Co-operation to Pollution Incidents by Hazardous and Noxious Substances, adopted 14 March 2000, 2003 ATNIF 9 (entered into force 14 June 2007).

[89] The 1996 International Convention on Liability and Compensation for Damage in Connection with Carriage of Hazardous and Noxious Substances by Sea (HNS), also modelled on the 1969 CLC, has not yet entered into force.

[90] Convention on the Protection of the Marine Environment of the Baltic Sea Area, adopted 9 April 1992, 2009 UNTS 197 (entered into force 17 January 2000).

In contrast, marine protection in areas beyond national jurisdiction is part of the remit of the Ad Hoc Open-ended Informal Working Group to study issues relating to the conservation and sustainable use of marine biological diversity beyond areas of national jurisdiction, established by UN General Assembly resolution in 2006.[91] The necessity for this gap-filling is occasioned by the failure of the UNCLOS to include the precautionary principle, marine protected areas management, or the ecosystem approach, gaps not surprising in a text reflecting the state of environmental law at the time. Amongst other things, marine genetic resources, and conservation and management tools such as area-based management and environmental impact assessment, have been under discussion by the Working Group.[92] In January 2015, it recommended that a Preparatory Commission be established with the view to drafting a text for a Diplomatic Conference to consider a further implementation agreement to fill this high seas 'marine biodiversity gap'.[93]

B. Nuclear Pollution[94]

The nuclear sector has been the subject of considerable regulatory activity at the international level.[95] Given that nuclear energy activities

---

91 See Ad Hoc Open-ended Informal Working Group to study issues relating to the conservation and sustainable use of marine biological diversity beyond areas of national jurisdiction, accessed 5 January 2015 at http://www.un.org/depts/los/biodiversityworkinggroup/biodiversityworkinggroup.htm.

92 *Ibid.*

93 See http://www.un.org/Depts/los/biodiversityworkinggroup/documents/AHWG_9_recommendations.pdf (accessed 10 February 2015); for background, see remarks by Nulifer Oral, '1982 UNCLOS +30: Confronting New Complexities in the Protection of Biodiversity and Marine Living Resources in the High Seas' in *Proceedings of the 106th Annual Meeting of the American Society of International Law* (Washington DC: ASIL, 2012) at 403; and David Freestone, 'Governance of Areas beyond National Jurisdiction: An Unfinished Agenda of the 1982 Convention?' in Jill Barrett and Richard Barnes (eds), *UNCLOS at Thirty* (London: BIICL, forthcoming in 2015) (working title).

94 This section draws upon Catherine Redgwell and Lavanya Rajamani, 'Energy Underground: What's International Law Got to Do with It?' in Donald N Zillman, Aileen McHarg, Adrian Bradbrook and Lila Barrera-Hernandez (eds), *The Law of Energy Underground: Understanding New Developments in Subsurface Production, Transmission, and Storage* (Oxford: Oxford University Press, 2014), Section C.

95 For a succinct overview, see Chapter 9, 'Nuclear Energy and the Environment' in Birnie, Boyle and Redgwell, *supra* note 47 at 488–534.

are ultra-hazardous in character with potentially devastating transboundary implications, this degree of international regulatory activity is unsurprising.[96] The first international agreements, at both global and regional level, were concerned with regulating liability and compensation for nuclear damage largely with a view to rendering a fledgling energy industry commercially viable. They also reflected the generally unsatisfactory nature of the general customary international law principles in respect of State responsibility as well as the magnitude of the potential harm to humans and the environment.

At the regional level, the 1960 Paris Convention on Third Party Liability in the Field of Nuclear Energy[97] and the 1963 Brussels Supplementary Convention,[98] were adopted under the auspices of the Nuclear Energy Agency of the Organisation for Economic Co-operation and Development (OECD). These are designed to elaborate and harmonise legislation relating to third-party liability to ensure that compensation is paid to persons suffering nuclear damage, whilst at the same time ensuring that the development and use of nuclear energy remains commercially feasible. Channelling and limiting liability achieves both of these aims, with the nuclear operator initially liable, then the State of operation and, finally, all contracting parties. Financial limits for liability, which is strict, are set for each level. In 2004, amendments to the Paris/Brussels regime (not yet in force) were made which increase

---

[96] Apart from the nuclear sector, there has also been considerable regulation of hazardous activities and substances, focusing on environmentally sound management and regulation of the transboundary movement of such substances. Three instruments are of particular importance in this area: (i) the 1989 Basel Convention on the Transboundary Movement of Hazardous Wastes and their Disposal was the first to occupy the field and influenced the scope and application of two other instruments with which it is closely linked; (ii) the 1998 Rotterdam Convention; and (iii) the 2001 Stockholm Convention. A fourth instrument, the 2000 Cartagena Protocol to the 1992 Convention on Biological Diversity, is addressed to the transboundary movement of living modified organisms, on the hazardous character of which there is as yet no international consensus. In 2013, the Minamata Convention on Mercury was concluded, and is brought within the Basel framework through, inter alia, the definition of mercury as 'waste' for the purposes of the latter. For further discussion, see Catherine Redgwell, 'International Environmental Law' in Malcolm Evans (ed), *International Law*, 3rd edn (Oxford: Oxford University Press, 2014) 688, at 710–14.

[97] Convention on Third Party Liability in the Field of Nuclear Energy, adopted 29 July 1960, 956 UNTS 251 (entered into force 1 April 1968).

[98] Convention Supplementary to the Paris Convention of 29th July 1960 on Third Party Liability in the Field of Nuclear Energy, adopted 31 January 1963 (entered into force 4 December 1974).

compensation levels, enhance operator liability, and enlarge their geographic scope.[99] The definition of nuclear damage thereunder includes not only personal injury and damage to property, but also economic losses such as the cost of preventive measures and of measures to reinstate an impaired environment, which may constitute the major portion of the damage caused by a nuclear incident.[100] However, there has not yet been a serious test of these provisions within Europe. The Chernobyl incident in 1986 did not fall under the Paris/Brussels regime since, as a non-OECD member, the Soviet Union was not a party to the Convention (nor is the Russian Federation now) and the Conventions do not apply to incidents outside States Parties. In 1988, a Joint Protocol was concluded which links this regional system with the similar global regime, the 1963 Vienna Convention on Civil Liability for Nuclear Damage,[101] with a further 1997 Supplemental Convention (in force 2003) increasing the amount of compensation available for nuclear damage.[102]

These global instruments have been concluded under the auspices of the International Atomic Energy Agency (IAEA), the UN body established in 1957 to facilitate the peaceful development of nuclear energy and adherence to non-proliferation safeguards. Following the Chernobyl accident in 1986, the IAEA concluded two further international treaties addressed to early notification (1986 Convention on Early Notification of a Nuclear Accident or Radiological Emergency) and to assistance in the event of an international nuclear incident (1986 Convention on Assistance in the Case of a Nuclear Accident or Radiological Emergency). The former embraces in treaty form the customary law obligation to notify in the event of significant transboundary harm.

In a more preventive vein, a further IAEA convention, the 1994 Convention on Nuclear Safety, is designed to ensure the safe operation of

---

[99] Protocol to Amend the Paris Convention, adopted 12 February 2004 (awaiting sufficient signatures to be entered into force) accessed 5 January 2015 at http://www.oecd-nea.org/law/paris_convention.pdf; Protocol to Amend the Brussels Supplementary Convention on Nuclear Third Party Liability, adopted 12 February 2004 (awaiting sufficient signatures to be entered into force) accessed 5 January 2015 at http://www.oecd-nea.org/law/brussels_supplementary_convention.pdf.

[100] *Ibid*, Art I.B (both Protocols).

[101] Joint Protocol relating to the Application of the Vienna Convention and the Paris Convention, adopted 21 September 1988, 1672 UNTS 293 (entered into force 27 April 1992).

[102] Convention on Supplementary Compensation for Nuclear Damage, adopted 12 September 1997, INFCIRC/567 (entered into force 4 October 2003).

land-based nuclear power plants.[103] The effectiveness of this instrument came under scrutiny following the Fukushima Daiichi nuclear plant incident in Japan in 2011.[104] An extraordinary meeting of the parties to the Convention was held in 2012 to consider the implications of the Fukushima incident for the Convention, which identified several key weaknesses including the non-binding peer review mechanism for national reports submitted under the Convention.[105] Procedural changes have been instituted to strengthen peer review and transparency, which unlike treaty amendment may have immediate effect, with a working group established to consider further strengthening through amendment of the Convention.[106]

Safety and security issues form an increasingly important area of the IAEA's activities with the object, inter alia, of protecting humans and the environment from harmful radiation exposure. In 1997, a further Joint Convention on the Safety of Spent Fuel and Radioactive Waste Management was concluded, modelled on the 1994 Convention, to ensure high safety standards and the prevention of accidents in the disposal of nuclear waste.[107] In many respects, these recent developments of both the OECD and IAEA regimes demonstrate parallels with the regulation of oil pollution liability: preventive concern to ensure that the activity is carried out to minimise or avert pollution consequences, emergency response and notification systems, and a liability regime which increasingly acknowledges the environmental costs of prevention and of reinstatement.

---

[103] Convention on Nuclear Safety, adopted 17 June 1994, 1963 UNTS 293 (entered into force 24 October 1996).

[104] See Selma Kus, 'International Nuclear Law in the 25 Years between Chernobyl to Fukushima and Beyond' (2011) 87 *Nuclear Law Bulletin* 7.

[105] See Second Extraordinary Meeting of the Contracting Parties to the Convention on Nuclear Safety, Final Summary Report, accessed 5 January 2015 at http://www.iaea.org/Publications/Documents/Conventions/cns-summaryreport 310812.pdf.

[106] For the outcome of the Second Extraordinary Meeting of the Contracting Parties to the Convention on Nuclear Safety, see http://www.iaea.org/newscenter/news/2012/cnsconcludes.html (accessed 5 January 2015). This agreement to further treaty revision was reinforced at the Sixth Review Meeting of the CNS in 2014 with the decision to hold a Diplomatic Conference in 2015: see https://www.iaea.org/newscenter/news/parties-nuclear-safety-convention-complete-sixth-review-meeting https://www.iaea.org/newscenter/news/parties-nuclear-safety-convention-complete-sixth-review-meeting (accessed 10 February 2015).

[107] Joint Convention on the Safety of Spent Fuel Management and on the Safety of Radioactive Waste Management, adopted 29 September 1997, INFCIRC/546 (entered into force 18 June 2001).

Despite the extent of legislative activity at the global and regional levels, a significant gap remains in the legal regime for disposal of radioactive waste. In addition to the recognition of the Antarctic continent as the first 'nuclear waste free zone',[108] it is widely recognised that the disposal of radioactive waste on or below the seabed is prohibited by international law. This flows either from specific treaty prohibition[109] or from the more general treaty and customary international law duty to protect the marine environment,[110] with which such disposal is regarded as inconsistent. This leaves States with the exclusive option of land-based disposal, though suitable land-based disposal sites may be at a premium or non-existent.[111] International law continues to play a role in the regulation of safe disposal of nuclear material, largely through the work of the IAEA. Amongst other issues, it is unclear whether the existing nuclear liability regime would extend to geological or subsurface disposal of high-level waste and spent nuclear waste.[112] The first geological repositories for high-level nuclear waste and spent fuel are expected to be

---

108 Antarctic Treaty, adopted 1 December 1959, 402 UNTS 71 (entered into force 23 June 1961), Art 5.

109 See e.g., 1972 London Dumping Convention, *supra* note 60, Art IV(1)(a); 1996 Protocol to the 1972 London Convention, *supra* note 49, Art 4(1); OSPAR, *supra* note 66, Art 3, Annex II.

110 See e.g., UNCLOS, *supra* note 3, Art 192.

111 For example, the state utility Taiwan Power Company has since 1982 stored nuclear waste from its three nuclear power plants at a facility on Orchid Island (Lanyu), a small volcanic island off the coast of the densely populated main island of Taiwan. There is considerable public opposition to the Lanyu facility, and efforts have been exerted by Taipower to locate a suitable offshore (land-based) facility, including reportedly storage in an abandoned coal mine in North Korea. This attracted strong international criticism from, inter alia, South Korea and the US. In March 2013, 16 years of negotiation pursuant to a 1996 agreement ended with, inter alia, the refusal by the Taiwanese Atomic Energy Council to grant Taipower an export licence. North Korea is now reportedly suing for breach of contract. 'North Korea suing Taipower', *Taipei Times* (2 March 2013) at 1.

112 For further discussion, see Catherine Redgwell and Lavanya Rajamani, 'Energy Underground: What's International Law Got to Do with It?' in Zillman, McHarg, Bradbrook, and Barrera-Hernandez (eds), *supra* note 94; and, generally, IAEA, 'The Long Term Storage of Radioactive Waste: Safety and Sustainability, A Position Paper of International Experts', (2003) at 5, accessed 5 January 2015 at http://www-pub.iaea.org/MTCD/Publications/PDF/LTS-RW_web.pdf.

commissioned by 2025 in the three countries closest to licencing, namely, Finland, France and Sweden.[113]

C. Air Pollution[114]

Unlike the marine context discussed above, there is no single instrument where the general 'law of the atmosphere' is codified, nor is this picture likely to change. In 2013, the ILC adopted the topic of 'Protection of the atmosphere' in its programme of work but with express caveats including that (i) this work would not interfere with political negotiations, including on climate change, ozone depletion and long-range transboundary air pollution; and (ii) the outcome of the work will be draft guidelines which do not seek to impose on existing treaty regimes legal principles not already contained therein.[115]

The *Trail Smelter* case referred to above was an early instance of an inter-State claim arising in respect of the harmful transboundary effects of air borne pollutants. Yet this case involved a single detectable source of air pollution (sulphur dioxide emissions from the smelter) causing quantified harm to health and property; today it would most likely be resolved through transnational litigation by access to national courts. What if the sources of air pollution are far more diffuse and its harmful effects upon the environment are widespread? These were the difficulties confronting States in negotiating the 1979 Convention on Long-Range Transboundary Air Pollution (LRTAP),[116] added to which was the initial scepticism of some parties as to the very existence and nature of the problem. In fact, it remains the only major regional agreement addressed to air pollution, a reflection both of the severity of the problem in Europe (though Canada and the US are also parties)[117] and the difficulty in

---

[113] European Commission and EURATOM, 'Implementing Geological Disposal of Radioactive Waste Technology Platform, Deployment Plan 2011–2016', (June 2012) at 5, accessed 5 January 2015 at http://www.igdtp.eu/index.php/component/docman/doc_details/65-deployment-plan-2011-2016?Itemid=134.

[114] See generally, Phoebe N Okowa, *State Responsibility for Transboundary Air Pollution in International Law* (Oxford: Oxford University Press, 2000).

[115] See http://legal.un.org/ilc/summaries/8_8.htm (accessed 5 January 2015).

[116] Convention on Long-Range Transboundary Air Pollution, adopted 13 November 1979, 1302 UNTS 217 (entered into force 16 March 1983) (LRTAP).

[117] LRTAP is one of five environmental treaties concluded under the auspices of the UN's Economic Commission for Europe, membership of which comprises 56 States within Europe, the CIS and North America. See http://www.unece.org (accessed 5 January 2015).

achieving international regulation of an activity impacting on sovereign energy choices.

The purpose of LRTAP is to prevent, reduce and control transboundary air pollution from both new and existing sources in order to protect human health and the environment from its adverse effects. 'Air pollution' is defined in terms reminiscent of the marine pollution definition in UNCLOS, and includes harm to living resources, ecosystems, and interference with amenities and legitimate uses of the environment.[118] 'Long-range transboundary air pollution' is defined as adverse effects in the jurisdiction of one State resulting from emissions originating in the jurisdiction of another State yet the individual source of which cannot necessarily be distinguished.[119] The framework character of LRTAP resulted in the absence of specific reduction targets in the treaty itself; rather, an environmental monitoring programme was first put in place pursuant to the first Protocol to LRTAP, in order to gather data to assess the extent of the problem, followed by the negotiation of further protocols to reduce emissions of specific air pollutants. In this respect, LRTAP is a good example of the interaction of monitoring and reporting obligations on the one hand and flexible treaty structures, easily adapted to changes in scientific knowledge, on the other. The framework approach of LRTAP has allowed a step-by-step approach by States allowing 'agreement' at the outset even where there is no consensus regarding the concrete steps to be taken to address a particular environmental problem. To date, eight protocols on monitoring and on specific air pollutants with specific reduction targets and timetables have been added to the initial framework convention, significantly expanding the scope of the parties' commitments.[120] Although there are provisions on notification and consultation in cases of significant risk of transboundary pollution, LRTAP does not itself contain provisions on liability nor on compliance. But in 1994 a second sulphur protocol was adopted with a non-compliance procedure based on the experience of the 1987 Montreal Protocol to the Ozone Convention. In 1997, this procedure was subsumed within a broader Implementation Committee established, inter alia, to

---

[118] LRTAP, *supra* note 116, Art 1(a). UNCLOS, *supra* note 3, Article 212 addresses pollution of the marine environment from or through the atmosphere, while MARPOL 73/78, *supra* note 62, Annex VI addresses the prevention and control of air pollution from ships.

[119] *Ibid*, Art 1(b).

[120] For the list and texts of the protocols, see http://www.unece.org/env/lrtap/status/lrtap_s.html (accessed 5 January 2015).

review compliance by parties with the reporting obligations of the convention and any non-compliance with the protocols.[121]

IV. CONCLUSION

One of the realities of the modern world is that States authorise a wide range of activities with the potential to cause significant transboundary harm. While there is no requirement of 'zero transboundary impact' at international law, which would be unrealistic and incapable of being met by any State, developed or developing, there are both substantive and procedural obligations which penetrate within the State in the exercise of control over activities which cause, or are likely to cause, significant transboundary harm and which in the exercise of due diligence must be subject to internal environmental controls, including EIA, to ensure the protection of humans and of the environment. These derive both from general principles of international environmental law and more specifically from the host of bilateral, regional and global instruments for pollution control one object of which is prevention of transboundary harm. Treaty-based rules and standards can provide concrete standards against which State compliance may be assessed, including obligations of result, and an institutional framework for cooperation. Marine and nuclear pollution standards and liability are amongst the most developed with multilateral regulatory regimes reducing reliance on general principles and the uncertainty (and cost and time) of international litigation. Air pollution – and protection of the atmosphere more generally – is less well regulated by treaty and in the absence of negotiated agreement is, for good or for ill, more apt to be subject to the general principles noted above.

[121] Decision 1997/2, 'The Implementation Committee, its structure and functions and procedures for review of compliance Annex: Structure and functions of the Implementation Committee and procedures for review of compliance (Annex III)', 1997 EB Decision: ECE/EB.AIR/53. The Implementation Committee's terms of reference were most recently updated in 2012 (Decision 2012/25). For reports of the non-compliance cases considered by the Implementation Committee, see http://www.unece.org/env/lrtap/ic/reports.html (accessed 5 January 2015).

2. Transboundary pollution in a global economy: general principles and problems in practice

Simon Tay*

I. INTRODUCTION

A central truism about international law and the environment is the tension between the ecological reality of our interconnected world and the political structures of international governance that focus on sovereign States. This surfaces vividly in concerns about cross-border pollution.

What one State does or allows to be done within its territory in exercise of its State sovereignty can and often does result in environmental consequences beyond its borders. These international 'externalities' often result in pollution that impacts other States even where there are insufficient international or regional systems to address the issue and 'internalise the externalities'.

In response to this, international law – while recognising the rights of State sovereignty – has evolved to declare the principle of responsibility for transboundary environmental harms. First codified as Principle 21 of the 1972 Stockholm Declaration,[1] this is often said by scholars to have achieved the status of customary international law, binding on all nations.[2]

There are, however, challenges and limits to this imposition of State responsibility. As discussed in this chapter, there are problems in the

* The author gratefully acknowledges the assistance of Ms Park Eunjeong, researcher at the Singapore Institute of International Affairs (SIIA).

[1] Stockholm Declaration on the Human Environment, in 'Report of the United Nations Conference on the Human Environment', UN Doc A/CONF.48/14, at 2 and Corr.1 (1972), 11 ILM 1416 (1972 Stockholm Declaration).

[2] See e.g., Catherine Redgwell's chapter in this book (Chapter 1).

record of practice of this principle – and practice remains the essential cornerstone of customary international law. There are, additionally, new concerns and complexities that emerge about transboundary pollution from the accelerating realities of economic interdependence.

The liberalisation of the world economy and the freer movement of goods and capital have been assisted not only by technology but changes in laws at the international, regional and national levels. The world economy is not fully borderless, but it has trended in that direction from the 1990s, albeit not always consistently or smoothly. While this has significant benefits, concerns have also arisen – most prominently about the social and environmental costs. Consequently, environmentally conscious political and social groups have been directing their concerns not only to environmental bodies but often demonstrating their concern and opposition to economic institutions and summits.

A considerable amount of academic literature has grown about the economics and politics of international trade and transboundary pollution. One branch of this literature, beginning from the mid-1990s, analyses trade and environment issues. With regard to transboundary environmental concerns, the studies on trade-environment issues go beyond the visible and immediate impacts of pollution and the traditional transboundary pollution analysis. For instance, questions of production in one country are raised, noting the flows of capital and goods within and among countries.

At the political level, statements of concern have been issued and some general parameters of the trade-environment agenda have been established. These concerns and studies have not, however, crystallised in specific reforms to the World Trade Organization (WTO) or other international bodies relevant to the trade-environment nexus. There is no trade-environment equivalent to Principle 21. Indeed, the issues have not been sufficiently conjoined so that we can see and appreciate how traditional transboundary pollution issues fit together with the newer trade-environment concerns.

That is the aim of this chapter. It argues that we can and should view transboundary pollution not only through the more simple and longer-established lens of environmental harms that are traceable physically from one country to another, but also through economic interdependencies.

This chapter will start by considering the traditional Principle 21 approach to cross-border pollution that has emphasised State responsibility. This will include not only the arguments that this Principle is now enshrined as customary international law, but also the problems of its practice, which countervail that status.

The chapter goes on to then consider whether the norm of State responsibility is and indeed should be transformed into one of cooperation. Where approaches to transboundary pollution have been more successful, this chapter argues that any effort to hold the polluting State responsible for the impact has been augmented and indeed supplanted by other means – including monitoring, funding, capacity-building and other means of assistance. The chapter postulates that we are in flux – moving from an emphasis on responsibility to one of cooperation.

Part IV will consider how trade-environment analyses and economic interdependence may necessitate new approaches and thinking about how international law can be used to combat transboundary pollution. The chapter argues that rather than focusing on the law between States, international legal approaches should also consider the national law of the States involved and the actors below the level of the State (for example, non-governmental organisations, corporations), are necessary. A few concluding comments will then be offered briefly about what these analyses may mean for the Association of Southeast Asian Nations (ASEAN) in dealing with the recurring issue of fires and haze.

II. PRINCIPLE 21 AND OTHER NEIGHBOURLY DUTIES: HONOURED IN NEGLECT

Principle 21 of the 1972 Stockholm Declaration proclaims that:

> States have, in accordance with the Charter of the United Nations and the principles of international law, the sovereign right to exploit their own resources pursuant to their own environmental policies, and the responsibility to ensure that activities within their jurisdiction or control do not cause damage to the environment of other States or of areas beyond the limits of national jurisdiction.[3]

The Principle reaffirms the sovereign rights of the State, but the flip side to this is to recognise State responsibility not to cause damage to the environment of other States. State responsibility for harm caused to another State is not without precedent in public international law generally and in the environmental domain.

It is trite law that the Declaration in itself does not establish international law. The venerable *Trail Smelter* case between the United States

[3] 1972 Stockholm Declaration, *supra* note 1, Principle 1.

and Canada in 1938 is often cited in support of arguments that this principle is observed in the practice of States.[4] This held that 'no state has the right to use or permit the use of its territory in such a manner as to cause injury by fumes in or to the territory of another or the properties or persons therein, when the case is of serious consequence and the injury is established by clear and convincing evidence'.[5]

Principle 21 has been reiterated (with some change) in the 1992 Rio Declaration, as Principle 2.[6] Additionally, Principle 21 has been embodied in some national laws and may, to this extent, be regarded as a general principle of law.[7] Academic writers also provide a number of general legal principles to buttress Principle 21, such as 'good neighbourliness' and the right to territorial integrity. Many emphasise that, arising from the norm of State responsibility, States also have the obligation to take preventive actions, to give notice and consult,[8] and to take measures to abate the harm. Principles 18 and 19 of the Rio Declaration encapsulate preventive and procedural 'soft law obligations'.

Principle 18 requires States to 'immediately notify other States of any natural disasters or other emergencies that are likely to produce sudden harmful effects on the environment of those States'.[9] Principle 19 states that 'States shall provide prior and timely notification and relevant information to potentially affected States on activities that may have a significant adverse transboundary environmental effect and shall consult with those States at an early stage and in good faith'.[10]

---

4 *Trail Smelter Case (United States of America, Canada)* (1938 and 1941) 3 RIAA 1905 *(Trail Smelter).*

5 *Ibid* at 1965.

6 Rio Declaration on Environment and Development, in 'Report of the United Nations Conference on Environment and Development', 12 August 1992, UN Doc A/CONF.151/26 (Vol I), 31 ILM 874 (1992 Rio Declaration). The 1972 Stockholm Declaration and the 1992 Rio Declaration, although declared by the majority of States at the conferences are not, by their nature, binding international treaties; they are 'soft' laws in that they set out stated general norms.

7 See Clean Air Act, USC § 7401 et seq (1963).

8 *Lake Lanoux Arbitration (France v Spain)* (1957) 12 RIAA 281 (French); with English translations at (1961) 24 ILR 101; (1959) 53 *American Journal of International Law* 156 (*Lake Lanoux Arbitration*). The case involved a dispute over French plans to develop the river on their side of the border, affecting the river flow to Spain. The Spanish argued that they had a right to veto the French plans. The finding, however, only emphasised a duty of France to consult Spain.

9 1992 Rio Declaration, *supra* note 6.

10 *Ibid.*

Notwithstanding these bases for the principle of State responsibility and ancillary duties, a number of doubts surround their certainty and sufficiency in application. For example, establishing legal responsibility would be difficult without a threshold for actionable transboundary damage, established by either custom or treaty.[11] Second, legal recourse becomes difficult because it depends on States to bring claims against each other and, because international jurisdiction over States is seldom compulsory, it requires those States to have access to a recognised court or tribunal.[12] Additionally, recourse for violations of the principle of State responsibility comes after damage has occurred, rather than preventively. It cannot, therefore, displace the need for a system of regulation that prevents harm from happening. Some observers dispute the remedies available under international law, with some believing that international law provides for monetary compensation between States or for an order to cease the wrongful act.[13] Some also argue that States bear strict liability for harm, while others suggest that a State need only take reasonable steps within its abilities to address the problem.[14]

Consequently, despite Principle 21 and other antecedents mentioned, subsequent practice by States has not consistently upheld and applied State responsibility to incidents of transboundary pollution. This has been the case even in high profile and controversial examples. Notably, no State brought suit in response to the Russian Chernobyl nuclear reactor incident in 1985 or to the Sandoz Spill in the Rhine in 1986.[15]

---

[11] For a critical overview of the principle of State responsibility for transboundary pollution, see Oscar Schachter, *International Law in Theory and Practice* (Dordrecht: Martinus Nijhoff Publishers, 1991) 362–87. See also Patricia Birnie and Alan Boyle, *International Law and the Environment* (Oxford: Oxford University Press, 1992) 158–60.

[12] The Statute of the International Court of Justice, Art 36, is the prime example of non-compulsory jurisdiction. In contrast, the World Trade Organization has established a compulsory and relatively effective system of dispute settlement over trade and trade-related issues. See Ernst-Ulrich Petersmann, *The GATT/WTO Dispute Settlement System: International Law, International Organizations and Dispute Settlement* (London: Kluwer Law International, 1997).

[13] Birnie and Boyle, *supra* note 11 at 152–3 (suggesting that fines can be levied).

[14] Principle 21 of the 1972 Stockholm Declaration and Principle 2 of the 1992 Rio Declaration do not use the term strict liability. They also recognise States' sovereignty and their right to set their own environmental and developmental policies.

[15] See generally Philippe Sands, *Chernobyl: Law and Communication* (Cambridge: Grotius, 1988) 1–47; Ellen B Moynagh, 'The Legacy of Chernobyl: Its

The haze pollution that has recurred across parts of Southeast Asia since the mid-1990s has not resulted in arbitration or other efforts to strictly uphold State responsibility – notwithstanding the acceptance of the Indonesian leader that his government accepted 'moral responsibility'.[16]

In the case of the nuclear incident at Fukushima in 2011, no country at present has sought to hold Japan legally responsible in international law for any transboundary impacts. This is despite the fact that many nearby residents and also some US Navy personnel are suing the company that operated the plant, the Tokyo Electric Power Co.[17]

Even as Principle 21 is upheld, some commentators note that there are gaps and a lack of detail in Principle 21. The principle acknowledges that States are liable only for significant harm; the level or type of pollution that is 'significant' may be open to debate, especially when the pollution arises from natural incidents or low-level and widespread activity, such as open burning, rather than from a single, man-made incident.[18]

If this is the position of customary international law, what of treaties and especially treaties between neighbours? Even if States are reluctant to accept responsibility to all others, it might be a different matter where it concerns neighbours. This is not only because of proximity and other ties, but because environmental harms are more likely to run in both

---

Significance for the Ukraine and the World' (1994) 21 *Boston College Environmental Affairs Law Review* 709; Aaron Schwabach, 'The Sandoz Spill: The Failure of International Law to Protect the Rhine from Pollution' (1989) 16 *Ecology Law Quarterly* 443; Devereaux F McClatchey, 'Chernobyl and Sandoz One Decade Later: The Evolution of State Responsibility for International Disasters, 1986–1996' (1996) 25 *Georgia Journal of International and Comparative Law* 659.

[16] Indonesian President Suharto issued a statement acknowledging moral responsibility in 1997. See 'Indonesia President Apologizes again for Haze', *The Associated Press* (5 October 1997) accessed 5 January 2015 at http://www.apnewsarchive.com/1997/Indonesia-president-apologizes-again-for-haze/id-7334f7140582dc0e796fb41e77e717f2. President Susilo Bambang Yudhoyono similarly apologised in 2013. See 'Indonesia formally apologizes for smoky haze', *The Associated Press* (25 June 2013) accessed 5 January 2015 at http://www.thejakartapost.com/news/2013/06/25/indonesia-formally-apologizes-smoky-haze.html.

[17] See 'Stakes high as ailing U.S. Navy sailors take on Tepco over Fukushima fallout', *The Japan Times* (10 March 2014) accessed 5 January 2015 at http://www.japantimes.co.jp/community/2014/03/10/issues/stakes-high-as-ailing-u-s-navy-sailors-take-on-tepco-over-fukushima-fallout/#.VB73-C6SzpA.

[18] Oscar Schachter, 'The Emergence of International Environmental Law' (1991) 44 *Journal of International Affairs* 457 at 463–8.

directions. In theory, therefore, regional or bilateral treaties might therefore be useful instruments by which the more general principles of State responsibility for transboundary pollution might be defined and applied to particular sets of relationships.

Yet, very few regional and bilateral treaties on transboundary pollution set out strict liability regimes in terms similar to the international law principles or provide for institutional means to enforce those legal principles of responsibility,[19] with the exception being those treaties that deal with 'ultra-hazardous' activities like oil spills and nuclear accidents.[20] Otherwise, and especially among developing countries, regional and bilateral treaties that are centred on State responsibility are not the norm.

From the above, even if we might wish otherwise, the necessary conclusion is that while Principle 21 and other neighbourly responsibilities for transboundary pollution are honoured on paper, they are largely neglected in practice.

III. FROM THE RHETORIC OF RESPONSIBILITY TO EFFECTIVE COOPERATION

Even if they do not hold fast to Principle 21, this does not mean that regional and bilateral agreements on transboundary pollution are nugatory. Rather than fixating on State responsibility, these efforts have generally centred on cooperative approaches. This is understandable from the perspective of political and international relations between neighbouring States, where there are many links and issues to be addressed, and not just environmental transboundary harms. While respecting and indeed giving some premium to State sovereignty such cooperative treaties generally are premised on goodwill for States to take actions such as:

---

[19] For a review of treaties that provide for strict liability, see Jan Schneider, *World Public Order of the Environment: Towards an International Ecological Law and Organization* (Toronto: University of Toronto Press, 1997) 168–71. For a review of international law on transboundary pollution, see generally Thomas Merrill, 'Golden Rules for Transboundary Pollution' (1997) 46 *Duke Law Journal* 931.

[20] See e.g., Convention on the Liability of Operators of Nuclear Ships, adopted 25 May 1962, 57 *American Journal of International Law* 268; Vienna Conference on Civil Liability for Nuclear Damage and Optional Protocol, adopted 21 May 1963, 1063 UNTS 265 (entered into force 12 November 1977); Convention on Third Party Liability in the Field of Nuclear Energy, adopted 29 July 1960, 956 UNTS 263 (entered into force 1 April 1968).

- advance notification and consultation;
- exchanges of information and research; or
- general undertakings to take 'appropriate measures'.[21]

Cooperative approaches however do not always work. On balance, most of such 'goodwill' cooperation treaties have been ineffective.[22]

Yet, other cooperation-focused treaties on transboundary pollution have proven effective over time, even though some of their efforts were initially flawed and ineffectual and later proved to be effective. For example, the treaties relating to cross-border acid rain included little effective regulation to address the problem at the beginning.[23] Subsequently, however, efforts have been more successful, with revisions to the European Long-Range Transboundary Air Pollution Convention (LRTAP)[24] and, in the United States, the enactment of the Clean Air Act. These have resulted in specific (if arbitrary) reductions in sulphur dioxide emissions.

What factors account for the difference in these cooperation-focused agreements? A look at LRTAP may assist understanding the cooperative approach taken in Europe as a region.[25] Signed in 1979, coming into force in 1984, LRTAP initially addressed acid rain but has become more complicated over time to include other environmental problems across European borders. The aim of the Convention is that Parties, 51 in total

---

[21] See e.g., Agreement for Co-operation in Dealing with Pollution of the North Sea by Oil (with annex), signed 9 June 1969, 704 UNTS 3 (entered into force 9 August 1969); Agreement for Co-operation in Dealing with Pollution of the North Sea by Oil and Other Harmful Substances, signed 13 September 1983, 1605 UNTS 33 (entered into force 1 September 1989).

[22] Merrill, *supra* note 19 at 967.

[23] See Jutta Brunnee, *Acid Rain and Ozone Layer Depletion: International Law and Regulation* (Ardsley, NY: Transnational Publishers, 1988) 81–141.

[24] See Marc Levy, 'European Acid Rain: The Power of Tote-Board Diplomacy', in Peter M Haas, Robert O Keohane and Marc A Levy (eds), *Institutions of the Earth: Sources of Effective International Environmental Protection* (Cambridge, MA: MIT Press, 1993); see also Gregory Wetstone and Armin Rosencranz, 'Transboundary Air Pollution: The Search for an International Response' (1984) 8 *Harvard Environmental Law Review* 89 at 100–7.

[25] LRTAP is one of the most studied international environmental regimes, and most policy evaluators have deemed it a success. The following draws on several sources as quoted elsewhere in these footnotes. However, the author wishes to especially acknowledge a recent academic work, Rolf Lidskog and Göran Sundqvist (eds), *Governing the Air: The Dynamics of Science, Policy, and Citizen Interaction* (Cambridge, MA: MIT Press, 2011).

as of October 2014,[26] shall endeavour to limit and, as far as possible, gradually reduce and prevent long-range transboundary pollution.

LRTAP outlines the general principles of international cooperation for the abatement of air pollution and provides an institutional framework linking science (as the basis for decision-making) and policy. It emphasises research, monitoring and information exchange on air pollution and its effects and developing emission reduction strategies.

The LRTAP abatement strategies were initially negotiated based on a flat-rate reduction basis, requiring all States to reduce emissions of pollutants (specifically, sulphur dioxides and nitrogen oxides), to the same quantum. These strategies were later called the 'first-generation protocols'. In the 1990s, the development of a second generation of protocols started, resulting in the Protocol on Further Reduction of Sulphur Emissions (Second Sulphur Protocol, 1994) and the Protocol to Abate Acidification, Eutrophication, and Ground-Level Ozone (multipollutant/multieffect protocol, 1999). These protocols focused on varying national reduction rates. These differentiations were based not on political bargaining, but on the approach of critical load – that is, effects in relation to what nature can withstand in different settings and ecologies – and cost effectiveness. These later protocols also broadened to include additional pollutive substances.[27]

LRTAP has been extended by eight protocols that identify specific measures to be taken by Parties to cut their emissions of air pollutants.[28]

---

[26] For status of LRTAP, see https://treaties.un.org/pages/ViewDetails.aspx?src=TREATY&mtdsg_no=XXVII-1&chapter=27&lang=en (accessed 5 January 2015).

[27] The multi-pollutant/multi-effect protocol was signed in Gothenburg in November 1999 and regulates four types of compounds (i.e., sulphur dioxide, nitrogen oxides, ammonia and volatile organic compounds) that affect human health, natural ecosystems, materials and crops through acidification, eutrophication and ground-level ozone. Protocol to the 1979 Convention on long-range transboundary air pollution to abate acidification, eutrophication and ground-level ozone, adopted 30 November 1999, 1593 UNTS 287 (entered into force 17 May 2005) (1999 Gothenburg Protocol), Art 2. The signatories have called the 1999 Gothenburg Protocol 'a smart protocol', one of the most sophisticated and most scientifically based protocols ever signed. See Jan Thompson, *Introduction to the Ministerial Declaration on Long-Range Transboundary Air Pollution* (mimeo, 1999).

[28] Within the framework of the 1979 LRTAP, eight protocols have been adopted. For the list and texts of the protocols, see http://www.unece.org/env/lrtap/status/lrtap_s.html (accessed 5 January 2015).

The Convention is supported by the United Nations Economic Commission for Europe (UNECE) as secretariat, with four professionals dedicated to LRTAP. To provide scientific support to the Convention, the Programme for Monitoring and Evaluation of the Long-Range Transmission of Air Pollutants in Europe (EMEP) has been established.

There is also an Executive Body, composed of environmental advisers to UNECE members, which meets annually to review the implementation of the Convention and assess the effectiveness of national policies. The Convention has its own compliance committee, called the Implementation Committee (IC). States are obliged to report annually to the IC. The secretariat can request explanations from countries that do not meet their targets.[29] Self-submission by countries not meeting their obligations is also possible.

It is significant to note that the compliance process is not akin to an arbitration or court proceeding: the aim is not to affix blame and seek recompense but rather to understand the non-compliance and discuss ways to improve the situation. As such, generally speaking, it is reported that States Parties remain open and even keen to explain what problems they face.

There are also factors below the level of the State that explain LRTAP's effectiveness. The 1980s, when LRTAP began, were the heyday of transboundary air pollution as a public issue. The term '*acid rain*' was widely known and of common concern among the public across Europe. Constituencies within key countries – green lobbies but also broader coalitions of citizens affected by the pollution – were aware of the issues and impacts, and moved to bring attention to what governments did, or failed to do.

When citizens recognise that their own health is at stake, they become more motivated to take action and support further abatement strategies. This is particularly true as attention centres on key cities with concentrations of people.

From the above, key elements that contributed to the effectiveness of LRTAP may be summarised as follows:

- **Specific Limits and Reliable Monitoring**: States agree on specific commitments and thresholds that indicate non-compliance. There

---

[29] See United Nations Economic Commission for Europe, 'Structure and functions of the implementation committee and procedures for review of compliance', accessed 5 January 2015 at http://www.unece.org/fileadmin/DAM/env/eia/documents/ImplementationCommittee/IC_functions_en.pdf.

are agreed methods to objectively assess whether or not a State has complied with the agreed commitment.

- **Scientific Understanding**: Research and expert discussions further the understanding of the models for transboundary movement of pollution since systems of dispersal are not always simple, and ecosystems differ in their sensitivity to pollutants.
- **Regional Institutions**: There is a body established outside of the individual States to undertake in monitoring, compliance process and to support further discussions and negotiations. This need not be large but competent and impartial.
- **Domestic Constituencies**: That there are constituencies and stakeholders within each State that are concerned with air pollution and have a voice in the public discussion of these issues that can impact on the government's actions.

These stand in contrast, on the one hand, to the focus on State responsibility in Principle 21 and on the other, to the ineffective 'goodwill' cooperation treaties. The former may be desired but are scant in State practice, especially because the consent of a State can be withheld. The latter is evidenced but ineffectual.

In contrast, the approach seen in LRTAP – where compliance is augmented by clear goals and measurements as well as supranational and sub-national elements of governance – represents something of a middle and workable path. While success may not be guaranteed, the 'institutionalised cooperation' seen in LRTAP would seem to improve the odds – whether or not State responsibility and liability are also included.

IV. TRADE-ENVIRONMENT, ECONOMIC INTERDEPENDENCE AND GLOBAL LAW APPROACHES

The concerns about transboundary pollution and the approaches and treaties surveyed above relate largely to environmental impact that is direct – whether in the air, water or land – and physically evidenced as pollution. The principles are fairly settled even if practice remains at issue. Cooperation is essential and this chapter has sought to distinguish between 'goodwill' cooperation and 'institutionalised' cooperation and make the case that the latter should have a higher likelihood of success. This section of the chapter considers an area that is usually seen to be quite separate: environmental concerns and their linkages to trade and investment.

Trade-environment concerns differ from transboundary pollution in that the environment impact is embodied in the product or in the process of making it, rather than as direct and distinct pollution. For example, the *Trail Smelter* case dealt with the direct and distinct environmental impact of fumes from a factory being carried across the border by wind and clouds and deposited on the other State as acid rain.[30] This fits the traditional matrix of transboundary pollution.

In contrast, the trade-environment matrix examines the product that issues from the factory. The process and method of its manufacture has resulted in the pollution and, as seen through the lens of trade-environment, must be internalised into the product itself to follow the adage of 'internalising the externality'. There is nothing pollutive in the product per se, however.

Using the trade-environment lens allows us to see transboundary pollution in new ways that are not only challenging to long-held precepts, but open up possibilities for dealing with a broader range of global environmental challenges – such as climate change. One important difference in perspective, for instance, is that the trade-environment lens allows us to consider the products and processes that relate to transboundary pollution. This can be especially significant where the pollution arises not from a one-off incident or natural calamity, but from ongoing activities in which humans and corporations are the key actors.

Another difference is that, in comparison to focusing on State responsibility, the trade-environment perspective allows us to look below the level of the State to determine what and who are involved in the activity that results in the transboundary pollution. In contrast, traditional views of State responsibility for transboundary pollution use international law operating at the level between States. In this respect, the trade-environment perspective can be most useful where the emphasis is given to institutionalised cooperative measures. Thus, in LRTAP, the domestic political process was key to its success, where segments of the public pushed their own government to address the issues.

The trade-environment lens also links with efforts to identify the economic activities and products that drive the transboundary pollution. Thus, in the case of LRTAP and acid rain, specific approaches were the focus, whereby different power generators and factories were compared and aligned so that efforts could be harmonised across borders. Less positively, the trade-environment lens would also lead some to consider imposing additional taxes and/or sanctions on the products involved in

---

30 *Trail Smelter*, *supra* note 4.

the transboundary pollution. While trade concerns may inveigh against such measures taken unilaterally, such considerations may have merit insofar as they seek to internalise the externality of pollution into the cost of the products.

In this regard, the trade-environment perspective seeks to match the paradigm of our ecological interdependence with a paradigm that recognises the increasingly borderless flow of trade and investment. In comparison, the traditional approach for transboundary pollution recognises and indeed depends upon State sovereignty and limits its concern to what happens beyond the borders of that State.

Trade-environment issues have, however, been controversial in the international community. In a number of cases, States unilaterally impose limits or bans on products from other States that are seeking to access their markets. Such attempts draw criticism and legal challenge vis-à-vis trade obligations. The *Tuna-Dolphin* case was an early and highly controversial example of this, which did not bode well for reconciling trade and environmental rules and concerns.[31]

The more recent *Shrimp-Turtle* case has been received more positively in terms of the basis for the decision as stated by the Appellate Body of the WTO.[32] Nevertheless, the controversy remains and the vast majority of trade-environment cases that have gone to the WTO for settlement have been resolved in favour of trade interests.

Despite this, it is important to recognise how the trade-environment perspective has continued and, indeed, evolved alongside the State responsibility approach. Take, for instance, the emerging regime for climate change.

The United Nations Framework Convention on Climate Change process is being negotiated in terms of State responsibility, whereby a State assumes responsibility for carbon dioxide and other climate change gases emitted from within its boundaries. At the same time, however, there are private sector efforts to track, monitor and limit the climate change impacts of their production, trade and commerce. For instance, the American supermarket chain, Wal-Mart, has begun to undertake this effort (although not without criticism) and thus to bring climate change questions to the many companies in its supply chain.[33]

---

[31] See WTO, 'Mexico etc versus US: "tuna-dolphin"', accessed 5 January 2015 at http://www.wto.org/english/tratop_e/envir_e/edis04_e.htm.

[32] See WTO, 'India etc versus US: "shrimp-turtle"', accessed 5 January 2015 at http://www.wto.org/english/tratop_e/envir_e/edis08_e.htm.

[33] See Jennifer Inez Ward, 'Missed targets: when companies fail to keep their key sustainability promises', *The Guardian* (21 July 2014) accessed 5 January

This emphasis on private supply chains cuts across boundaries, without much concern for the sovereign territory in which those suppliers are based. Rather, the carbon embedded in the products – wherever produced – is the focus. Such an approach seeks to accommodate both the global nature of the environmental challenge of climate change and the ability of producers to shift from one State to another.

V. CONCLUDING OBSERVATIONS

Chapters 11 and 12 of this book will address in some detail specific issues relating to the Southeast Asian haze and fires. The focus of this chapter is instead on the more general principles that apply to transboundary pollution. It is nevertheless necessary to offer some brief observations in conclusion on how these general principles may be seen to apply to the ASEAN region's challenge of haze and fires.

Since the fires of 1994 and even more after 1997–98, ASEAN has made efforts to address the haze and fires. At the start, these efforts centred on Regional Haze Action Plans and the Sub-Regional Ministerial Steering Committee. While not without effort or any impact, these plans can be described, in terms of the general principles discussed in this chapter, as 'goodwill cooperation'. Accordingly, while these have been acceptable to Indonesia, they have not been as effective as hoped, so that the problems have continued and recurred over these past 15 years.

A second stage of the regional effort was the 2002 ASEAN Agreement on Transboundary Haze Pollution (2002 ASEAN Haze Agreement), which has been ratified by all ASEAN Member States and has come into force.[34] This treaty has been criticised for its lack of 'teeth'.[35] However, it does iterate Principle 21 and the principle of State responsibility. It is for this reason that the Indonesian Parliament delayed its agreement to ratify the 2002 ASEAN Haze Agreement for 12 years. In this regard, we may observe that while Principle 21 may seek to balance sovereignty with responsibility, the emphasis of some governments remains on sovereignty.

2015 at http://www.theguardian.com/sustainable-business/blog/2014/jul/21/sustainability-goals-promise-broken-failure-target-walmart-disney.

[34] See Adella Anjani Putri, 'Indonesia Ratifies Haze Treaty after 12 Years of Stalling', *The Jakarta Globe* (17 September 2014) accessed 5 January 2015 at http://www.thejakartaglobe.com/news/indonesia-ratifies-haze-treaty-12-years-stalling/.

[35] See Alan Khee-Jin Tan's chapter in this book (Chapter 12).

As ASEAN governments continue to try to address the haze challenge, it remains to be seen if methods of 'institutionalised' cooperation will be adopted. One step in this direction was the agreement among environment ministers to share satellite information and overlay these on maps that show the landowners. This step increases the monitoring shared among the States. However, other elements – such as clear limits – have not been discussed and agreed.

The time might be ripe to consider the emergent trade-environment perspective. Actors below the level of the State have become more evident. Environmental activists in Indonesia have been active on the issue, increasingly over the last decade. The concerns of citizens and residents in Singapore and Malaysia were reignited when the pollution levels hit a record high of 401 PSI (Pollution Standards Index) in Singapore in June 2013.[36] Much of this has led to concerns and accusations not against the government of Indonesia (which apologised for the pollution) but against companies operating in Indonesia and accused of using fire to clear land or owning those lands affected by fires. The focus on corporations moreover was not overly swayed by whether the owners were Indonesian nationals or Malaysian or Singaporean. This suggests an emerging approach that aligns with the trade-environment perspective in which concerns include actors below the level of the State, and what happens within the State.

Many decision-makers still ponder how to reconcile the growing complexity of an issue with the need for public awareness, inclusion and engagement.

---

[36] See Charissa Yong, 'Haze update: PSI 401 at noon; many pharmacies still out of masks', *The Straits Times* (21 June 2013) accessed 5 January 2015 at http://www.straitstimes.com/breaking-news/singapore/story/singapore-haze-update-psi-401-noon-many-pharmacies-still-out-masks-201#sthash.qbIuovSs.dpuf.

3. Unpacking the elements of a state responsibility claim for transboundary pollution

Jacqueline Peel

I. INTRODUCTION

The legal principles governing State responsibility for harm, and their application to situations of transboundary pollution, are now relatively well-settled as a matter of international law. The International Law Commission's (ILC) Draft Articles on Responsibility of States for Internationally Wrongful Acts (Draft Articles on State Responsibility),[1] while they have not given rise to a treaty, are widely regarded as codifying customary international law on the topic.[2] Moreover, transboundary pollution disputes, such as the *Trail Smelter* arbitration[3] and the *Gabčíkovo-Nagymaros* case,[4] form some of the main precedents from

---

1 'State Responsibility' in 'Report of the International Law Commission covering the work of its fifty-third session, with commentaries, 2001' (UN Doc A/56/10), Ch IV, in *Yearbook of the International Law Commission 2001*, vol II, Part Two (UN, 2001) (Draft Articles on State Responsibility).

2 Alain Pellet, 'The ILC's Articles on State Responsibility for Internationally Wrongful Acts and Related Texts' in James Crawford, Alain Pellet and Simon Olleson (eds), *The Law of International Responsibility* (Oxford: Oxford University Press, 2010) 75 at 87. See also UNGA Report of the Secretary-General, 'Compilation of Decisions of International Courts, Tribunals and other Bodies', A/62/62 (2007) in which Annex I references 129 cases in which the Draft Articles on State Responsibility (or the draft on first reading) and the commentaries have been cited.

3 *Trail Smelter Case (United States of America, Canada)* (1938 and 1941) 3 RIAA 1905 (*Trail Smelter*).

4 *Case concerning the Gabčíkovo-Nagymaros Project (Hungary/Slovakia)*, [1997] ICJ Rep 7 (*Gabčíkovo-Nagymaros*).

which the ILC derived its Draft Articles.[5] Nonetheless, substantial uncertainty continues to surround the application of State responsibility principles in practice to instances of transboundary pollution involving States. The limited number of transboundary pollution cases brought or decided on the basis of State responsibility doctrines leaves open key questions such as whether the polluting actions of private entities are attributable to States; the nature and extent of a State's duty to prevent transboundary pollution and how to demonstrate a causal link between a State's activities and injuries suffered as a result of transboundary pollution.[6]

This chapter seeks to unpack the elements of a State responsibility claim for transboundary pollution in order to focus on these questions pertaining to practical application of the relevant international rules. The chapter first outlines the general elements of a State responsibility claim for transboundary pollution before turning to three particular aspects that pose the most complex issues of proof and application: (1) attribution of private polluting conduct to a State; (2) the standard of care required of a State in acting to prevent transboundary pollution; and (3) proving causation of harm. Analysis of these elements of a State responsibility transboundary pollution claim is grounded in the specific example of greenhouse gas (GHG) pollution leading to harm through climate change. While GHG pollution is not a conventional situation of transboundary pollution (which more often involves a bilateral dispute where one State's

---

[5] James Crawford, *The International Law Commission's Articles on State Responsibility: Introduction, Text and Commentaries* (Cambridge: Cambridge University Press, 2002).

[6] This chapter focuses on some of the key practical difficulties encountered in applying State responsibility rules in transboundary pollution scenarios. For more detailed analysis of State responsibility rules and transboundary harm, see Rebecca M Bratspies and Russell A Miller (eds), *Transboundary Harm in International Law: Lessons from the Trail Smelter Arbitration* (New York: Cambridge University Press, 2006); Günther Handl, 'International Accountability for Transboundary Environmental Harm Revisited: What Role for "State Liability"?' (2007) 37(2–3) *Environmental Policy and Law* 116; Phoebe N Okowa, *State Responsibility for Transboundary Air Pollution in International Law* (Oxford: Oxford University Press, 2000); Roda Verheyen, *Climate Change Damage and International Law: Prevention Duties and State Responsibility* (Leiden: Martinus Nijhoff Publishers, 2005); Xue Hanquin, *Transboundary Damage in International Law* (Cambridge: Cambridge University Press, 2003).

actions cause harm across borders to another State(s)),[7] climate change amplifies the practical difficulties that will be encountered in applying State responsibility rules to instances of environmental harm.[8] As such, it provides a useful case study for testing the capacity of State responsibility rules to be applied in a range of 'real life' situations of transboundary pollution.

II. ELEMENTS OF A STATE RESPONSIBILITY CLAIM FOR TRANSBOUNDARY POLLUTION

The Draft Articles on State Responsibility establish a straightforward schema of the legal elements necessary to found a State responsibility claim for transboundary pollution. In the first instance, the complainant State(s) must be able to point to an 'internationally wrongful act' on the part of another State.[9] According to Article 2 of the Draft Articles, an internationally wrongful act of a State consists of conduct, whether by way of an action or omission, that (1) is attributable to the State under international law, and (2) constitutes a breach of an international obligation of that State.

A. Attribution

In many cases of transboundary pollution, the harm results not from actions of the State per se or its authorities but from the activities of private individuals, groups or entities, including corporations. For example, GHG pollutants are released from a wide variety of private and commercial activities such as farming, forestry, transportation and energy production.[10] Under State responsibility rules, purely private actions cannot be attributed to the State and hence cannot form the basis of a

---

[7] For this reason, some authors argue that the 'no-harm' rule prohibiting transboundary environmental damage is inapplicable to climate change. Verheyen, *supra* note 6 at 166–7, citing Meinhard Schröder, 'Klimaschutz als Problem des internationalen Rechts', in Rüdiger Breuer, Michael Kloepfer, Peter Marburger and Meinhard Schröder (eds), *Jahrbuch des Umwelt und Technikrechts* (Heidelberg: Decker, 1993) 191 at 196.

[8] For an extensive discussion of these issues, see Verheyen, *supra* note 6; see also Christina Voigt, 'State Responsibility for Climate Change Damages' (2008) 77 *Nordic Journal of International Law* 1.

[9] Draft Articles on State Responsibility, *supra* note 1, Art 1.

[10] Data by country and sector can be found on the United Nations Framework Convention on Climate Change website, accessed 6 January 2015 at

State responsibility claim.[11] States may only be held accountable for private conduct in circumstances where private actors are 'under the direction or control' of the State in carrying out the conduct concerned.[12] Whether private polluting conduct is attributable to the State in which it takes place, or in other words, whether it takes place under the State's direction or control, is thus often an important practical question in establishing a State responsibility claim for transboundary pollution.

B. Breach of an International Obligation

The second element of an internationally wrongful act – establishing breach of an international obligation of the State – can also pose difficulties of application in practice. Identifying a relevant international obligation is generally straightforward. A State may be under certain international obligations as a result of a treaty to which it is party. For instance, under the Kyoto Protocol, parties included in Annex I to the UN Convention on Climate Change were required to achieve specific emissions reductions from 1990 levels over the first commitment period running from 2008–2012.[13] Provided that other requirements are met, '[a] breach of these very specific quantified emission limitation and reduction commitments could thus be considered a breach of a treaty obligation that potentially could give rise to State liability'.[14] Alternatively, in the absence of suitable treaty obligations, a complainant State may fall back on the general customary international law duty of States to prevent transboundary harm to the environment of other States or to areas of the global commons.[15]

---

http://unfccc.int/ghg_data/items/3800.php. Adding to the complexity, emissions of GHG are both spread out in time and space.

[11] Draft Articles on State Responsibility, *supra* note 1, Commentary to Art 8, para 1, reproduced in Crawford, *supra* note 5 at 110.

[12] Draft Articles on State Responsibility, *supra* note 1, Art 8.

[13] Kyoto Protocol to the UN Framework Convention on Climate Change, adopted 10 December 1997, 37 ILM 22 (entered into force 16 February 2005) (Kyoto Protocol), Art 3.1 and Annex B.

[14] Michael G Faure and André Nollkaemper, 'International Liability as an Instrument to Prevent and Compensate for Climate Change' (2007) 43A *Stanford Journal of International Law* 123 at 144.

[15] The customary law status of this obligation was confirmed by the ICJ in *Legality of the Threat or Use of Nuclear Weapons*, Advisory Opinion, [1996] ICJ Rep 226 (*Legality of Nuclear Weapons*) at 242. For further discussion of this obligation, see Catherine Redgwell's chapter in this book (Chapter 1).

In most cases alleging a breach of environmental obligations, States will not be held to a standard of strict liability.[16] More often, the required standard of care to be exercised by States in pursuance of their environmental obligations is one of 'due diligence'.[17] The due diligence requirement is an obligation of conduct, that is, for a State to 'take all appropriate measures'[18] or 'do the best it can'[19] to prevent harm and minimise transboundary pollution risks. As such, compliance with the due diligence standard can only be assessed in light of the circumstances of each particular case and taking into account the capacities and capabilities of the States concerned. Consideration of the relative technological and economic capacity of the State involved, whether the responsible State had the opportunity to take preventive action, the foreseeability of harm and the proportionality of the measures chosen to reduce the harm, will all be factors relevant to this determination.[20] For States that are developing countries, issues surrounding their capacity to enact, implement and enforce environmental standards to prevent transboundary pollution are likely to be critical in the assessment of due

---

[16] Limited exceptions apply in the case of ultra-hazardous activities. See Wilfred Jenks, 'The Scope and Nature of Ultra-Hazardous Liability in International Law' (1966) 117 *Recueil des Cours* 99. The relevant standard required is determined by the primary obligation with which the State has failed to comply.

[17] Organisation for Economic Co-operation and Development (OECD), Environment Committee, 'Responsibility and Liability of States in Relation to Transfrontier Pollution' (1984) at 4; see also 'International Liability for Injurious Consequences Arising out of Acts Not Prohibited by International Law (Prevention of Transboundary Harm from Hazardous Activities)' in 'Report of the International Law Commission to the General Assembly covering the work of its fifty-third session, with commentaries, 2001' (UN Doc A/56/10), Ch V, in *Yearbook of the International Law Commission 2001*, vol II, Part Two (UN, 2001) (Draft Articles on Prevention), Art 3, para 7; Riccardo Pisillo-Mazzeschi, 'Forms of International Responsibility for Environmental Harm' in Francesco Francioni and Tullio Scovazzi (eds), *International Responsibility for Environmental Harm* (London: Graham & Trotman, 1991) 15.

[18] Draft Articles on Prevention, *supra* note 17, Art 3. See also *Case concerning Pulp Mills on the River Uruguay (Argentina v Uruguay)*, [2010] ICJ Rep 14 (*Pulp Mills*) at para 101 ('use all the means at its disposal').

[19] Voigt, *supra* note 8 at 11.

[20] See Draft Articles on Prevention, *supra* note 17, Commentary to Art 3. For discussion of the no-harm rule in the climate change context, see Verheyen, *supra* note 6 at 145–92.

diligence.[21] In this respect, the international environmental principle of 'common but differentiated responsibilities' may also weigh into the assessment of whether particular conduct breaches an international obligation of the State.[22] Therefore, determining whether a State has acted with due diligence to prevent transboundary harm in any particular case will frequently involve a complicated factual evaluation, which will depend heavily upon context.

C. Reparation for Injury Caused by the Breach

In the event that a complainant State(s) is able to establish an internationally wrongful act on the part of another State, that other State will bear responsibility for the act under international law,[23] barring the existence of any defences, such as necessity or *force majeure*.[24] As a consequence, the responsible State will be under an obligation to make 'full reparation' for the injury 'caused' by the internationally wrongful

---

[21] See e.g., Alan Khee-Jin Tan, 'Forest Fires of Indonesia: State Responsibility and International Liability' (1999) 48 *International and Comparative Law Quarterly* 826.

[22] Principle 7 of the Rio Declaration on Environment and Development states 'States shall cooperate in a spirit of global partnership to conserve, protect and restore the health and integrity of the Earth's ecosystem. In view of the different contributions to global environmental degradation, States have common but differentiated responsibilities. The developed countries acknowledge the responsibility that they bear in the international pursuit to sustainable development in view of the pressures their societies place on the global environment and of the technologies and financial resources they command'. Rio Declaration on Environment and Development, in 'Report of the United Nations Conference on Environment and Development', 12 August 1992, UN Doc A/CONF.151/26 (Vol I), 31 ILM 874 (1992 Rio Declaration). For discussion of the principle, including its application in the context of climate change, see Lavanya Rajamani, 'The Principle of Common but Differentiated Responsibility and the Balance of Commitments under the Climate Regime' (2000) 9 *Review of European Community and International Environmental Law* 120; Christopher D Stone, 'Common but Differentiated Responsibilities in International Law' (2004) 98 *American Journal of International Law* 276.

[23] Draft Articles on State Responsibility, *supra* note 1, Art 1.

[24] In Chapter V of the Draft Articles on State Responsibility, such defences are known as 'circumstances precluding wrongfulness' and include consent, self-defence, countermeasures, *force majeure*, distress, necessity and compliance with peremptory norms. If a defence can be established by a defendant State, then no internationally wrongful act occurs. A defence of *force majeure* may be relevant in instances such as the Fukushima nuclear disaster where an earthquake and consequent tsunami caused flooding of the plant's nuclear reactors.

act.[25] Most commonly the reparation sought in respect of injury due to transboundary pollution is financial compensation given the difficulties involved in restoring polluted environments to ecological health.[26] However, for the injured State, or other State(s) seeking to invoke the responsibility of the responsible State,[27] a more complex issue may be whether the pollution injury complained of was 'caused' by the responsible State's wrongful conduct.

In the commentaries to its Draft Articles on State Responsibility, the ILC makes clear that causation in this sense is specific causation requiring a showing that the wrongful action of the defendant State was the 'proximate' cause of the harm.[28] This task is often complicated in transboundary pollution scenarios by the potential for alternate contributing causes of the harm, for example, natural sources of emissions or pollution from the injured State. Climate change harms provide an apt illustration of the difficulties here given multiple GHG emitters, natural and anthropogenic sources of GHG emissions, and long time lags between the emission of GHG and the manifestation of changes in the

---

[25] Draft Articles on State Responsibility, *supra* note 1, Art 31, codifying the customary law principle established in the *Case concerning Factory at Chorzów* (*Germany v Poland*) (1927), 9 PCIJ (Ser A) No 12 at 21. Reparation can take the form of restitution, compensation and satisfaction, either singly or in combination. Draft Articles on State Responsibility, *supra* note 1, Art 34.

[26] While restitution of the status quo is an alternative remedy to financial compensation, it will not be required of a responsible State where it is materially impossible and imposes burdens disproportionate to the benefit derived. Draft Articles on State Responsibility, *supra* note 1, Art 35. Calculating appropriate compensation for environmental damage is also far from straightforward (see Philippe Sands and Richard B Stewart, 'Valuation of Environmental Damage – US and International Approaches' (1996) 5 *Review of European Community and International Environmental Law* 290) although the practice of the UN Compensation Commission offers valuable experience in this regard. See further Peter H Sand, 'Compensation for Environmental Damage from the 1991 Gulf War' (2005) 35 *Environmental Policy and Law* 244.

[27] The Draft Articles on State Responsibility make provision for the invocation of responsibility by either the injured State or a third State. In the latter instance, Article 48 limits the circumstances of invocation to situations of breaches of collective obligations or obligations *erga omnes*. The third party invoking responsibility may only claim from the responsible State cessation of the internationally wrongful act and performance of the obligation of reparation in the interest of the injured State or beneficiaries of the obligation breached.

[28] Crawford, *supra* note 5, commentary to Art 31, para 10, at 204–5.

climate.[29] As a general matter, detailed scientific evidence will be necessary in order to prove a link between activities in the responsible State and transboundary harms in other States. Uncertainties in relevant scientific knowledge or the potential for injury to manifest only over the long term may raise additional issues as to the role and relevance of international environmental principles, such as the precautionary principle and the principle of inter-generational equity.[30]

III. ATTRIBUTING POLLUTING CONDUCT TO THE STATE

Transboundary pollution often involves the accumulated harms of a multitude of actions taken by both State and non-State actors. This is certainly the case for GHG pollution but also for other air, land and water pollution problems that stem from emissions from various point and non-point sources. Indeed, unless there is a very large pollution event originating from a single source (for example, the Fukushima nuclear accident), individual instances of pollution are unlikely to reach the requisite threshold of 'seriousness' or 'significance' of harm necessary to give rise to a breach of international environmental obligations.[31] This means that a complainant State(s) will often be faced with the task of seeking to attribute cumulative harms – due to the actions of multiple polluters – to a single responsible State.

---

[29] As Faure and Nollkaemper note, this also raises the issue of the time point at which emissions become unlawful, which is connected to the issue of responsibility for historical GHG emissions. Faure and Nollkaemper, *supra* note 14 at 171–2.

[30] See Part III, below. See also Richard J Lazarus, 'Super Wicked Problems and Climate Change: Restraining the Present to Liberate the Future' (2009) 94 *Cornell Law Review* 1153; Richard Lord, Silke Goldberg, Lavanya Rajamani and Jutta Brunnée (eds), *Climate Change Liability* (Cambridge: Cambridge University Press, 2012); Philippe Sands and Jacqueline Peel, *Principles of International Environmental Law,* 3rd edn (Cambridge: Cambridge University Press, 2012), Ch 6.

[31] For discussion of the threshold at which environmental damage becomes actionable at international law see Sands and Peel, *ibid* at 708–11; Patricia Birnie, Alan Boyle and Catherine Redgwell, *International Law & the Environment*, 3rd edn (Oxford: Oxford University Press, 2009) at 186–8.

A. Attributing Private Conduct to a State

As discussed above, the Draft Articles on State Responsibility generally exclude the actions of private actors from conduct considered to be 'attributable to the State under international law'.[32] Where transboundary pollution derives principally from the actions of private entities with no direct connection to a State, questions arise as to how far responsibility can be attributed to that State for injuries caused to other States. In the case of climate change, for instance, this poses a problem as the GHG that give rise to harm are generally emitted by private entities rather than States (although in some countries there remains significant State ownership of GHG-emitting facilities such as power stations). In addition, emissions of GHG are diffuse and not limited to stationary sources such as industrial and electric facilities that are more readily identifiable and controllable by States.

One approach to dealing with the attribution problem in a State responsibility transboundary pollution claim could lie in drawing an indirect link between the polluting activities of private actors and a State. Article 8 of the Draft Articles on State Responsibility extends the concept of attributable conduct from situations where individuals or organisations are acting on behalf of the State or exercising elements of the governmental authority,[33] to situations where persons or entities are acting 'under the direction and control of that State in carrying out the conduct'. In general, States use licencing or permitting systems to regulate the release of environmental pollutants, arguably bringing such activities under the 'control' of the State.[34] Accordingly then, 'the failure to stop, reduce or regulate emitting activities with due care' could be the basis for attributing private emitting conduct to States.[35] However, in the commentaries to Article 8, the ILC stressed that the necessary degree of 'control' exercised by the State in order for private conduct to be attributable to it must be such that the State 'directed or controlled the specific operation and the conduct complained of was an integral part of that operation'.[36]

---

[32] Draft Articles on State Responsibility, *supra* note 1, Art 2.

[33] See *ibid*, Arts 4 and 5.

[34] Richard S J Tol and Roda Verheyen, 'State Responsibility and Compensation for Climate Change Damages – a Legal and Economic Assessment' (2004) 32 *Energy Policy* 1109 at 1111–12.

[35] *Ibid*.

[36] Crawford, *supra* note 5, Commentary to Art 8, para 3. No environmental examples are cited by the ILC in its commentary, which instead deals with cases arising in the fields of international humanitarian law and diplomatic protection.

This test may be difficult to satisfy in circumstances where the only link to a State consists of the issue of a pollution permit by the latter.

A more plausible strategy may be to shift analysis from the attribution issue to the conduct of the State itself. Here the question is not whether the State (indirectly) controlled the actions of a private polluter but rather whether the State's responsibility is engaged by its failure to control adequately the activities of private entities that give rise to harm beyond borders.[37] In this case, the question becomes one not of whether private conduct is attributable to the State but instead whether the State discharged its duty of due diligence, a matter considered further in Part IV below.

The most difficult problems will arise in those scenarios where polluting conduct is the result of illegal actions by private entities. For instance, smoke haze caused by illegal logging and burning of forested areas by individual farmers or corporate plantation owners is generally the result of purely private conduct with no direct link to the State.[38] On the other hand, it could be argued that widespread illegal logging is the result of failures on the part of the State to implement and enforce biodiversity protection and air pollution prevention measures as required under international treaties, such as the Convention on Biological Diversity,[39] the ASEAN Agreement on Transboundary Haze Pollution,[40] and customary international law. In these scenarios, assessment of the attribution question bleeds into an evaluation of whether a breach has occurred through the failure of the State to exercise due diligence in carrying out and enforcing the environmental laws on its statute books. Questions regarding the technical capacity and financial resources of the

---

[37] See Alan Boyle's chapter in this book (Chapter 9).

[38] Tan, *supra* note 21. See also Alan Khee-Jin Tan's chapter in this book (Chapter 12).

[39] For an overview of this issue, including discussion of illegal logging practices in Indonesia, see Luca Tacconi (ed), *Illegal Logging: Law Enforcement, Livelihoods and the Timber Trade* (Abingdon: Earthscan, 2007).

[40] ASEAN Agreement on Transboundary Haze Pollution, signed 10 June 2002 (entered into force 25 November 2003) accessed 6 January 2015 at http://agreement.asean.org/media/download/20140119170340.pdf (2002 ASEAN Haze Agreement). Indonesia – one of the main countries contributing to transboundary haze in Southeast Asia, recently agreed to ratify the Agreement. See Wahyudi Soeriaatmadja, 'Indonesia's parliament agrees to ratify Asean haze pact', *The Straits Times* (16 September 2014) accessed 6 January 2015 at http://www.straitstimes.com/news/asia/south-east-asia/story/indonesias-parliament-agrees-ratify-asean-haze-pact-20140916.

State concerned will be key in determining whether the requisite standard of due diligence has been met in such circumstances.

B. Shared Responsibility

Transboundary pollution problems frequently raise questions of shared responsibility where the pollution in question results from the actions of more than one State or from the actions of both States and non-State entities.[41] Where large, private entities, such as multinational energy, chemical or other manufacturing companies, are major contributors to the transboundary pollution issue in question, complainant States may wish to consider alternative (or parallel) avenues to a State responsibility claim. For instance, in contemplating claims for climate change damage against major GHG emitters in 2002, the Pacific Island nation of Tuvalu considered separate actions in the International Court of Justice (ICJ) against the United States and Australia, and in US civil courts against US based oil companies.[42] In many cases, the activities of multinational corporations will make a more significant contribution to the pollution problem concerned than any one State(s). For instance, the GHG contribution of large, often multinational emitters (for example, oil companies, power stations, automobile manufacturers) significantly exceeds the collective contribution of many smaller States.[43]

Despite the horizontally and vertically shared nature of responsibility in many instances of transboundary pollution,[44] as a matter of practice, actions against States and private entities seeking to attribute responsibility for transboundary pollution are likely to be brought in different fora. This reflects the jurisdictional limitations of existing venues to address

---

41 André Nollkaemper and Dov Jacobs, 'Shared Responsibility in International Law: A Conceptual Framework' (2013) 34 *Michigan Journal of International Law* 359.

42 Richard C Paddock, 'Tuvalu's Sinking Feeling', *Los Angeles Times* (4 October 2002) accessed 6 January 2015 at http://articles.latimes.com/2002/oct/04/world/fg-tuvalu4.

43 See Matthew F Pawa and Benjamin A Krass, 'Global Warming as a Public Nuisance: Connecticut v. American Electric Power' (2004–2005) 16 *Fordham Environmental Law Review* 407 at 420.

44 Ilias Plakokelfalos, 'Liability for Transboundary Harm', in André Nollkaemper and Ilias Plakokelfalos (eds), *The Practice of Shared Responsibility in International Law* (Cambridge: Cambridge University Press, forthcoming).

all aspects of a shared responsibility claim.[45] In the ICJ, for example, only States may be parties in cases brought before the Court in its contentious dispute settlement jurisdiction.[46] Even in a case where responsibility is shared amongst several States (for example, upriver States responsible for a significant diminution in water quality or quantity in an international watercourse), it may not be possible to bring all responsible States before a single international adjudicator. This situation raises difficult jurisdictional questions for the tribunal involved if 'it would have to rule, as a prerequisite, on the lawfulness'[47] of another State's conduct in the latter's absence and without its consent.[48]

IV. ESTABLISHING A FAILURE OF DUE DILIGENCE

The obligation of States to prevent transboundary harm to the environments of other States or to global commons areas has been repeatedly reiterated in international environmental declarations,[49] endorsed as custom by international courts and tribunals,[50] and adopted as a guiding

---

[45] See André Nollkaemper, 'Procedural Aspects of Shared Responsibility in International Adjudication: Introduction' (2013) 4(2) *Journal of International Dispute Settlement* 277.

[46] Statute of the International Court of Justice, adopted 26 June 1945, 33 UNTS 993 (entered into force 24 October 1945), Art 34(1) (ICJ Statute).

[47] *Case concerning East Timor (Portugal v Australia)*, [2010] ICJ Rep 105 at para 35.

[48] This is known as the *Monetary Gold* principle. *Case of the Monetary Gold Removed from Rome in 1943,* [1954] ICJ Rep 19 at 32. Debate exists in the literature as to whether this principle acts as a jurisdictional bar in cases involving a plurality of responsible States. In the case of *Phosphate Lands in Nauru*, cited by the ILC in the commentary to its Draft Articles on State Responsibility, the ICJ found it had jurisdiction over Australia in a claim brought in relation to the administration of a trusteeship agreement to which the United Kingdom and New Zealand (not before the Court) were also parties. See, however, Faure and Nollkaemper, *supra* note 14 at 167, who reject the argument that this line of jurisprudence will necessarily provide a way around the *Monetary Gold* principle in complex responsibility cases.

[49] E.g., 1992 Rio Declaration, *supra* note 22, Principle 2; Stockholm Declaration on the Human Environment, in 'Report of the United Nations Conference on the Human Environment', UN Doc A/CONF.48/14, at 2 and Corr.1 (1972), 11 ILM 1416 (1972 Stockholm Declaration), Principle 21.

[50] *Legality of Nuclear Weapons*, supra note 15 at para 29; *Pulp Mills*, supra note 18.

objective in a number of multilateral environmental treaties.[51] Despite the centrality of the so-called 'no-harm rule' to international environmental law, the standard of care required by States to fulfil this duty remains exceptionally vague. As the Seabed Disputes Chamber of the International Tribunal for the Law of the Sea (ITLOS) put it recently, due diligence at international law is 'a variable concept'.[52]

A. Requisite Standard of Care

As an obligation of conduct rather than of result, due diligence does not require the attainment of a specific standard in all circumstances. Taking climate change as an example, it is not possible to specify a given level of emissions reduction that would equate to duly diligent action on the part of an individual State to prevent climate change harms.[53] Instead, much will turn on the circumstances of each particular case and the capacities and capabilities of the States concerned.[54] Invoking the principle of 'common but differentiated responsibilities', defendant States

---

[51] See e.g., Convention on Long-Range Transboundary Air Pollution, adopted 13 November 1979, 1302 UNTS 217 (entered into force 16 March 1983); Montreal Protocol on Substances that Deplete the Ozone Layer to the Vienna Convention for the Protection of the Ozone Layer, adopted 22 March 1985, 1513 UNTS 293 (entered into force 22 September 1988).

[52] *Responsibilities and Obligations of States Sponsoring Persons and Entities with respect to Activities in the Area*, Advisory Opinion, [2011] ITLOS Rep 10 *(Advisory Opinion on Seabed Activities)* at para 117.

[53] That said, there have been numerous attempts made to come up with formulae for sharing the burden of emissions reduction between countries. See e.g., Michel G J den Elzen, Nicklas Höhne, Markus Hagemann, Jasper van Vliet and Detlef P van Vuuren, *Sharing Developed Countries' Post-2012 Greenhouse Gas Emission Reductions Based on Comparable Efforts* (Netherlands Environmental Assessment Agency, December 2009); Kelly Levin and Rob Bradley, 'Comparability of Annex I Emission Reduction Pledges' WRI Working Paper (World Resources Institute, 2010) accessed 6 January 2015 at http://pdf.wri.org/working_papers/comparability_of_annex1_emission_reduction_pledges_2009-11-05.pdf; Malte Meinshausen, Nicolai Meinshausen, William Hare, Sarah C B Raper, Katja Frieler, Reto Knutti, David J Frame and Myles R Allen, 'Greenhouse-Gas Emission Targets for Limiting Global Warming to 2°C' (2009) 458 *Nature* 1158.

[54] In its work on the topic of International Liability for Injurious Consequences Arising out of Acts not Prohibited by International Law, the ILC sought to deal with situations of liability where there is no breach of an international obligation of the State giving rise to an internationally wrongful act. For a critique of the Commission's work on this topic see Alan Boyle, 'Globalizing

that are developing countries may also seek to argue that they should be subject to a less stringent standard of care than wealthier States. For instance, in a claim involving injury from transboundary smoke haze, Indonesia might argue that it has done all that could be reasonably expected of it as a developing State to prevent illegal logging practices in its forests given its particular capacities and capabilities. On the other hand, in the event of an oil spill from an installation in the Timor Sea which was authorised under Australian law, Indonesia might well contend that Australia, as a rich developed country, should be held to a higher standard.[55]

It should be noted, however, that formulations of the common but differentiated responsibilities principle, set out in instruments such as the 1992 Rio Declaration[56] and the United Nations Framework Convention on Climate Change,[57] were developed in the context of attributing accountability for responding to environmental problems, with a particular burden of action placed on developed States.[58] How this principle applies in situations of State responsibility for transboundary harm remains unclear. In its Advisory Opinion on the *Responsibilities and*

Environmental Liability: The Interplay of National and International Law' (2005) 17 *Journal of Environmental Law* 3. Work on the liability arm of this topic resulted in the ILC adopting 'Draft Principles on the Allocation of Loss in the Case of Transboundary Harm Arising out of Hazardous Activities, with commentaries, 2006'. 'Report of the International Law Commission at its fifty-eighth session, in 2006, and submitted to the General Assembly as a part of the Commission's report covering the work of that session' (UN Doc A/61/10) in *Yearbook of the International Law Commission, 2006*, vol II, Part Two (UN, 2006). The principles apply in relation to transboundary damage caused by activities 'not prohibited by international law which involve a risk of causing significant transboundary harm through their physical consequences'. *Ibid*, Principle 1. These principles do not give rise to a regime of State liability for such activities but rather encourage States to channel liability to the operator of the relevant activity, which will often be a private entity.

55 See e.g., the 2009 Montara Oil Spill discussed in the Australian Embassy Indonesia Media Release: Montara Oil Spill (2 November 2009) accessed 6 January 2015 at http://www.indonesia.embassy.gov.au/jakt/MR09_086.html.

56 1992 Rio Declaration, *supra* note 22, Principle 7.

57 UN Framework Convention on Climate Change, 29 May 1992, 1771 UNTS 107, reprinted in (1992) 31 ILM 849, Art 3.1.

58 The concept of common but differentiated responsibilities is undergoing rapid evolution, particularly in the context of the international climate change regime. See International Law Association, 'Legal Principles Relating to Climate Change', Washington Conference (2014) accessed 6 January 2015 at http://www.ila-hq.org/en/committees/index.cfm/cid/1029.

Obligations of States Sponsoring Persons and Entities with respect to Activities in the Area (*Advisory Opinion on Seabed Activities*), the Seabed Disputes Chamber of ITLOS (Chamber) found that the general provisions of the UN Convention on the Law of the Sea (UNCLOS) concerning the responsibilities and liabilities of a State sponsoring entities conducting activities in the deep seabed applied equally to all sponsoring States, whether developed or developing.[59] The Chamber saw '[e]quality of treatment between developing and developed sponsoring States' as being:

> consistent with the need to prevent commercial enterprises based in developed States from setting up companies in developing States, acquiring their nationality and obtaining their sponsorship in the hope of being subjected to less burdensome regulations and controls. The spread of sponsoring States 'of convenience' would jeopardise uniform application of the highest standards of protection of the marine environment, the safe development of activities in the [deep seabed] Area and protection of the common heritage of mankind.[60]

In considering the application of the precautionary approach under Principle 15 of the 1992 Rio Declaration, which refers to the approach being applied by States 'according to their capabilities', the Chamber appeared to endorse a differentiated standard, commenting that 'the requirements for complying with the obligation to apply the precautionary approach [under the Nodules and Sulphides Regulations] may be stricter for the developed than for the developing sponsoring States'.[61] Although ITLOS indicated that the reference to different capabilities in Principle 15 of the 1992 Rio Declaration did not apply to State obligations to follow 'best environmental practices' in the deep seabed area,[62] it concluded that 'the reference to "capabilities" is only a broad and imprecise reference to the differences in developed and developing States. What counts in a specific situation is the level of scientific knowledge and technical capacity available to a given State in the relevant scientific and technical fields'.[63]

While the legal hedge – 'it all depends on the circumstances' – may be an accurate representation of the applicable international law on due diligence, it is a frustrating response for States facing substantial injuries from transboundary pollution. Recent case law dealing with State

---

59 *Advisory Opinion on Seabed Activities*, supra note 52 at para 158.
60 *Ibid* at para 159.
61 *Ibid* at para 161.
62 *Ibid.*
63 *Ibid* at para 162.

responsibility in an environmental context offers some basis for refining the general notion of due diligence as a 'variable concept'. For instance, in its Judgment in the *Pulp Mills* case, the ICJ indicated that having laws on the books to prevent environmental harm is, of itself, insufficient to discharge a standard of due diligence. Rather the obligation 'entails not only the adoption of appropriate rules and measures, but also a certain level of vigilance in their enforcement and the exercise of administrative control applicable to public and private operators, such as the monitoring of activities undertaken by such operators'.[64] The context for this statement by the Court was a transboundary water pollution dispute between two developing country neighbours in South America. It suggests that if a similar dispute were to arise in a Southeast Asian context the Court would look to be satisfied as to the robustness of the entire regime a State had in place to prevent environmental harms to other States, including aspects relating to administration, monitoring and enforcement. What level of 'robustness' would be required is more difficult to determine from the statements of the Court. As Professor Alan Boyle discusses in Chapter 9 of this book, insights can be drawn from instances where States have been found to fail the due diligence standard. For instance, if a State regulatory agency responsible for carrying out environmental requirements was systematically underfunded, officers were poorly trained and lacked experience, procedures for inspection and monitoring were rarely carried out, and few enforcement actions were brought, this could provide indications of inadequate 'vigilance' in enforcement and administration by the State.

In its *Advisory Opinion on Seabed Activities,* the Chamber also provided some guidance on interpretation of the obligation of due diligence. In particular, the Chamber drew a strong link between acting with due diligence and acting with precaution. Even where scientific uncertainty exists as to the extent of likely transboundary damage, the Chamber suggested that ignoring 'plausible indications of potential risks' would amount to a failure of due diligence.[65] The Chamber also highlighted the role of evolving scientific knowledge in assessing what amounts to due diligence at a particular point in time. It noted that 'measures considered sufficiently diligent at a certain moment may

---

64 *Pulp Mills, supra* note 18 at para 197.

65 *Advisory Opinion on Seabed Activities*, *supra* note 52 at para 131.

become not diligent enough in light, for instance, of new scientific or technological knowledge'.[66]

B. Due Diligence and Procedural Obligations

Fulfilling the due diligence standard relevant to the prevention of transboundary harm may also entail the adoption of certain procedures by the State authorising an activity with a risk of significant transboundary harm. In its Draft Articles on Prevention of Transboundary Harm from Hazardous Activities, the ILC articulated a number of procedural duties for States contemplating activities 'which involve a risk of causing significant transboundary harm through their physical consequences'.[67] These include requirements of cooperation, notification, consultation and prior risk assessment.[68] The endorsement of such procedural standards in numerous international environmental treaties and other instruments,[69] in addition to the Draft Articles on Prevention, provides strong evidence supporting the relevance of compliance with these standards in assessing the requirement of due diligence in the prevention of transboundary harm. Moreover, if the due diligence standard is treated as an evolving requirement as in the *Advisory Opinion on Seabed Activities*, this could suggest the need for ongoing processes of risk assessment by States to ensure measures to prevent transboundary pollution are, and remain, adequate over time.

Emerging as an essential procedural measure for demonstrating due diligence to prevent transboundary harm is the carrying out of an environmental impact assessment. In the *Pulp Mills* case, the ICJ found that the practice of environmental impact assessment had gained in recent years 'so much acceptance amongst States that it may now be considered a requirement under general international law to undertake an environmental impact assessment where there is a risk that the proposed

66 *Ibid* at para 117, also suggesting the application of a stricter standard for 'riskier' activities.

67 Draft Articles on Prevention, *supra* note 17, Art 1.

68 *Ibid*, Arts 4, 7–9.

69 See e.g., the *1975 Statute of the River Uruguay* which was central to the *Pulp Mills* case. See *supra* note 18; see also 1992 Rio Declaration, *supra* note 22, Principle 17; Convention on Environmental Impact Assessment in a Transboundary Context, adopted 25 February 1991, 1989 UNTS 309 (entered into force 10 September 1997).

industrial facility may have a significant adverse impact in a transboundary context, in particular, on a shared resource'.[70] The Court went on to tie this requirement to the obligation of due diligence, stating 'due diligence, and the duty of vigilance and prevention which it implies, would not be considered to have been exercised if a party planning works liable to affect … the quality of [the river's] waters did not undertake an environmental impact assessment on the potential effects of such works'.[71]

In sum, while the standard of care entailed by due diligence remains highly dependent on contextual factors in a transboundary pollution scenario, a claim alleging breach of that standard would be considerably strengthened by evidence showing plausible risks of harm were ignored, protective laws were on the books but not adequately enforced, or that facilities carrying a significant risk of transboundary harm were authorised without a prior environmental impact assessment.

V. CAUSATION AND THE PROOF OF HARM

Proving a causal link exists between the breach of an international obligation attributable to a State and injury suffered by another State or States presents one of the greatest practical challenges in any State responsibility claim involving transboundary pollution. This problem is especially acute for transboundary pollution problems, such as climate change, with multiple contributors and diffuse harms. However, even in the case of a bilateral transboundary pollution issue, such as arose in the famous *Trail Smelter* arbitration between the United States and Canada, proof of injury that is attributable to the actions of a State often involves the production and assessment of complex scientific evidence.

In *Trail Smelter*, the Arbitral Tribunal, convened in the late 1930s, considered US claims for compensation for damages from atmospheric pollution deriving from the Trail zinc smelter operating across the border in British Columbia, Canada. The Arbitral Tribunal received advice from independent scientists and heard from a variety of experts regarding the links between sulphur dioxide emissions from the smelter and damage to crops, forests, soil and waterways in Washington State.[72] On this basis,

---

70 *Pulp Mills*, *supra* note 18 at para 204.

71 *Ibid.*

72 The two awards issued by the Arbitral Tribunal are reproduced in (1939) 33 *American Journal of International Law* 182 (First Award) and (1941) 35 *American Journal of International Law* 684. Interestingly, the Arbitral Tribunal

the Arbitral Tribunal made findings about the mechanism of atmospheric distribution of the sulphur dioxide gas and the extent of its detrimental effects on property and the environment in Washington State.

More recently, in the *Pulp Mills* case, the parties placed before the ICJ a 'vast amount' of scientific and technical information regarding potential transboundary pollution of the River Uruguay as a result of pollutants emitted from pulp and paper mills on the banks of the River that had been authorised by Uruguay.[73] The Court saw its role as being to give 'careful consideration' to all the evidence placed before it by the parties and on that basis 'to determine which facts must be considered relevant, to assess their probative value, and to draw conclusions from them as appropriate'.[74]

In drawing conclusions about State responsibility based on the available scientific evidence, key questions facing decision-makers will include: (1) how to deal with uncertainties in the scientific information, particularly regarding pollution harms that will only manifest over the long-term; and (2) whether the internationally wrongful act committed by the defendant State can be considered sufficiently proximate to the transboundary pollution injury suffered so as to give rise to an obligation of reparation. The following sections consider each of these issues in turn.

A. Dealing with Uncertain, Long-term Harms

Scientific evidence linking an activity with particular polluting effects may often suffer from information gaps, methodology deficiencies or

was little impressed by the evidence presented by the parties' experts given their conflicting views and the potential for bias in an adversarial setting. First Award, *ibid* at 195–6.

73 *Pulp Mills*, *supra* note 18 at para 165.

74 *Ibid* at para 168. Dissenting Judges, Al Khasawneh and Simma criticised the majority in this regard for not seeking additional expert evidence to supplement the scientific evidence presented by the parties. More recently, the Court has revisited the issue of the standard of review to be applied in evaluating scientific evidence in the case of *Whaling in the Antarctic (Australia v Japan; New Zealand intervening)*, Judgment of 31 March 2014, ICJ, accessed 6 January 2015 at http://www.icj-cij.org/docket/files/148/18136.pdf. In reviewing the grant of permits by Japan to allow whaling 'for purposes of scientific research', the Court indicated that the standard of review applied is an objective one and that the Court is not called upon to resolve issues of scientific policy, such as the appropriate policy of the international community in respect of whales or whaling. *Ibid* at paras 62–69.

other forms of uncertainty that complicate the establishment of cause-and-effect relationships.[75] For example, radioactive contaminants released following a nuclear disaster are known to have ill effects for human health and the environment. But greater scientific uncertainty surrounds the question of whether small amounts of such contaminants released from nuclear facilities pose a significant risk of transboundary harm in the short or longer-term.[76] Uncertainties of this kind might be seen to invite the application of international environmental principles, such as the precautionary principle or the principle of inter-generational equity. The precautionary principle is expressly directed to situations '[w]here there are threats of serious or irreversible damage' and 'lack of full scientific certainty' and requires that uncertainty 'should not be used as a reason for postponing measures to prevent environmental degradation'.[77] The inter-generational equity principle requires consideration of the interests of future generations in decision-making on the allocation of the environmental burdens or risks associated with present day activities.[78]

The relevance of the precautionary principle in assessing a claim of potential transboundary radioactive pollution was at issue in the *MOX Plant* case decided by ITLOS in 2001. There Ireland claimed that the United Kingdom had failed to apply a precautionary approach to the protection of the Irish Sea in the exercise of its decision-making authority in relation to the direct and indirect consequences of the operation of the Sellafield mixed oxide (MOX) nuclear materials reprocessing plant.[79] The precautionary principle was invoked by Ireland at the provisional measures phase to support its claim that the United Kingdom had the burden of demonstrating that no harm would arise from radioactive discharges and other consequences of the operation of the MOX plant. It argued that 'this principle might usefully inform the assessment by the Tribunal of the urgency of the measures it was required to take in respect

---

[75] See Vern Walker, 'The Siren Songs of Science: Toward a Taxonomy of Scientific Uncertainty for Decisionmakers' (1991) 23 *Connecticut Law Review* 567.

[76] This was at issue in the *MOX Plant case (Ireland v United Kingdom)*, Provisional Measures, Order of 3 December 2001, ITLOS Case No 10, 41 ILM 405 (2002) (*MOX Plant*).

[77] 1992 Rio Declaration, *supra* note 22, Principle 15.

[78] See e.g., *ibid*, Principle 3.

[79] *MOX Plant*, *supra* note 76; see Ireland's Statement of Claim, 25 October 2001, at para 34 ('the precautionary principle is a rule of customary international law which is binding upon the United Kingdom and relevant to the assessment of the United Kingdom's actions by reference to [UNCLOS]').

of the operation of the MOX plant'.[80] While ITLOS ultimately did not require the suspension of the operation of the plant, it ordered the parties to cooperate and enter into consultations to exchange further information on possible consequences for the Irish Sea arising out of the commissioning of the MOX plant and to devise, as appropriate, measures to prevent pollution of the marine environment which might result from the operation of the MOX plant.[81] That Order was premised on considerations of 'prudence and caution',[82] which suggests a certain precautionary character.

Seemingly then, the precautionary principle would be relevant to the assessment of injury causation in circumstances where scientific evidence regarding a causal link is subject to uncertainty. Overall though, international courts have been reluctant to move far beyond general calls for 'prudence and caution' when faced with uncertain scientific evidence concerning cause-and-effect relationships.[83] An exception is the *Advisory Opinion on Seabed Activities* of ITLOS's Seabed Chamber, mentioned above, which adopted a relatively robust approach to interpretation and application of the precautionary principle. By contrast, other international courts, including the ICJ, have insisted on strong scientific evidence in environmental cases to support a finding of harm causation, notwithstanding the existence of uncertainties.[84]

For instance in the *Gabčíkovo-Nagymaros* case, involving a jointly operated dams project on a shared river, Hungary claimed that its failure to carry out obligations regarding the project under a treaty with Slovakia was justified by the ecological 'peril' the proposed dams posed to freshwater supplies and biodiversity.[85] Despite finding there were serious

---

[80] *MOX Plant*, *supra* note 76 at para 71.

[81] *Ibid* at para 89(1).

[82] *Ibid* at para 84. *C.f.* the Separate Opinion of Ad Hoc Judge Szekely (the Arbitral Tribunal 'should have been responsive, in the face of such uncertainty, to the Irish demands regarding the application of the precautionary principle (see paragraphs 96 to 101 of the Request, at 43–46). It is regrettable that it did not do so, since acting otherwise would have led to granting the provisional measure requested by Ireland regarding the suspension of the commissioning of the plant.').

[83] See Jacqueline Peel, 'Precaution: A Matter of Principle, Approach or Process?' (2004) 5 *Melbourne Journal of International Law* 483.

[84] Sands and Peel, *supra* note 30 at 223–8.

[85] *Gabčíkovo-Nagymaros*, *supra* note 4 at paras 53–54. To preclude the wrongfulness of its breach of treaty obligations, Hungary raised a defence of ecological necessity. The ICJ accepted that an environmental crisis could be the

uncertainties concerning future environmental harms from the dams project, the ICJ found:

> [these uncertainties] could not, alone, establish the objective existence of a 'peril' in the sense of a component element of the State of necessity. The word 'peril' certainly evokes the idea of 'risk'; that is precisely what distinguishes 'peril' from material damage. But a State of necessity could not exist without a 'peril' duly established at the relevant point in time; the mere apprehension of a possible 'peril' could not suffice in that respect.[86]

This is not precautionary language, premised as it is on the need to establish the certainty and inevitability of serious harm.[87]

In its judgment some ten years later in the *Pulp Mills* case, the ICJ also rejected application of the precautionary principle in order to find a causal link between serious pollution of the River Uruguay and proposed pulp mills in circumstances of scientific uncertainty. In response to arguments by Argentina calling for a 'precautionary approach', the Court ruled that, 'while a precautionary approach may be relevant in the interpretation and application of the provisions of the [1975 Uruguay River] Statute, it does not follow that it operates as a reversal of the burden of proof'.[88] These precedents suggest that establishing causation of pollution injury will continue to provide a significant hurdle for transboundary State responsibility claims in the absence of compelling scientific evidence linking a responsible State's conduct with the harm suffered in another State(s).

B. Alternative Approaches for Establishing Causation

The difficulties encountered in establishing specific causation in instances of transboundary pollution, particularly those involving multiple and/or diffuse sources, have sparked discussion in the literature of alternative approaches for establishing causation in responsibility

---

basis for a necessity claim subject to the other elements of the defence being made out, including the existence of 'a grave and imminent peril'.

[86] *Ibid* at para 54.

[87] Sands and Peel, *supra* note 30 at 224, noting however that the Court's assessment was concerned with the application of the law as it stood in 1989 at a time when the precautionary principle had not fully emerged as a principle of international law. See also Afshin A-Khavari and Don Rothwell, 'The ICJ and the Danube Dam Case: A Missed Opportunity for International Environmental Law?' (1998) 22 *Melbourne University Law Review* 507.

[88] *Pulp Mills*, *supra* note 18 at para 164.

claims.[89] These alternative approaches often draw on national law precedents, making an argument for their extension to international law as part of 'the general principles of law recognised by civilised nations'.[90]

One option in this regard is to base a finding of a sufficient causal link on the fact that actions attributable to a State make a material or meaningful contribution to the risk of transboundary pollution harm. For instance, where the activities of multiple upstream States affect sediment and pollutant levels in a shared river, a material risk approach to assessing causation would look to see if a particular State's activities made a meaningful contribution to the overall diminution in river water quality rather than seeking to establish that those activities were the proximate and substantial cause of the harm. Each individual State's responsibility for the injured State's damage could then be assumed to be proportional to its share of water use from the river.[91] This strategy has been pursued in domestic liability cases in the United States alleging damage from corporate activities contributing to climate change. These cases, albeit unsuccessful to date, have targeted the largest emitters within the most GHG-intensive industries in order to buttress the plaintiffs' claim that the defendants' activities are causally linked to global warming and climate change harms.[92]

In domestic liability law, some support exists for the material risk approach, although as yet the practice is probably not sufficiently widespread to amount to a general principle of law recognised by civilised nations. For example, the Canadian Supreme Court in a 2012 decision, *Clements v Clements*, ruled that in negligence cases involving multiple tortfeasors, the standard 'but for' test of causation should be replaced with a standard whereby the plaintiff need only establish that the defendant materially contributed to the risk of injury.[93] The Canadian Supreme Court ruled that the 'material contribution to risk' approach,

---

89 See e.g., Phillip Barton, 'State Responsibility and Climate Change: Could Canada be Liable to Small Island States?' (2002) 11 *Dalhousie Journal of Legal Studies* 65.

90 ICJ Statute, *supra* note 46, Art 38(1)(c) (specifying sources of law to be applied by the ICJ in resolving disputes before it).

91 Faure and Nollkaemper, *supra* note 14 at 174.

92 For discussion, see Matthew F Pawa, 'Global Warming Litigation Heats Up' (2008) April *Trial* 18.

93 *Clements v Clements* 2012 SCC 32, [2012] 2 SCR 181. The 'but for' test applied in many common law jurisdictions requires the plaintiff to demonstrate that the harm/injury would not have occurred but for the defendant's wrongful conduct.

which has also been applied in cases in the United Kingdom,[94] 'applies where "but for" causation cannot be proven against any of multiple defendants, all negligent in a manner that might have in fact caused the plaintiff's injury, because each can use a "point the finger" strategy to preclude a finding of causation on a balance of probabilities'.[95] The Court noted that this is not a test of factual causation but that its approach was justified on the basis of considerations of fairness and justice.[96]

In a different context, the US Supreme Court in *Massachusetts v Environmental Protection Agency* (EPA), considered whether the State of Massachusetts had standing to challenge the federal Environmental Protection Agency's failure to regulate GHG emissions from motor vehicles under the Clean Air Act. The applicable test of standing required the State to demonstrate 'a concrete and particularised injury that … is fairly traceable to the defendant'.[97] On the issue of causation, the Supreme Court rejected the Agency's argument that its decision not to regulate GHG from new motor vehicles would not contribute to climate change damage in Massachusetts, noting that US motor vehicle emissions (at around 6 per cent of worldwide carbon dioxide emissions) make a 'meaningful contribution' to GHG concentrations and global warming.[98]

These causation tests leave open the question of what amounts to a material or meaningful contribution to the risk, which is likely to turn on the perceived seriousness of the harm. On one view, since many transboundary pollution problems are due to the accumulated environmental effects of multiple pollutant releases, every additional quantity of a pollutant released into the environment increases the cumulative risk of damage and hence could be viewed as material.[99] In the *Nuclear Tests* case before the ICJ concerning French atmospheric nuclear testing in the Pacific, an analogous argument was pursued by Australia. Australia alleged that the nuclear tests carried out by France had, inter alia, 'resulted in additional radiation doses to persons living in [the southern]

---

[94] See e.g., *Fairchild v Glenhaven Funeral Services Ltd*, [2002] UKHL 22, [2002] 3 All ER 305; and *Barker v Corus UK Ltd*, [2006] UKHL 20, [2006] 2 AC 572.

[95] *Clements v Clements*, *supra* note 93 at para 43.

[96] *Ibid.*

[97] *Massachusetts v Environmental Protection Agency*, 549 US 497 (2007) at 14.

[98] *Ibid* at 525.

[99] Keely Boom, 'See You in Court: the Rising Tide of International Climate Litigation' (2011) *The Conversation* (28 September 2011) accessed 6 January 2015 at http://theconversation.com/see-you-in-court-the-rising-tide-of-international-climate-litigation-3542.

hemisphere and in Australia in particular'.[100] The ICJ did not ultimately rule on the merits of this submission. In any event, this interpretation of the material risk approach comes very close to calling for imposition of a strict liability standard, which would seem to be at odds with the due diligence requirements applicable under treaty and customary law obligations to prevent transboundary environmental harm.

A more acceptable approach may instead be to lower the standard of proof demanded of an injured State in establishing that its injury is linked to the internationally wrongful act. Some international tribunals have relied upon the precautionary principle in order to lower the required standard of proof where the complexity of facts leads to a degree of scientific uncertainty.[101] For example, in the *Southern Bluefin Tuna Provisional Measures* case, decided by ITLOS in 1999, 'scientific uncertainty regarding measures to be taken to conserve the stock of southern bluefin tuna'[102] did not deter the Tribunal from ordering the parties to the dispute to 'act with prudence and caution to ensure that effective conservation measures are taken to prevent serious harm to the stock of the southern bluefin tuna'.[103] The Tribunal ruled that although it could not conclusively assess the scientific evidence presented by the parties regarding the status of the tuna stock, 'measures should be taken as a matter of urgency to preserve the rights of the parties and to avert further deterioration of the southern bluefin tuna stock'.[104]

VI. CONCLUSION

State responsibility claims involving transboundary pollution continue to be a rarity in international law.[105] The analysis in this chapter of the key

---

100 *Nuclear Tests Cases (Australia v France)*, [1973] ICJ Pleadings, vol I at 13, para 46.

101 Voigt, *supra* note 8 at 16, citing *Southern Bluefin Tuna,* Request for Provisional Measures, Order of 7 August 1999, ITLOS Case Nos 3 and 4 (*Bluefin Tuna*).

102 *Ibid* at para 79.

103 *Ibid* at para 77.

104 *Ibid* at para 80. The Separate Opinion of Judge Treves in the case recognised that ITLOS was plainly taking a precautionary approach in this regard as 'it would seem to me that the requirement of urgency is satisfied only in the light of such precautionary approach'. *Ibid* at para 8.

105 This is certainly the situation in terms of decided case law, albeit that this may not be an accurate reflection of the true extent of broader international practice raising claims of State responsibility for transboundary harms.

practical difficulties facing the establishment of such claims underlines just why this is the case. Apart from the fact that States often seem reluctant to bring claims against other States for environmental and other damage, even where there might be good legal grounds for doing so,[106] claimants face numerous hurdles and uncertainties in seeking to make out the elements of a State responsibility claim. These include difficulties in establishing attribution of wrongful conduct, breach of an international obligation and causation of injury.[107] Successive calls in international instruments for States 'to develop further international law regarding liability and compensation for adverse effects of environmental damage caused by activities within their jurisdiction or control to areas beyond their jurisdiction'[108] also seem to have fallen on deaf ears. Indeed, States may well regard the ambiguities respecting the application of general international law principles of responsibility to transboundary pollution as a convenient buffer against State responsibility claims.[109]

Nonetheless, improving understanding of the legal difficulties likely to be encountered in bringing a State responsibility claim for transboundary pollution is not a futile exercise. On the one hand, identification of the existing legal hurdles is an important prerequisite for further development of the law in this field and could spark law reform efforts. On the other hand, an awareness of the difficulties involved in applying general responsibility rules in instances of transboundary pollution claims may encourage States to pursue alternative avenues to ensure the prevention of transboundary harms. This might include bilateral treaty efforts agreeing on the measures necessary to protect a shared resource or an agreement on a specific State liability or civil liability regime applicable within a given pollution field.

It is also important to appreciate that the international law of State responsibility and transboundary pollution is not static. While the number

---

106 The Chernobyl accident was a case in point. Despite a clearly identifiable link between maintenance and regulatory failures at the Chernobyl plant in Ukraine and radioactive pollution damage in neighbouring States no State responsibility claims were publicly brought following the incident. See further, Philippe Sands (ed), *Chernobyl, Law and Communication: Transboundary Nuclear Air Pollution, the Legal Materials* (Cambridge: Grotius, 1988).

107 See also the discussion in Birnie, Boyle and Redgwell, *supra* note 31 at 216–37.

108 1992 Rio Declaration, *supra* note 22, Principle 13; see also 1972 Stockholm Declaration, *supra* note 49, Principle 22.

109 See Jutta Brunnée, 'Of Sense and Sensibility: Reflections on International Liability Regimes as Tools for Environmental Protection' (2004) 53 *International Law and Comparative Law Quarterly* 351.

of cases coming before international courts and tribunals may represent more of a trickle than a flood, each case adds to the existing body of law. In its *Gabčíkovo-Nagymaros* decision, the ICJ lent support to the idea of an 'evolutionary' approach to the interpretation of environmental treaty obligations. Considering a 1977 bilateral treaty for the joint dams project at issue, it held that its provisions were not static but were 'open to adapt to emerging norms of international law'.[110] This statement could provide an opening for the injection of new environmental norms – including the environmental principles of precaution, inter-generational equity and common but differentiated responsibilities – into the consideration of existing treaty or customary obligations with important implications for State responsibility claims. It is also noteworthy that a small but increasing proportion of recent State responsibility claims in the environmental field have been brought by developing States which, in time, may lead to greater clarification of standards of due diligence, including the procedural obligations that attach to them.[111]

Just as international law on State responsibility for transboundary pollution continues to evolve, our knowledge about the links between human activities and environmental damage is also continually improving. The 'science of causation' is an area receiving increasing attention in the face of complex environmental problems such as climate change. For instance, some scientists have begun using statistical analysis to assess how anthropogenic GHG emissions may make an extreme weather event more likely.[112] At the same time, recent studies such as the Climate Accountability Institute's peer-reviewed report, 'Tracing anthropogenic

---

110 *Gabčíkovo-Nagymaros, supra* note 4 at paras 67–68, 112.

111 Examples include the *Gabčíkovo-Nagymaros* and *Pulp Mills* cases, and the *Case concerning Land Reclamation by Singapore in and around the Straits of Johor (Malaysia v Singapore)*, Provisional Measures, Order of 8 October 2003, ITLOS Case No 12, at para 99. A further case submitted by Ecuador to the ICJ concerning responsibility for the environmental effects of aerial herbicide spraying by Colombia was withdrawn following a settlement reached by the parties: see further, http://www.icj-cij.org/docket/index.php?p1=3&code=ecol&case=138 (accessed 6 January 2015). Most recently ITLOS has been requested by the Sub-Regional Fisheries Commission in West Africa to issue an advisory opinion on questions pertaining to flag State responsibilities in respect of illegal, unreported and unregulated fishing and the management of shared fish stocks by coastal States. See ITLOS Press Release, 'The Tribunal receives a request for an Advisory Opinion from the Sub-Regional Fisheries Commission' (28 March 2013) ITLOS/Press 190.

112 Pardeep Pall, 'Anthropogenic Greenhouse Gas Contribution to Flood Risk in England and Wales in Autumn 2000' (2011) 470 *Nature* 382.

carbon dioxide and methane emissions to fossil fuel and cement producers 1854–2010', provide greater clarity around the contribution of multiple emitting entities to the global GHG pollution problem.[113] These developments may in turn inspire innovations in legal causation tests to better match existing law to the complexities of transboundary pollution problems.

Ultimately, the significance of State responsibility rules in addressing transboundary pollution depends upon the extent to which they exercise a deterrent effect in order to promote preventative efforts to avoid or minimise risks of transboundary environmental harm. Even if rarely used in practice, a robust and effective regime of State responsibility for transboundary pollution can stimulate more concerted efforts to prevent environmental harm or to better mitigate damage through improved systems of risk management, risk transfer, rehabilitation and compensation mechanisms.

---

[113] Richard Heede, 'Tracing Anthropogenic Carbon Dioxide and Methane Emissions to Fossil Fuel and Cement Producers 1854–2010' (2014) 122 *Climatic Change* 229. The paper finds 90 major corporate entities are responsible for around two-thirds of global greenhouse gas emissions from industrial sources.

PART II

Transboundary pollution: problems and prospects

4. Pollution of shared freshwater resources in international law

Stephen C McCaffrey

I. INTRODUCTION

In a world of shrinking per capita water supplies,[1] pollution of fresh water poses increasing challenges. And since much of the globe's fresh water is shared by two or more countries,[2] factors such as pollution that affect availability present a growing potential for igniting international disputes. Three cases involving transboundary freshwater pollution issues have already reached the International Court of Justice (ICJ)[3] and a

[1] While the quantity of fresh water on Earth remains constant, the growing human population means there will be less per capita on a daily basis until the population tops out, probably at some 9 billion people around 2050 under the United Nations (UN) Population Division's medium scenario: 'Under the assumptions made in the medium scenario projection, world population will not vary greatly after reaching 8.92 billion in 2050 … [W]orld population growth beyond 2050, at least for the following 250 years, is expected to be minimal.' UN Department of Economic and Social Affairs/Population Division, *World Population to 2300* (UN, 2004) at 12.

[2] There are over 260 international river basins, and additional transboundary aquifers that are not connected to those surface-water basins. Some 40 per cent of the world's population lives in international river basins. See Oregon State University, United Nations Environment Programme (UNEP), and Food and Agriculture Organization of the UN, *Atlas of International Freshwater Agreements* (2002) at 2.

[3] *Case concerning the Gabčíkovo-Nagymaros Project (Hungary/Slovakia)*, [1997] ICJ Rep 7 (*Gabčíkovo-Nagymaros*); *Case concerning Pulp Mills on the River Uruguay (Argentina v Uruguay)*, [2010] ICJ Rep 14 (*Pulp Mills*); *Construction of a Road in Costa Rica along the San Juan River (Nicaragua v Costa Rica)*, 'Application Instituting Proceedings Submitted by Nicaragua' (22 December 2011) accessed 6 January 2015 at http://www.icj-cij.org/docket/files/152/16917.pdf *(Construction of a Road along the San Juan)*.

number of other interstate water pollution disputes have been resolved through negotiations leading to agreements.[4]

This chapter will offer an overview of treaty and case law setting forth generally applicable rules of international law governing transboundary pollution of freshwater resources. The chapter will not dwell upon the general obligation to avoid transboundary pollution harm, since the general obligation will be examined by others.[5] It will also avoid for the most part extended analytical or theoretical discussions of the issues involved in this topic, as these have already been the subject of in-depth, book-length treatment, to which the interested reader is referred.[6]

It should be noted at the outset that while transboundary water pollution problems have by no means gone away, the focus today is more on protecting the environment and ecosystems of shared freshwater resources than on pollution of those resources alone.[7] That is, the concern is the broader one of protecting aquatic ecosystems from threats, of which pollution is one, to be sure, but which also include such things as exotic species, sedimentation, riverbed erosion and reduced flow.

After taking a brief look at the nature of the problem, this chapter will consider how the problem of transboundary water pollution is dealt with

---

[4] See e.g., the '*Rhine Salt Case*' discussed in Johan G Lammers, *Pollution of International Watercourses* (The Hague: Martinus Nijhoff Publishers, 1984) at 171–3, 196–206; International Commission for the Protection of the Rhine (ICPR) accessed 6 January 2015 at http://www.iksr.org/. See also the '*Colorado River Salinity Case*', resolved in Minute 242 to the Treaty between the United States and Mexico, 3 February 1944, Treaty Series 994, accessed 6 January 2015 at http://www.ibwc.gov/Files/1944Treaty.pdf, and discussed at the International Symposium on the Salinity of the Colorado River set out in Volume 15 of the *National Resources Journal* (1975).

[5] See Catherine Redgwell and Jacqueline Peel's chapters in this book (Chapter 1 and Chapter 3 respectively).

[6] Perhaps the classic in the field is Lammers, *supra* note 4. Another valuable work is also by a Dutch scholar: André Nollkaemper, *The Legal Regime for Transboundary Water Pollution: Between Discretion and Constraint* (London: Martinus Nijhoff/Graham & Trotman, 1993). See also, e.g., José Sette-Camara, 'Pollution of International Rivers' (1984) 186 *Recueil des Cours del'Académie de Droit International* 119; Anthony P Lester, 'River Pollution in International Law' (1963) 57 *American Journal of International Law* 828; Anthony P Lester, 'Pollution' in Albert Henry Garretson, Robert D Hayton and Cecil J Olmstead (eds), *The Law of International Drainage Basins* (New York: Oceana Publications, 1967) 89.

[7] See e.g., Owen McIntyre, *Environmental Protection of International Watercourses under International Law* (Farnham: Ashgate, 2007).

in the two global treaties relating to the subject, and then describe how it has been treated in case law.

II. WATER STRESS, WATER POLLUTION AND INCREASING CONFLICT

As recognised in a recent United Nations (UN) report, '[w]ater has long ceased to be solely a local issue', in part because 'many river basins and aquifers transcend national boundaries …'.[8] The 1997 UN Comprehensive Assessment of the Freshwater Resources of the World found that:

> Water use has been growing at more than twice the rate of the population increase during this century, and already a number of regions are chronically water-short. About one-third of the world's population lives in countries that are experiencing moderate-to-high water stress partly resulting from increasing demands from a growing population and human activities. By the year 2025, as much as two-thirds of the world population could be under stress conditions.[9]

These two facts – that much of the world's fresh water is shared by two or more States, and that water use continues to grow rapidly – combine to produce a worrisome picture of potential conflict as water stress increases. In addition, in a report released in 2008, the Intergovernmental Panel on Climate Change observed that '[a]ccording to many experts, water and its availability and quality will be the main pressures on, and issues for, societies and the environment under climate change …'.[10]

While it does not affect the quantity of available water, pollution can make water unusable for some or virtually all purposes. And, as the UN noted in its Comprehensive Assessment, human use of rivers, especially, to dispose of waste is a venerable practice:

> For millennia, people have used water as a convenient sink into which to dump wastes. The pollution comes from many sources, including untreated sewage, chemical discharges, petroleum leaks and spills, dumping in old

[8] UNESCO, Managing Water under Uncertainty and Risk, World Water Development Report 4 (UNESCO, 2012) at 40.

[9] Comprehensive Assessment of the Freshwater Resources of the World, UN Doc E/CN.17/1997/9 (4 February 1997) (Comprehensive Assessment) at 3, para 2.

[10] Bryson C Bates, Zbigniew W Kundzewicz, Shaohong Wu and Jean P Palutikof (eds), *Climate Change and Water, Technical Paper of the Intergovernmental Panel on Climate Change* (IPCC) (IPCC Secretariat, 2008) at 7.

> mines and pits, and agricultural chemicals that are washed off or seep downward from farm fields. In one area after another, the amounts and types of waste discharged have outstripped nature's ability to break them down into less harmful elements. Pollution spoils large quantities of water which then cannot be used or, at best, can be used for restricted purposes only.[11]

Moreover, '[c]ontaminated water that people drink without adequate treatment is one of the major causes of human illness'.[12] It has been estimated that 1.4 million children die each year from a lack of access to clean water and sanitation, which is the equivalent of a jumbo jet full of children crashing every four hours.[13]

Both because surface water is often polluted and because there is not always readily available access to surface water, human populations are relying increasingly on groundwater. This water source was historically more pure than surface water because of its filtration by the soil, sand and geologic formations that contain it. And, some 97 per cent of the fresh water potentially available for human use is stored underground.[14] However, even this source has come under threat:

> Nitrates from fertilizers, human and cattle wastes are polluting groundwater in many regions. High nitrate levels in drinking water decrease the oxygen-carrying capacity of haemoglobin in blood, which can threaten the health of infants. A United Nations study has stated that nitrate pollution will likely be one of the most pressing water quality problems in Europe and North America in the coming decade, and will become a serious problem in other countries, such as India and Brazil, if present trends continue …[15]

The UN Special Rapporteur on the human right to safe drinking water and sanitation is reported to have found on a visit to the State of California in the United States (US) that over 250,000 California residents, most of whom are poor agricultural workers, lacked access to safe water due largely to nitrate pollution of well water and were forced

---

[11] Comprehensive Assessment, *supra* note 9 at 10, para 51.

[12] *Ibid* at para 53.

[13] Water.org, 'Water Facts', accessed 6 January 2015 at http://water.org/water-crisis/water-facts/water, citing UNICEF/WHO, 'Diarhhoea: Why children are still dying and what can be done' (2009).

[14] UNEP, *Vital Water Graphics* (UNEP, 2008) Chapter 1, accessed 6 January 2015 at http://www.unep.org/dewa/vitalwater/article5.html.

[15] Comprehensive Assessment, *supra* note 9 at 10, para 53(c).

to purchase bottled water.[16] This is the largest state in the US with an economy that ranks in the top ten of those in the world.[17]

In light of these developments, it is perhaps not surprising that instances of conflict over water in general are increasing. According to the latest volume of *The World's Water*, '[i]n recent years, there has been an increase in reported cases of water-related disputes and violence'.[18] This work also states that there has been 'a shift in the nature of these conflicts, away from water disputes between nations and toward sub-national and local violence over water access'.[19] The volume also notes, however, that this shift is 'almost certainly' due in part to 'improved local reporting of water conflicts'.[20]

In addition, the proposition that there is such a shift may not take fully into account the evidence of an increase in water disputes between States, in the form of filings of cases before judicial and arbitral bodies. Two cases involving disputes over shared fresh water were heard by the Permanent Court of International Justice (PCIJ) in the 1920s and 1930s (the *River Oder*[21] and *Meuse*[22] cases). No freshwater disputes were brought before the ICJ thereafter until the 1990s, when the *Gabčíkovo-Nagymaros* case[23] was brought by Hungary and Slovakia. But beginning with that case, a number of disputes concerning international water-courses were brought to the ICJ in rather rapid succession over the next decade and a half: *Dispute regarding Navigational and Related Rights (Costa Rica v Nicaragua)* (Judgment of 2009),[24] *Pulp Mills on the River*

16 See Dan Bacher, 'Governor Brown Signs Human Right to Water Bill' *Indybay.og* (26 September 2012) accessed 6 January 2015 at http://www.indybay.org/newsitems/2012/09/26/18722446.php.

17 See e.g., Dan Walters, 'California Poised to Regain No. 8 Ranking in Global Economy' *The Sacramento Bee* (11 July 2013) accessed 6 January 2015 at http://blogs.sacbee.com/capitolalertlatest/2013/07/california-poised-to-regain-eighth-8-rank-in-global-economy.html.

18 Peter H Gleick (ed), *The World's Water*, *Volume 8* (Washington, DC: Island Press, 2014) at 159.

19 *Ibid* at 161.

20 *Ibid.*

21 *Case Relating to Territorial Jurisdiction of the International Commission of the River Oder (Czechoslovakia, Denmark, France, Germany, Great Britain and Sweden/Poland)* (1929), PCIJ (Ser A) No 23 (*River Oder*).

22 *Diversion of Water from Meuse (Netherlands v Belgium)* (1937), PCIJ (Ser A/B) No 70 (*Meuse*).

23 *Gabčíkovo-Nagymaros*, *supra* note 3.

24 *Dispute regarding Navigational and Related Rights (Costa Rica v Nicaragua)*, [2009] ICJ Rep 213 (*Navigational and Related Rights*).

Uruguay (Argentina v Uruguay) (Judgment of 2010),[25] *Certain Activities carried out by Nicaragua in the Border Area (Costa Rica v Nicaragua)* (Application of 18 November 2010),[26] and *Construction of a Road in Costa Rica along the San Juan River (Nicaragua v Costa Rica)* (Application of 22 December 2011).[27]

As far as arbitrations are concerned, until well into the 21st century there had been only one that could be characterised as major: the *Lake Lanoux Arbitration* of 1957.[28] But, seemingly as part of the trend toward increased filings by governments of disputes with international tribunals, on 17 May 2010, Pakistan brought a dispute with India to arbitration under the 1960 Indus Waters Treaty. The final award in the case, the *Indus Waters Kishenganga Arbitration,* was issued in December 2013.[29]

Thus, while only three major water disputes (the *River Oder, Meuse* and *Lake Lanoux Arbitration* cases) were brought by States before international courts and tribunals until the 1990s, since then six cases (the *Gabčíkovo-Nagymaros, Navigational and Related Rights, Pulp Mills, Certain Activities, Construction of a Road along the San Juan* and *Indus Waters Kishenganga Arbitration* cases) have been filed. Three of those cases have involved allegations of transboundary water pollution, broadly defined: the *Gabčíkovo-Nagymaros, Pulp Mills* and *Construction of a Road along the San Juan* cases. Other recent disputes, however, implicate the protection of freshwater ecosystems more generally, in particular the *Certain Activities* and *Indus Waters Kishenganga Arbitration* cases. Because of the current concern with protection of the ecosystems of international watercourses and their environment mentioned above, the latter two cases will be discussed herein.

Before moving to a review of treaty and case law on the subject, however, it is appropriate to indicate how the term 'pollution' will be understood in this chapter. Guidance is provided by the 1997 UN

---

25 *Pulp Mills, supra* note 3.

26 *Certain Activities carried out by Nicaragua in the Border Area (Costa Rica v Nicaragua)* 'Application Instituting Proceedings Submitted by Costa Rica' (18 November 2010) accessed 6 January 2015 at http://www.icj-cij.org/docket/files/150/16279.pdf (*Certain Activities*).

27 *Construction of a Road along the San Juan, supra* note 3.

28 *Lake Lanoux Arbitration (France v Spain)* (1957) 12 RIAA 281 (French); with English translations at (1961) 24 ILR 101; reprinted in (1959) 53 *American Journal of International Law* 156 (*Lake Lanoux Arbitration*).

29 *Indus Waters Kishenganga Arbitration (Pakistan v India)*, Partial Award of 18 February 2013 (*Kishenganga*, Partial Award), Final Award of 20 December 2013 (*Kishenganga*, Final Award) accessed 6 January 2015 at http://www.pca-cpa.org/showpage.asp?pag_id=1392 (*Indus Waters Kishenganga Arbitration*).

Convention on the Law of the Non-Navigational Uses of International Watercourses[30] (1997 Convention), a universally-applicable treaty whose negotiation within the UN was based on a set of draft articles prepared by the International Law Commission (ILC).[31] The expression 'pollution of an international watercourse' is defined in Article 21(1) of that agreement as 'any detrimental alteration in the composition or quality of the waters of an international watercourse which results directly or indirectly from human conduct'.[32] This definition is thus sufficiently broad to cover everything from chemicals discharged into a watercourse (as alleged in the *Pulp Mills* case) to sediment washed into one as a result of construction activities on land in the vicinity (as in the *Construction of a Road along the San Juan* case).

Also needed is an indication of what is understood in this chapter by the expression 'international watercourse'. Again, the 1997 Convention provides useful guidance. That instrument defines 'watercourse' as 'a system of surface waters and groundwaters constituting by virtue of their

---

[30] Convention on the Law of the Non-navigational Uses of International Watercourses, adopted 21 May 1997, 36 ILM 700 (entered into force 17 August 2014) (1997 Convention).

[31] See 'The Law of the Non-Navigational Uses of the International Watercourses' in 'Report of the International Law Commission on the work of its forty-sixth session' (UN Doc A/CN.4/L.503 and Add.1 and 2), Ch III, in *Yearbook of the International Law Commission 1994*, vol II, part Two (UN, 1994) (1994 Draft Articles).

[32] 1997 Convention, *supra* note 30, Art 21(1). Compare the definition in the 1979 Athens Resolution of the Institute of International Law: 'For the purpose of this Resolution, "pollution" means any physical, chemical or biological alteration in the composition or quality of waters which results directly or indirectly from human action and affects the legitimate uses of such waters, thereby causing injury.' Athens Resolution on the Pollution of Rivers and Lakes and International Law, Institute of International Law (IIL), 12 September 1979, Art 1(1), English translation accessed 6 January 2015 at http://www.idi-iil.org/idiE/resolutionsE/1979_ath_02_en.PDF (IIL Athens Resolution). Similarly, the Helsinki Rules, adopted by the International Law Association in 1966, contain the following definition: 'the term "water pollution" refers to any detrimental change resulting from human conduct in the natural composition, content, or quality of the waters of an international drainage basin.' Committee on the Uses of the Waters of International Rivers, in 'The Helsinki Rules on the Uses of the Waters of International Rivers' (1966) (International Law Association, 1967), Art IX, accessed 6 January 2015 at http://webworld.unesco.org/water/wwap/pccp/cd/pdf/educational_tools/course_modules/reference_documents/internationalregionconventions/helsinkirules.pdf (1966 Helsinki Rules).

physical relationship a unitary whole and normally flowing into a common terminus'.[33]

The 1997 Convention then defines 'international watercourse' simply as 'a watercourse, parts of which are situated in different States ...'.[34] It is important to emphasise here that the concept of an international watercourse under the 1997 Convention embraces not only surface water but also groundwater that is connected to surface water. It should also be borne in mind that most groundwater is related to surface water, either because it is recharged from the surface or because it contributes to surface water flows, or both (which is often the case).[35]

Against this background we may turn to consider two treaties, both open to all States, which treat the subject of transboundary water pollution.

III. TRANSBOUNDARY WATER POLLUTION IN GLOBALLY APPLICABLE TREATIES

This section will focus upon two treaties on international watercourses that are open to participation by all members of the international community: the 1997 Convention mentioned above and the 1992 UN Economic Commission for Europe (UNECE) Convention on the Protection and Use of Transboundary Watercourses and International Lakes[36] (1992 Convention). Originally conceived of as a regional agreement, the latter instrument became open to accession by all UN Member States with the entry into force on 6 February 2013 of amendments added in 2003 to convert it into a global instrument.[37] Both agreements contain provisions on transboundary water pollution.

---

[33] 1997 Convention, *supra* note 30, Art 2(a).

[34] *Ibid*, Art 2(b).

[35] See e.g., Foreword to Thomas C Winter, Judson W Harvey, O Lehn Franke and William M Alley, 'Ground Water and Surface Water: A Single Resource' (1998) US Geological Survey Circular 1139 ('Nearly all surface-water features ... interact with ground water.').

[36] Convention on the Protection and Use of Transboundary Watercourses and International Lakes, adopted 17 March 1992, 31 ILM 1312 (entered into force 6 February 2013) (1992 Convention).

[37] However, due to a decision adopted by the Meeting of the Parties on 30 November 2012 (decision VI/3), the amendments do not become operational until all Parties to the Convention that were Parties in 2003 ratify them. Meeting of the Parties to the Convention on the Protection and Use of Transboundary Watercourses and International Lakes (Sixth session), 'Report of the Meeting of

The two Conventions deal with the same subject matter, international watercourses, but their provenance, focus and approaches are quite different. While the 1992 Convention was drafted in less than two years on the basis of an initial draft of possible elements prepared by the UNECE secretariat,[38] the preparation of the 1997 Convention took place over a period of 20 years, from 1974 to 1994, in the ILC alone. The latter agreement was the result of a process initiated by the UN General Assembly in 1970, when it tasked the ILC with taking up the study of the law of the non-navigational uses of international watercourses with a view to its progressive development and codification.[39] This followed a resolution adopted just over a decade earlier, in which the General Assembly requested the Secretary-General to prepare a report on legal problems relating to the utilisation and use of international rivers 'with a view to determining whether the subject is appropriate for codification'.[40]

It should also be borne in mind that the 'object' of the ILC is 'the promotion of the progressive development of international law and its codification'.[41] Thus, in preparing sets of draft articles, the ILC takes into account the practice of States, as revealed in treaty and case law, as well as in other forms of practice. The ILC also interacts with States on a regular basis through its annual reports to the General Assembly and the comments of governments on those reports. Thus, the ILC's work takes into account not only historic practice of States, but also governments' current views concerning the topics on which the ILC is working. For

---

the Parties on its sixth session, Addendum – Decisions and vision for the future of the Convention' (28–30 November 2012) ECE/MP.WAT/37/Add.2. As of February 2015, Belgium, Kazakhstan and Ukraine, Parties in 2003, had not yet ratified the amendments.

38 UNECE 'Note by the Secretariat, Possible Elements for a Draft Framework Convention on the Protection and Use of Transboundary Watercourses and International Lakes' (15 March 1990) ECE Doc ENVWA/WP.3/R.17.

39 Progressive Development and Codification of the Rules of International Law relating to International Watercourses, GA Res 2669 (XXV), UNGAOR, 25th Sess, UN Doc A/CN.4/244/Rev.1 (1970).

40 *Preliminary Studies on the Legal Problems relating to the Utilization and Use of International Rivers,* GA Res 1401 (XIV), UNGAOR, 14th Sess, UN Doc A/RES/1401 (XIV) (1959), Preamble. The report was later published, together with a supplementary report, in the 1974 *Yearbook of the International Law Commission*, vol 2, part 2 at 33.

41 Statute of the International Law Commission, adopted by GA Res 174 (II) (21 November 1947), as amended by GA Res 485 (V) (12 December 1950), GA Res 984 (X) (3 December 1955), GA Res 985 (X) (3 December 1955) and GA Res 36/39 (18 November 1981) UN Doc A/CN.4/4/Rev.2 (1982), Art 1(1).

these reasons, the ILC's work on a given topic will almost inevitably be of longer duration than that of States negotiating directly on a treaty. But the product of the ILC's work is likely to reflect, at least in part, rules of customary international law on the topic in question. These rules would thus be applicable to States generally, regardless of whether the ILC's draft articles ever took the form of a treaty or, even if they did, whether that treaty entered into force or a given State were a party to it. One need only think of the ILC's work on treaty[42] and diplomatic[43] law, as well as its articles on State Responsibility,[44] for examples of this phenomenon.

The general focus, or emphasis, of these two Conventions is also somewhat different, as mentioned above, though this does not make them in any way incompatible. While the 1992 Convention was originally envisaged as a regional agreement between States many of which were well-watered and were more concerned with pollution than allocation, the 1997 Convention was always seen as a global instrument that would have to take into account the situations of both arid States and those with plentiful rainfall. This led to an emphasis in the 1992 Convention on the avoidance of 'transboundary impact', while the key principle of the 1997 Convention may be said to be equitable and reasonable utilisation – that is, a principle relating largely to allocation rather than to avoidance of impact. Nevertheless, as mentioned, the two Conventions are complementary despite their different approaches.

A. The 1997 Convention

The 1997 Convention contains one principal article relating to transboundary pollution, although it is one of the most detailed provisions of the agreement. Article 21, 'Prevention, reduction and control of pollution', reads as follows:

> 1. For the purpose of this article, 'pollution of an international watercourse' means any detrimental alteration in the composition or quality of the waters of an international watercourse which results directly or indirectly from human conduct.

---

[42] Vienna Convention on the Law of Treaties, adopted 22 May 1969, 1155 UNTS 331 (entered into force 27 January 1980).

[43] Vienna Convention on Diplomatic Relations, adopted 18 April 1961, 500 UNTS 95 (entered into force 24 April 1964); Vienna Convention on Consular Relations, adopted 24 April 1963, 596 UNTS 261 (entered into force 19 March 1967).

[44] UN Legislative Series, Materials on the Responsibility of States for Internationally Wrongful Acts, UN Doc ST/LEG/SER.B/25 (UN, 2012).

2. Watercourse States shall, individually and, where appropriate, jointly, prevent, reduce and control the pollution of an international watercourse that may cause significant harm to other watercourse States or to their environment, including harm to human health or safety, to the use of the waters for any beneficial purpose or to the living resources of the watercourse. Watercourse States shall take steps to harmonize their policies in this connection.
3. Watercourse States shall, at the request of any of them, consult with a view to arriving at mutually agreeable measures and methods to prevent, reduce and control pollution of an international watercourse, such as:
   (a) Setting joint water quality objectives and criteria;
   (b) Establishing techniques and practices to address pollution from point and non-point sources;
   (c) Establishing lists of substances the introduction of which into the waters of an international watercourse is to be prohibited, limited, investigated or monitored.[45]

This article thus establishes obligations of prevention, reduction and control of transboundary water pollution. Article 3 of the 1992 Convention also covers these three kinds of measures, and fleshes out in greater detail their meaning and how they are to be achieved. This is an example of the complementarity of the two conventions. It will be noted that the specific obligation in paragraph 2 of the above-quoted provision is to prevent, reduce and control water pollution that 'may cause significant harm' to other riparian States or their environment. This underscores the preventive nature of the obligation. It is not necessary that significant harm has actually occurred for the obligation to be triggered, but only that it 'may' occur.

Paragraph 2 goes on to specify three entities or interests that are protected by this obligation: (i) other watercourse States or their environment, including human health or safety; (ii) the use of the waters for any beneficial purpose; and (iii) the living resources of the watercourse. These three categories comprising the 'protected class' are quite broad and call for brief comment. Concerning the first category, while the disjunctive 'or' following 'other watercourse States' could seem puzzling, since the environment of those States is an integral part of their territory, the fact that the environment is singled out serves to emphasise its importance as the basic, foundational, system that sustains life. The words 'including harm to human health or safety' recognise that watercourse States have a primary interest in the health and safety of their populations, and resonate with Article 10 of the 1997 Convention, to be

[45] 1997 Convention, *supra* note 30, Art 21.

discussed below. The second category, protecting from significant pollution harm 'the use of the waters for any beneficial purpose', is virtually all-encompassing in terms of uses that could be made of a watercourse, short of malicious or hostile uses that would be prohibited by other rules of watercourse law or by the law of armed conflict. The third category could be seen as duplicative of the term 'environment' referred to above, but serves to focus attention on the 'living' features of the environment, the biota, rather than those that are non-living or abiotic, but may still support life, such as material in riverbeds.

Paragraph 2 of Article 21 concludes by stating an obligation of riparian States to 'take steps to harmonize their policies in this connection'. This is important to ensure that protective measures adopted by one State are not frustrated by measures adopted, or not adopted, by another riparian State, and generally that the water pollution regimes of the States concerned operate synergistically. The obligation is to 'take steps' toward achieving harmony in this regard, not to achieve it, but since this is a due diligence obligation, as will be seen below, it is clearly foreseen that the steps that are taken should ultimately lead to harmony among pollution-control policies.

Paragraph 3 of Article 21 establishes an obligation of consultation on the request of any of the States sharing the same watercourse regarding measures and methods to be taken to achieve water pollution prevention, reduction and control. An indicative list of the types of measures that may be considered is provided in sub-paragraphs (a) to (c).

Article 21 of the 1997 Convention may be compared with corresponding provisions adopted by the Institute of International Law (IIL) in 1979:

> Article II
>
> In the exercise of their sovereign right to exploit their own resources pursuant to their own environmental policies, and without prejudice to their contractual obligations, States shall be under a duty to ensure that their activities or those conducted within their jurisdiction or under their control cause no pollution in the waters of international rivers and lakes beyond their boundaries.
>
> Article III
>
> 1. For the purpose of fulfilling their obligation under Article II, States shall take, and adapt to the circumstances, all measures required to:
>   (a) prevent any new form of pollution or any increase in the existing degree of pollution; and
>   (b) abate existing pollution within the best possible time limits.

2. Such measures shall be particularly strict in the case of ultra-hazardous activities or activities which pose a danger to highly exposed areas or environments.[46]

Despite the rather strict language in Article II ('States shall be under a duty to ensure'), Article III makes clear that these are in the nature of due diligence obligations ('States shall take … all measures required to').

For their part, the International Law Association's (ILA) 1966 Helsinki Rules treat the issue of transboundary water pollution in the context of equitable utilisation, the *leitmotif* of the rules:

Article X

1. Consistent with the principle of equitable utilization of the waters of an international drainage basin, a State:
   (a) Must prevent any new form of water pollution or any increase in the degree of existing water pollution in an international drainage basin which would cause substantial injury in the territory of a co-basin State;
   (b) Should take all reasonable measures to abate existing water pollution in an international drainage basin to such an extent that no substantial damage is caused in the territory of a co-basin State.[47]

This approach, like that of the IIL Athens Resolution, is consistent with the requirement of Article 21 of the 1997 Convention that pollution be prevented, reduced and controlled. We will see that the 1992 Convention takes the same general approach.

There are two other articles in the 1997 Convention that bear directly on transboundary water pollution. First and, perhaps, foremost, is Article 20, 'Protection and preservation of ecosystems', which provides in its entirety as follows:

Watercourse States shall, individually and, where appropriate, jointly, protect and preserve the ecosystems of international watercourses.[48]

This provision establishes a broad and all-encompassing obligation to protect the 'ecosystems' of international watercourses, with no requirement of actual or threatened harm to other watercourse States. These ecosystems include both those in the water and those adjoining or otherwise influencing watercourses. Thus, for example, logging in the

---

46 IIL Athens Resolution, *supra* note 32, Arts II and III.
47 1966 Helsinki Rules, *supra* note 32, Art X.
48 1997 Convention, *supra* note 30, Art 20.

neighbourhood of or upslope from a watercourse can lead to sedimentation of its bed, which can in turn have adverse effects on aquatic species not adapted to the changed conditions. The obligation set forth in Article 20 is an umbrella for those that follow in Articles 21–23.

The second provision that bears on transboundary water pollution is Article 23, which concerns protection of the marine environment from such pollution and provides as follows:

> Watercourse States shall, individually and, where appropriate, in cooperation with other States, take all measures with respect to an international watercourse that are necessary to protect and preserve the marine environment, including estuaries, taking into account generally accepted international rules and standards.[49]

This article addresses the well-known problem of land-based sources of marine pollution, which account for some 70 per cent of all marine pollution. The vastness of the seas fuels the human tendency, alluded to earlier, to dispose of wastes in rivers on the assumption that they will eventually be carried out into the ocean and swallowed up in its 'infinite' depths. We now know better, but the problem has grown, not diminished. The harm to the marine environment from waste carried by rivers to the sea begins in estuaries, which are nurseries for marine life. Much of this waste is contained in agricultural runoff, which carries residues from fertilisers, herbicides and pesticides into watercourses and thence to estuaries and the sea.

The obligation is to 'take all measures with respect to an international watercourse that are necessary to protect and preserve the marine environment', indicating that the obligation is one of due diligence, as discussed further below. In discharging this obligation, States are to take into account 'generally accepted international rules and standards', which is an allusion to those reflected in the 1982 UN Convention on the Law of the Sea[50] and in various regional sea conventions.[51]

The obligations set forth in Articles 21 and 23, and probably that contained in Article 20 as well, are likely ones of due diligence rather than strict obligations of result. 'Statements of Understanding' adopted by the Working Group of the Whole, in which the 1997 Convention was

---

[49] *Ibid*, Art 23.

[50] UN Convention on the Law of the Sea, adopted 10 December 1982, 1833 UNTS 3 (entered into force 16 November 1994) (UNCLOS), Arts 194(3)(a) and 207.

[51] See 1994 Draft Articles, *supra* note 31 at 124, commentary to Art 23, on which Art 23 of the 1997 Convention was based.

negotiated, include the following statement relating to Articles 21 and 23, inter alia: 'As regards *articles 21, 22 and 23*: As reflected in the commentary of the International Law Commission, these articles impose a due diligence standard on watercourse States.'[52] While Article 20 was not included in the list, it seems inevitable that it would also be interpreted as reflecting a due diligence obligation. This is especially the case in light of what the ICJ said in its judgment in the *Pulp Mills* case, discussed below, concerning the nature of similar obligations involved there.

One final provision of the 1997 Convention that bears mention here is Article 10, 'Relationship between different kinds of uses', which provides as follows:

> 1. In the absence of agreement or custom to the contrary, no use of an international watercourse enjoys inherent priority over other uses.
> 2. In the event of a conflict between uses of an international watercourse, it shall be resolved with reference to articles 5 to 7, with special regard being given to the requirements of vital human needs.[53]

The primary purpose of this article is to make clear that no kind of use – navigational, domestic, agricultural or industrial – enjoys inherent priority over other kinds of uses. It may once have been the case that navigational uses enjoyed such a priority,[54] but this is no longer true, in general. If different kinds of uses conflict with each other – for example, navigation and filling a reservoir for hydroelectric power production – the conflict is to be resolved according to the principles of equitable and reasonable utilisation (Articles 5 and 6) and prevention of significant harm (Article 7). But in resolving such a conflict, the last clause in paragraph 2 instructs that 'special regard' is to be given to 'the requirements of vital human needs'.[55]

---

[52] UNGA Sixth Committee, 'Statements of Understanding' contained in the 'Report of the Sixth Committee convening as the Working Group of the Whole' (11 April 1997) UN Doc A/51/869 (Statements of Understanding) (emphasis in original). Article 22 is entitled 'Introduction of alien or new species'. It was not unusual for the negotiating governments to seek guidance from the ILC's commentary. A general statement of understanding at the end of the document reads as follows: 'Throughout the elaboration of the draft Convention, reference had been made to the commentaries to the draft articles prepared by the International Law Commission to clarify the contents of the articles.' *Ibid.*

[53] 1997 Convention, *supra* note 30, Art 10.

[54] See 1994 Draft Articles, *supra* note 31 at 110, Commentary to Art 10.

[55] 1997 Convention, *supra* note 30, Art 10(2).

This final clause was the subject of significant discussion in the Working Group in which the 1997 Convention was negotiated. In the end, no changes were made to the ILC's text, but a Statement of Understanding was adopted, which reads as follows: 'In determining "vital human needs," special attention is to be paid to providing sufficient water to sustain human life, including both drinking water and water required for production of food in order to prevent starvation.'[56] This standard sounds very much like the human right to water, as understood by the UN Committee on Economic, Social and Cultural Rights (ESC Committee) in its General Comment 15, adopted in 2002:[57] 'The human right to water entitles everyone to sufficient, safe, acceptable and affordable water for personal and domestic uses.'[58] But to suggest that State A could violate the right to water of an individual living in State B would be going quite far, as it would give extraterritorial effect to a human right. However, the 'special regard' to be given to 'vital human needs' under Article 10(2) presumably implicates only the rights and duties of States, not those of individuals, since it is States that have rights and obligations under the 1997 Convention. Still, the meaning of 'vital human needs' could usefully be informed not only by the Statement of Understanding quoted above, but also by the content of the human right to water as explained by the ESC Committee, also quoted above.

Having reviewed the relevant aspects of the 1997 Convention, we may now turn to the 1992 Convention, focusing again on its approach to transboundary water pollution.

B. The 1992 Convention

As noted earlier, the apparent *leitmotif* of the 1992 Convention is the avoidance of 'transboundary impact'. This expression is defined in Article 1 of the Convention as follows:

> 'Transboundary impact' means any significant adverse effect on the environment resulting from a change in the conditions of transboundary waters caused by a human activity, the physical origin of which is situated wholly or in part within an area under the jurisdiction of a Party, within an area under the jurisdiction of another Party. Such effects on the environment include

---

[56] Statements of Understanding, *supra* note 52.

[57] UN Committee on Economic, Social and Cultural Rights, 'General Comment No. 15' in 'The Right to Water (Arts 11 and 12 of the International Covenant on Economic, Social and Cultural Rights)' UN Doc E/C.12/2002/11 (26 November 2002).

[58] *Ibid* at 2, para 2.

> effects on human health and safety, flora, fauna, soil, air, water, climate, landscape and historical monuments or other physical structures or the interaction among these factors; they also include effects on the cultural heritage or socio-economic conditions resulting from alterations to those factors ...[59]

This is quite a broad definition, but the fact remains that it is framed in terms of environmental harm. It is not incompatible with the definition of 'pollution' in the 1997 Convention, set forth above, but goes well beyond it.

The first operative provision of the 1992 Convention is Article 2, entitled 'General Provisions'. The first two paragraphs of this provision, which consists of eight paragraphs, are set forth here because they provide the basic framework for the provisions that follow:

> 1. The Parties shall take all appropriate measures to prevent, control and reduce any transboundary impact.
> 2. The Parties shall, in particular, take all appropriate measures:
>   (a) To prevent, control and reduce pollution of waters causing or likely to cause transboundary impact;
>   (b) To ensure that transboundary waters are used with the aim of ecologically sound and rational water management, conservation of water resources and environmental protection;
>   (c) To ensure that transboundary waters are used in a reasonable and equitable way, taking into particular account their transboundary character, in the case of activities which cause or are likely to cause transboundary impact;
>   (d) To ensure conservation and, where necessary, restoration of eco-systems.[60]

Thus, the general theme of the agreement, avoidance of transboundary impact via international watercourses, is established in paragraph 1. Paragraph 2(c) does refer to use of transboundary waters 'in a reasonable and equitable way', but this is in the context of 'activities which cause or are likely to cause transboundary impact'.

Pollution is dealt with specifically in Article 3 of the 1992 Convention, entitled 'Prevention, Control and Reduction'. Only the chapeau of the first paragraph and the first two sub-paragraphs will be set forth here because of the length of the remainder of paragraph 1 and of the article as a whole, and because in any event the provisions quoted indicate the approach of the 1992 Convention:

---

59 1992 Convention, *supra* note 36, Art 1(2).

60 *Ibid*, Art 2(1) and (2).

> 1. To prevent, control and reduce transboundary impact, the Parties shall develop, adopt, implement and, as far as possible, render compatible relevant legal, administrative, economic, financial and technical measures, in order to ensure, inter alia, that:
>   (a) The emission of pollutants is prevented, controlled and reduced at source through the application of, inter alia, low- and non-waste technology;
>   (b) Transboundary waters are protected against pollution from point sources through the prior licensing of waste-water discharges by the competent national authorities, and that the authorized discharges are monitored and controlled.[61]

It is clear even from this brief excerpt that the obligations set forth in the 1992 Convention are far more specific and detailed than those in Article 21 of the 1997 Convention with regard to how parties are to implement the basic obligation to prevent, control and reduce the transboundary impact of water pollution. But the end objective of the two conventions is the same: to avoid transboundary pollution harm and to protect the environment of international watercourses.

The 1992 Convention goes on to provide for, inter alia, monitoring (Article 4), research and development (Article 5), exchange of information (Article 6), and responsibility and liability (Article 7).[62] Up to this point, all provisions of the 1992 Convention have been in Part I, entitled 'Provisions relating to All Parties'. Part II, entitled 'Provisions relating to Riparian Parties', contains provisions on bilateral and multilateral cooperation (Article 9), consultations (Article 10), joint monitoring and assessment (Article 11), common research and development (Article 12), exchange of information between riparian parties (Article 13), emergencies (Articles 14 and 15) and public information (Article 16). Part III contains 'Institutional and Final Provisions', including articles that provide for a meeting of the parties (Article 17) and establish a secretariat (Article 19).

The 1992 Convention thus contains a detailed programme for the management and progressive reduction of transboundary water pollution and institutional mechanisms for the implementation of this programme. While the agreement was concluded five years earlier than the 1997 Convention, its preparatory work took into account the work that the ILC

---

61 *Ibid*, Art 3(1)(a) and (b).

62 This article does not actually contain rules on responsibility and liability, but only requires parties to 'support appropriate international efforts to elaborate rules, criteria and procedures in the field of responsibility and liability'. *Ibid*, Art 7.

had done on its draft articles on international watercourses. Indeed, by 1991, when work on a draft of what became the 1992 Convention was completed, the ILC had itself adopted on first reading a complete set of draft articles on the law of international watercourses. Moreover, the terms of reference for the elaboration of a draft framework convention within the Economic Commission for Europe provided that 'the ongoing work in the United Nations International Law Commission' should be taken into account.[63] Thus, from the beginning the 1992 and 1997 Conventions were seen as being compatible and mutually reinforcing. The 1992 Convention may thus be regarded as further refining the provisions of the 1997 Convention concerning transboundary water pollution.

We may now turn to decisional law to ascertain how international courts and tribunals have treated the subject of transboundary water pollution.

IV. TRANSBOUNDARY WATER POLLUTION IN SELECTED EXAMPLES OF CASE LAW

As noted at the outset of this chapter, recent years have seen an increase in cases concerning international watercourses brought by States before international courts and arbitral tribunals. Four of these cases have involved transboundary water pollution: the *Gabčíkovo-Nagymaros*,[64] *Pulp Mills*,[65] *Construction of a Road along the San Juan*[66] and *Indus Waters Kishenganga Arbitration*[67] cases. All of these, except the *Construction of a Road along the San Juan* case, have been finally decided. While this is not the place for a full treatment of these cases, it is useful to highlight the ways in which they bear upon the problem of transboundary water pollution.

---

[63] Senior Advisers to ECE Governments on Environmental and Water Problems, Working Party on Water Problems, 'Provisional Agenda for the First Special Session' (Geneva, 7–10 May 1990) (12 March 1990) ECE Doc ENVWA/WP.3/6 at 2.

[64] *Gabčíkovo-Nagymaros*, *supra* note 3.

[65] *Pulp Mills*, *supra* note 3.

[66] *Construction of a Road along the San Juan*, *supra* note 3.

[67] *Indus Waters Kishenganga Arbitration*, *supra* note 29.

A. The *Gabčíkovo-Nagymaros* Case

In 1977, Hungary and Czechoslovakia concluded a treaty providing for the construction of dams and related works along a stretch of the Danube River some 200 kilometres in length, most of which (142 kilometres) forms the border between them. In the words of the ICJ, the project was:

> essentially aimed at the production of hydroelectricity, the improvement of navigation on the relevant section of the Danube and the protection of the areas along the banks against flooding. At the same time, by the terms of the Treaty, the contracting parties undertook to ensure that the quality of water in the Danube was not impaired as a result of the Project, and that compliance with the obligations for the protection of nature arising in connection with the construction and operation of the System of Locks would be observed.[68]

The upstream portion of the project involved diverting most of the water of the Danube through a 31-kilometre bypass canal in Czechoslovakia running parallel to the river on its north side. The Gabčíkovo dam is on this canal and includes both ship locks and turbines for the production of hydroelectric power.

In May 1992, Hungary issued a declaration purporting to terminate the treaty unilaterally, citing a number of justifications. One of these was that water in the reservoir behind the main dam would become stagnant, resulting in serious impairment of the quality of groundwater on which it relied.[69] This, said Hungary, along with other effects of the project, gave rise to a 'state of ecological necessity' that permitted it to terminate the treaty.[70]

The Court ruled against Hungary, holding that it had not established a state of necessity, ecological or otherwise. However, in its judgment, the Court did make statements about water pollution and environmental protection in general that are instructive for present purposes. The Court indicated its attitude concerning the importance to humans of the environment in the following passage:

> The Court recalls that it has recently had occasion to stress, in the following terms, the great significance that it attaches to respect for the environment, not only for States but also for the whole of mankind:

---

[68] *Gabčíkovo-Nagymaros*, *supra* note 3 at 18, para 15. Disclosure: The author served as counsel to Slovakia in this case.

[69] *Ibid* at 35, para 40.

[70] *Ibid*.

'the environment is not an abstraction but represents the living space, the quality of life and the very health of human beings, including generations unborn. The existence of the general obligation of States to ensure that activities within their jurisdiction and control respect the environment of other States or of areas beyond national control is now part of the corpus of international law relating to the environment.' *(Legality of the Threat or Use of Nuclear Weapons, Advisory Opinion, ICJ Reports 1996,* pp. 241–242, para 29)[71]

The Court thus confirmed that the obligation not to cause transboundary environmental harm, whether by pollution or otherwise, is part of customary international law. However, because Hungary had not satisfied the rigorous requirements for making out a case for a 'state of necessity', the Court found that Hungary's argument for treaty termination[72] on this ground failed. The Court 'note[d] that the dangers ascribed to the upstream reservoir were mostly of a long-term nature and, above all, that they remained uncertain'.[73]

Hungary also argued that:

if the works at [the downstream dam site at] Nagymaros had been carried out as planned, the environment – and in particular the drinking water resources – in the area would have been exposed to serious dangers on account of problems linked to the upstream reservoir on the one hand and, on the other, the risks of erosion of the riverbed downstream.[74]

But the Court also rejected this argument on the ground that the availability to Hungary of alternative methods of solving the problem vitiated its state of necessity argument.[75]

Slovakia[76] had argued that its having put the upstream portion of the project into operation unilaterally via its 'Variant C' of the original project was lawful in that it was essentially what Hungary had agreed to

---

71 *Ibid* at 41, para 53.

72 The Court in fact found that even if Hungary had been able to establish a state of necessity, this would only preclude the wrongfulness of Hungary's unilateral termination of the treaty while the facts giving rise to the necessity existed. Therefore, the effect of the treaty would only be suspended, not terminated.

73 *Gabčíkovo-Nagymaros, supra* note 3 at 42, para 55.

74 *Ibid.*

75 *Ibid* at 42–43, para 55.

76 While the Court used 'Czechoslovakia' throughout its opinion to refer to the state on the northern bank of the river, Slovakia became an independent state on 1 January 1993 and by virtue of an agreement with the Czech Republic took over responsibility for the project and the case against Hungary.

in the treaty. The Court disagreed. It noted that Variant C resulted in Czechoslovakia's 'appropriat[ing], essentially for its use and benefit, between 80 and 90 per cent of the waters of the Danube before returning them to the main bed of the river, despite the fact that the Danube is not only a shared international watercourse but also an international boundary river'.[77] Further, Hungary's withdrawal of its consent to the treaty,

> constituted a violation of Hungary's legal obligations, demonstrating, as it did, the refusal by Hungary of joint operation; but that cannot mean that Hungary forfeited its basic right to an equitable and reasonable sharing of the resources of an international watercourse.[78]

This 'basic right' therefore exists with or without a treaty providing for it, and even survives breaches by the right holder of other obligations relating to an international watercourse. In addition, the 'resources of an international watercourse' clearly go beyond mere water and include the entire spectrum of things water can be used for – everything from irrigating crops, producing electricity and fishing to domestic uses, swimming and boating.

In its analysis of whether Variant C could be justified as a countermeasure, the Court harked back to a concept articulated by its predecessor in the *River Oder* case:

> In 1929, the Permanent Court of International Justice, with regard to navigation on the River Oder, stated as follows:
>
> > '[the] community of interest in a navigable river becomes the basis of a common legal right, the essential features of which are the perfect equality of all riparian States in the user of the whole course of the river and the exclusion of any preferential privilege of any one riparian State in relation to the others.' *(Territorial Jurisdiction of the International Commission of the River Oder, Judgment No. 16, 1929, P. C. I. J., Series A, No. 23,* p. 27)
>
> Modern development of international law has strengthened this principle for non-navigational uses of international watercourses as well, as evidenced by the adoption of the Convention of 21 May 1997 on the Law of the Non-Navigational Uses of International Watercourses by the United Nations General Assembly.
>
> The Court considers that Czechoslovakia, by unilaterally assuming control of a shared resource, and thereby depriving Hungary of its right to an equitable and reasonable share of the natural resources of the Danube – with the continuing effects of the diversion of these waters on the ecology of the

---

77 *Gabčíkovo-Nagymaros, supra* note 3 at 54, para 78.

78 *Ibid.*

> riparian area of the Szigetkoz – failed to respect the proportionality which is required by international law.[79]

In this remarkable passage, the Court applies the concept of community of interest to non-navigational uses of international watercourses, with its features of 'the perfect equality of all riparian States in the use of the whole course of the river and the exclusion of any preferential privilege of any one riparian State in relation to the others'. These principles are given concrete shape with regard to transboundary water pollution in the form of specific obligations in the 1992 Convention, and general ones in the 1997 Convention. The status of the latter treaty as a codification of the basic principles of customary international law in the field is reinforced by the Court's reference to it, a mere four months after it was concluded, as support for the modern vitality of the principle of community of interest in international watercourses.[80]

The *Gabčíkovo-Nagymaros* case thus contains a wealth of statements by the ICJ that apply specifically or generally to transboundary water pollution. The next case to be considered involved a dispute that focused on that problem.

B. The *Pulp Mills* Case

Argentina brought an action against Uruguay before the ICJ in 2006 alleging that Uruguay had breached the 1975 Statute of the River Uruguay,[81] a treaty between the two countries, by authorising the construction of two pulp mills on the river, which would allegedly affect the quality of its waters. In the end, only one of these plants, the Orion (Botnia) mill, was constructed and put into operation, the latter having occurred during the proceedings in the case. The Court held that Uruguay had breached its procedural obligations under the treaty by failing to notify Argentina of plans for the plant through the CARU, the joint commission established by the 1975 Statute. However, it also held that Uruguay had not breached its substantive obligations under the 1975 Statute. The main rules concerning transboundary water pollution applicable in the case were contained in the 1975 Statute and are thus not

---

[79] *Ibid* at 53, para 85.

[80] The Court also quoted from Article 5, para 2, on equitable participation, in support of its finding that the parties had to re-establish the joint regime under the 1997 Treaty. *Ibid* at 80, para 147.

[81] Statute of the River Uruguay, signed 26 February 1975, 1295 UNTS 340 (entered into force 18 September 1976) (1975 Statute).

generally applicable, per se. However, those rules, contained chiefly in Chapter X of the 1975 Statute, are consistent with the provisions on the subject in the 1992 and 1997 Conventions. Thus the Court's statements about the rules in the 1975 Statute on pollution may be taken to apply generally to the corresponding provisions of the two conventions. In addition, in its judgment, the Court made several general statements concerning pollution and transboundary environmental impact assessment that are of present interest.

First, the Court had recourse to the principle of community of interest in discussing the duty of the parties to cooperate in ensuring the fulfilment of the object and purpose of the 1975 Statute. It stated: 'By acting jointly through CARU, the Parties have established a real community of interests and rights in the management of the River Uruguay and in the protection of its environment.'[82] Second, the Court expanded on the content of what it called the 'principle of prevention':

> The Court points out that the principle of prevention, as a customary rule, has its origins in the due diligence that is required of a State in its territory. It is 'every State's obligation not to allow knowingly its territory to be used for acts contrary to the rights of other States.' (*Corfu Channel (United Kingdom* v. *Albania), Merits, Judgment, ICJ Reports 1949*, p. 22)
>
> A State is thus obliged to use all the means at its disposal in order to avoid activities which take place in its territory, or in any area under its jurisdiction, causing significant damage to the environment of another State. This Court has established that this obligation 'is now part of the corpus of international law relating to the environment.' (*Legality of the Threat or Use of Nuclear Weapons, Advisory Opinion, ICJ Reports 1996 (I)*, p. 242, para. 29)[83]

Thus, States are required to exercise due diligence to prevent transboundary environmental harm, including but not limited to harm from pollution of shared freshwater resources.

Third, the Court drew on the principle of equitable and reasonable utilisation in elaborating on the obligation to contribute to the optimum and rational utilisation of the river under Article 1 of the 1975 Statute:

> The Court wishes to add that such utilization could not be considered to be equitable and reasonable if the interests of the other riparian State in the shared resource and the environmental protection of the latter were not taken into account. Consequently, it is the opinion of the Court that Article 27

---

[82] *Pulp Mills*, *supra* note 3 at 105, para 281. Disclosure: The author served as counsel to Uruguay in this case.

[83] *Ibid* at 55, para 101.

embodies this interconnectedness between equitable and reasonable utilization of a shared resource and the balance between economic development and environmental protection that is the essence of sustainable development.[84]

Here the Court seems to imply that use of an international watercourse must be equitable and reasonable, independently of an applicable treaty, unless, of course, the parties agreed to the contrary. It also sheds light on the role of this principle in the context of treaty and customary regimes.

Fourth, the Court in its judgment repeatedly characterised obligations under the 1975 Statute as ones of due diligence. For example, Article 36 of the 1975 Statute provides: 'The parties shall coordinate, through the Commission, the necessary measures to avoid any change in the ecological balance and to control pests and other harmful factors in the river and the areas affected by it.'[85] The Court found that '[b]oth Parties are … called upon, under Article 36, to exercise due diligence in acting through the Commission for the necessary measures to preserve the ecological balance of the river'.[86] The Court elaborated on the meaning of the obligation of due diligence in the context of a specific obligation as follows:

> [T]he obligation [under Article 41(a) of the Statute] to 'preserve the aquatic environment, and in particular to prevent pollution by prescribing appropriate rules and measures' is an obligation to act with due diligence in respect of all activities which take place under the jurisdiction and control of each party. It is an obligation, which entails not only the adoption of appropriate rules and measures, but also a certain level of vigilance in their enforcement and the exercise of administrative control applicable to public and private operators, such as the monitoring of activities undertaken by such operators, to safeguard the rights of the other party. The responsibility of a party to the 1975 Statute would therefore be engaged if it was shown that it had failed to act diligently and thus take all appropriate measures to enforce its relevant regulations on a public or private operator under its jurisdiction.[87]

The Court thus goes well beyond the actual language of the provision in question in explaining the significance of its finding that the provision is a due diligence obligation. It seems likely that the Court would likewise require States to take 'all appropriate measures' toward the fulfilment of other, similar obligations, whether based in a treaty or in customary international law. In addition, the Court, in addressing environmental impact assessment, again referred to due diligence:

---

84 *Ibid* at 75, para 177.
85 1975 Statute, *supra* note 81, Art 36.
86 *Pulp Mills*, *supra* note 3 at 77, para 187.
87 *Ibid* at 79, para 197.

> [d]ue diligence, and the duty of vigilance and prevention which it implies, would not be considered to have been exercised, if a party planning works liable to affect the régime of the river or the quality of its waters did not undertake an environmental impact assessment on the potential effects of such works.[88]

This obligation to conduct an environmental impact assessment was addressed directly by the Court in the context of the parties' obligations under the 1975 Statute:

> there are situations in which the parties' intent upon conclusion of the treaty was, or may be presumed to have been, to give the terms used – or some of them – a meaning or content capable of evolving, not one fixed once and for all, so as to make allowance for, among other things, developments in international law. (*Dispute Regarding Navigational and Related Rights (Costa Rica* v. *Nicaragua), Judgment, ICJ Reports 2009*, p. 242, para. 64)
>
> In this sense, the obligation to protect and preserve, under Article 41 *(a)* of the Statute, has to be interpreted in accordance with a practice, which in recent years has gained so much acceptance among States that it may now be considered a requirement under general international law to undertake an environmental impact assessment where there is a risk that the proposed industrial activity may have a significant adverse impact in a transboundary context, in particular, on a shared resource. Moreover, due diligence, and the duty of vigilance and prevention which it implies, would not be considered to have been exercised, if a party planning works liable to affect the régime of the river or the quality of its waters did not undertake an environmental impact assessment on the potential effects of such works.[89]

In this passage, the Court begins by quoting its earlier finding that the parties may have intended to give certain terms of a treaty a meaning that evolves so that they will keep pace with developments in international law. It then interprets the obligation to protect and preserve to include an obligation, which it finds to be part of customary international law, to conduct a transboundary environmental impact assessment. The Court also finds that conducting such an assessment would be part of the due diligence required where planned measures could have a significant transboundary impact. These findings are highly significant and should be noted and implemented by all States that are riparian to international watercourses.

---

88 *Ibid* at 83, para 204.
89 *Ibid.*

C. The *Indus Waters Kishenganga* Arbitration

On 19 September 1960, India and Pakistan concluded the Indus Waters Treaty.[90] Pursuant to Article IX of the Treaty, Pakistan initiated arbitral proceedings against India on 17 May 2010. In its Request for Arbitration, Pakistan stated that the parties had failed to resolve a dispute concerning the Kishenganga Hydro-Electric Project (KHEP) and identified two specific questions to be resolved by the Court of Arbitration concerning that project. The first question was whether India's planned KHEP was consistent with the Indus Waters Treaty. The second question, which will not be considered here, was whether India's proposed method of controlling sediment in the reservoir behind the dam, known as draw-down flushing, was permissible under the Indus Waters Treaty. The Court of Arbitration held that the KHEP was consistent with the Indus Waters Treaty, but interpreted the agreement to require India to release a minimum environmental flow through the dam year around. It is this latter feature of the Court's Award that will be the focus of discussion here.

The Kishenganga River is a tributary of the Jhelum, one of the six rivers divided between the parties by the 1960 Treaty (India is upstream on all six rivers, including the Jhelum, and Pakistan downstream). The Indus Waters Treaty literally allocates three rivers to each country, with carefully controlled exceptions for certain uses by the other. The Jhelum is one of the rivers allocated to Pakistan, but the treaty permits India to use the river and its tributaries for the production of hydroelectric power, under detailed conditions. As planned by India, the KHEP would divert water from a dam on the Kishenganga through a 23.5-kilometre tunnel to a power plant in the catchment of another tributary of the Jhelum, 666 metres below the dam.[91] From there, the water would pass into the Jhelum, before the latter flows from India-administered Kashmir into Pakistan-administered Kashmir, and on into Pakistan. Meanwhile, Pakistan was in the process of constructing its own hydroelectric plant downstream on the Kishenganga in Pakistan-administered Kashmir, the Neelum-Jhelum Hydroelectric Plant (NJHEP). It was concerned that the KHEP would reduce its ability to produce hydroelectric power at the NJHEP.

---

[90] Indus Waters Treaty, signed 19 September 1960, 419 UNTS 126 (entered into force 12 January 1961) (Indus Waters Treaty).

[91] *Kishenganga*, Partial Award, *supra* note 29 at 50, para 155. Disclosure: The author served as counsel to India in this arbitration.

As is the case with many Himalayan rivers, the flow of the Kishenganga varies widely with the seasons. Monsoon rains bring high flows from May to August, while the flows from early October to mid-March are considerably lower.[92] Pakistan contended that the KHEP was designed to be operated in such a way that it would divert the entirety of the flow of the Kishenganga through the tunnel during the low-flow period, so that 'there will be no flow immediately below the dam site for six months of the year'.[93] India contested this assertion on the ground that it 'is in conflict with the river flow data and fails to take into account the waters contributed to the Kishenganga/Neelum from tributaries below the [dam] site'.[94] The Court of Arbitration also took note of 'India's commitment to ensure a minimum environmental flow downstream of the KHEP at all times',[95] something that was in fact required by the law of India.

The Court of Arbitration found that India was required to preserve downstream flows for two purposes: to protect Pakistan's rights under the Indus Waters Treaty to use the Kishenganga for agricultural and hydroelectric power generation purposes[96] and to protect the environment as required by customary international law.[97] It is the latter purpose that I will focus upon here.

The term 'environment' appears nowhere in the Indus Waters Treaty, perhaps understandably since the agreement was concluded before the expression had taken on the connotation it has today. However, the Indus Waters Treaty does employ the term 'pollution' in the context of providing for the prevention of harm to uses of the rivers, rather than protection of their ecosystems or the environment, per se.[98] This is in line with the Indus Waters Treaty's principal focus, which is the allocation of

---

92 *Ibid* at 42–43, para 129.
93 *Ibid* at 87, para 249.
94 *Ibid* at 71, para 206.
95 *Ibid* at 171, para 453.
96 See especially, *ibid* at 161, para 436.
97 *Ibid* at 168–72, paras 447–454.
98 Indus Waters Treaty, *supra* note 90, Art IV, para 10:

> Each Party declares its intention to prevent, as far as practicable, undue pollution of the waters of the Rivers which might affect adversely uses similar in nature to those to which the waters were put on the Effective Date, and agrees to take all reasonable measures to ensure that, before any sewage or industrial waste is allowed to flow into the Rivers, it will be treated, where necessary, in such manner as not materially to affect those uses: Provided that the criterion of reasonableness shall be the customary practice in similar situations on the Rivers.

water for purposes of irrigation and hydroelectric power production. Despite the Indus Waters Treaty's silence on the environment, in its pleadings Pakistan had raised environmental concerns relating to the alleged lack of flow below the Kishenganga dam for six months. Of course, any order by the Court of Arbitration to release water through the Kishenganga dam would not only be of possible benefit to the aquatic ecosystems below the dam but would also increase Pakistan's hydroelectric generation capacity.

The Court of Arbitration found that India's 'KHEP preceded the NJHEP, such that India's right to divert the waters of the Kishenganga/Neelum for power generation by the KHEP is protected under the Treaty'.[99] However, the Court also concluded that:

> the constraints specified by the Treaty … and, in addition, by the relevant principles of customary international law to be applied by the Court … when interpreting the Treaty … require India to operate the KHEP in a manner that ensures a minimum flow of water in the riverbed of the Kishenganga/Neelum downstream of the Plant.[100]

But the Court of Arbitration found that it had insufficient data to determine a precise minimum flow. The Court therefore 'defer[red] its determination of the appropriate minimum flow downstream of the KHEP to a further, Final Award, to be issued after it has had the benefit of considering further written submissions on the matter from the Parties'.[101] The Court therefore requested that the parties submit further data and information, which they did.[102] In its Final Award, issued in December 2013,[103] the Court set the minimum flow to be released by the Kishenganga dam. The following discussion will focus upon the rationale given by the Court in its Partial Award for requiring the release of a minimum environmental flow, rather than the quantum of that flow.

The Court first pointed out that '[w]ell before the Treaty was negotiated, a foundational principle of customary international environmental law had already been enunciated in the *Trail Smelter* arbitration'.[104] After

---

[99] *Ibid* at 161, para 435.

[100] *Ibid* at 168, para 445. The Indus Waters Treaty provides in para 29(b) of Annexure G that in interpreting the treaty the Court may apply, 'whenever necessary for its interpretation or application, but only to the extent necessary for that purpose … (b) Customary international law'.

[101] *Ibid* at 172, para 456.

[102] *Ibid* at 173–4, paras 458–462.

[103] *Kishenganga*, Final Award, *supra* note 29.

[104] *Ibid* at 169, para 448.

setting forth the rule stated by the tribunal in that case, which involved transboundary air pollution,[105] the Court moved on to the 'broader restatement of the duty to avoid transboundary harm' contained in Principle 21 of the 1972 Stockholm Declaration.[106] The Court then drew the following conclusion from these precedents: 'There is no doubt that States are required under contemporary customary international law to take environmental protection into consideration when planning and developing projects that may cause injury to a bordering State.'[107] It will be noted that this finding is not limited to projects on international watercourses, but applies to the environment generally. In support of the requirement, the Court referred to 'a series of international conventions, declarations and judicial and arbitral decisions [that] have addressed the need to manage natural resources in a sustainable manner'.[108] It proceeded to draw upon and quote from the decisions in the *Gabčíkovo-Nagymaros*,[109] *Pulp Mills*[110] and *Iron Rhine*[111] cases, including the finding of the *Iron Rhine* tribunal that according to a '"principle of general international law" … States have "a duty to prevent, or at least mitigate" significant harm to the environment when pursuing large-scale construction activities'.[112] The Court then invoked what it called the 'principle' of sustainable development, citing the ICJ's explanation that this 'concept' expresses the 'need to reconcile economic development with protection of the environment'.[113]

Explaining its rationale for applying current principles of environmental protection to a case governed by a treaty concluded in 1960, the Court of Arbitration declared that '[i]t is established that principles of international environmental law must be taken into account even when

---

105 *Ibid*, quoting from the award in the *Trail Smelter Case (United States of America, Canada)* (1941) 3 RIAA 1905 at 1965.

106 *Kishenganga,* Partial Award, *supra* note 29 at 169, para 448, citing the 1972 Stockholm Declaration.

107 *Ibid* at 169, para 449.

108 *Ibid* at 169–70, para 449 (footnotes omitted).

109 *Gabčíkovo-Nagymaros*, *supra* note 3.

110 *Pulp Mills*, *supra* note 3.

111 *In the Arbitration Regarding the Iron Rhine (Ijzeren Rijn) Railway (Kingdom of Belgium v Kingdom of the Netherlands)*, Award (24 May 2005) ICGJ 373 (Permanent Court of Arbitration 2005) (*Iron Rhine*).

112 *Kishenganga,* Partial Award, *supra* note 29 at 170, para 451, quoting from *Iron Rhine*, *supra* note 111 at para 59.

113 *Ibid* at 170, para 449, quoting from *Gabčíkovo-Nagymaros*, *supra* note 3 at 78, para 141.

(unlike the present case[114]) interpreting treaties concluded before the development of that body of law'.[115] As support for this proposition, the Court referred to the ICJ's declaration in the *Gabčíkovo-Nagymaros* case that '"new norms have to be taken into consideration, and … new standards given proper weight"',[116] 'whenever necessary for the application of a treaty'.[117] Thus, the Court concluded: 'It is therefore incumbent upon this Court to interpret and apply this 1960 Treaty in light of the customary international principles for the protection of the environment in force today.'[118]

These findings concerning treaty interpretation will not escape the attention of governments. There are now three interstate cases[119] in which a treaty has been interpreted, not in light of the law existing when it was concluded, but in accordance with the law existing when the case was decided. This will likely not be the case with every treaty, but these decisions display a strong tendency on the part of international tribunals to interpret treaty provisions in accordance with modern principles of international environmental law. And the Court of Arbitration indicated that the new environmental principles apply equally to the upstream and downstream States (India and Pakistan, respectively): 'hydro-electric projects (including Pakistan's projects) must be planned, built and operated with environmental sustainability in mind'.[120]

To sum up, the Court of Arbitration interpreted the Indus Waters Treaty to require protection of the environment in all phases of a project's life: planning, construction and operation. In particular, the Court interpreted the Indus Waters Treaty to mean that while India is permitted to divert the flow of the Kishenganga into another basin, principles of contemporary international environmental law require that India release a sufficient flow from the dam to protect the environment – despite the fact that nothing in the Indus Waters Treaty itself required this.

---

114 Here the Court was presumably referring to the *Trail Smelter* award rendered in 1941. However, while that decision is often referred to as the font of international environmental law, in the author's view it cannot be said that the field was 'developed' in 1941, or even before 1960 (Author's footnote).

115 *Kishenganga,* Partial Award, *supra* note 29 at 171, para 452.

116 *Ibid*, quoting from *Gabčíkovo-Nagymaros*, *supra* note 3 at 78, para 140.

117 *Kishenganga,* Partial Award, *supra* note 29 at 171, para 452.

118 *Ibid.*

119 The three cases are *Gabčíkovo-Nagymaros*, *supra* note 3, *Navigational and Related Rights*, *supra* note 24, and *Indus Waters Kishenganga Project*, *supra* note 29.

120 *Kishenganga,* Partial Award, *supra* note 29 at 172, para 454.

D. Other Recent Cases

Finally, brief mention may be made of two cases between Costa Rica and Nicaragua that are currently before the ICJ, the *Certain Activities* case[121] and the case concerning *Construction of a Road along the San Juan.*[122] In *Certain Activities*, on 18 November 2010 Costa Rica instituted proceedings against Nicaragua in the ICJ regarding Nicaragua's alleged incursion into Costa Rica's territory, resulting in claimed damage to a Costa Rican Ramsar site. Nicaragua denied having entered Costa Rica's territory, contending that the area in question was in fact Nicaraguan territory. The case is worthy of note here because of the environmental character of many of the parties' arguments.

The case concerning *Construction of a Road along the San Juan*, filed by Nicaragua on 22 December 2011, is closer to the subject of the present chapter inasmuch as it involves allegations by Nicaragua of pollution of the San Juan River by Costa Rica as a result of the construction of a road along the Costa Rican bank of the river. The pollution takes the form, primarily, of increased sedimentation of the river, which already carries a rather heavy sediment load. Nicaragua is particularly concerned because, unlike most boundary rivers, the San Juan is wholly within Nicaragua; the border runs along the Costa Rican bank of the river. This case was joined with the *Certain Activities* case by the ICJ, which has indicated that it intends to hold hearings in the two cases in the early spring of 2015.

V. CONCLUSION

The foregoing discussion has shown that contemporary international law places emphasis on the importance of the prevention of pollution of international watercourses and the protection of aquatic ecosystems. This is evident from the two global treaties on international freshwater law and the decisions of international courts and tribunals concerning the subject discussed above. Moreover, the ICJ has indicated in two cases that it will take contemporary principles of international environmental law into account in interpreting earlier treaties concerning international watercourses, and the recent decision of an arbitral tribunal in a significant interstate case was to the same effect.

---

121 *Certain Activities, supra* note 26.

122 *Construction of a Road along the San Juan, supra* note 3. Disclosure: The author is counsel to Nicaragua in both of these cases.

Countries sharing international watercourses – including, of course, those in Asia – would be well advised to take these developments into account in their relations with other riparian States and in the management of the portions of international watercourses and their watersheds located within their territories. One of the great rivers of Asia is the Mekong, which since 1995 has been governed by a treaty between the four lower-riparian countries: the Lao PDR, Thailand, Cambodia and Vietnam.[123] While the Agreement is open to the two upper riparian States, China and Myanmar, they have not yet joined it. China, however, has been participating as an observer in the Mekong River Commission (MRC) established by the treaty since 1996.[124] In view of the location of the conference for which this chapter was prepared and the regional perspective of this book, it is perhaps appropriate to conclude by considering briefly the approach of the Mekong regime to transboundary pollution and related problems.

The 1995 Mekong Agreement provides generally that the Parties agree to protect the Mekong River Basin from 'pollution or other harmful effects resulting from any development plans and uses of water … in the Basin'.[125] It also provides, separately, for 'prevention and cessation of harmful effects' including harm to water quality and the ecological balance of the river system.[126] The inclusion of these provisions is no doubt a positive feature of the treaty, but their generality could give rise to issues as to their implementation in specific cases.

A function of the Joint Committee, one of the three permanent bodies comprising the MRC, is to 'conduct appropriate studies and assessments for the protection of the environment and maintenance of the ecological balance of the Mekong River Basin'.[127] While commendable, this provision falls short of requiring the Parties themselves to conduct an environmental impact assessment for their planned projects in the Mekong River Basin.[128]

---

[123] Agreement on the Cooperation for the Sustainable Development of the Mekong River Basin, concluded 5 April 1995, 34 ILM 864 (1995 Mekong Agreement).

[124] Mekong News, Newsletter of the Mekong River Commission (April–June 2002/2) at 1.

[125] 1995 Mekong Agreement, *supra* note 123, Art 3.

[126] *Ibid*, Art 7.

[127] *Ibid*, Art 24(D).

[128] *Ibid*, Art 2. In Art 2 of the 1995 Mekong Agreement, the parties agree to protect the environment and ecological balance of the basin from pollution or

With respect to planned projects, always a potential source of contention on international watercourses, the treaty provides only for the provision of notification to the Joint Committee for 'intra-basin uses and inter-basin diversions' on Mekong tributaries,[129] and for the mainstream, 'prior consultation which aims at arriving at an agreement by the Joint Committee' regarding intra-basin use.[130] These provisions have led to uncontrolled hydroelectric development of Mekong tributaries in Laos and Vietnam,[131] and sharp controversy concerning Laos's Xayaburi Hydropower Project, an 820 metre-long dam under construction on the main stem of the Mekong where both banks of the river are in Lao territory. Laos had submitted prior consultation documents to the MRC and a number of groups were established to support the consultation process. However, after the expiration of the six-month consultation period provided for in guidelines adopted by the MRC,[132] Laos indicated that any concerns of the other three parties would be accommodated and there was no need to extend consultations beyond the six-month period. The other three lower riparians disagreed. Construction at Xayaburi is reportedly continuing at the time of this writing though the controversy has not been resolved. While the primary concerns regarding this project appear to relate to flow, the same process would apply in the case of possible pollution by a planned project on or near the river or one of its tributaries.[133]

It is noteworthy that Vietnam has recently ratified the 1997 Convention (its ratification was the 35th, the required number for entry into force of the agreement, which occurred on 17 August 2014). Since Vietnam is already a party to the 1995 Mekong Agreement, it might be wondered why it saw fit to join the 1997 Convention. One possibility is that the 1997 Convention's provisions on prior notification of planned measures are far more detailed and stringent than those under the 1995 Mekong

other harmful effects, but there are no specific provisions for the implementation of this obligation.

[129] *Ibid*, Art 5, para A.

[130] *Ibid*, Art 5(B)(2).

[131] See Bennett L Bearden, 'Following the Proper Channels: Tributaries in the Mekong Legal Regime' (2012) 14 *Water Policy* 991.

[132] Mekong River Commission, Guidelines on Implementation of the Procedures for Notification, Prior Consultation and Agreement, Lao PDR, 31 August 2005, accessed 6 January 2015 at http://www.mrcmekong.org/assets/Publications/policies/Guidelines-on-implementation-of-the-PNPCA.pdf.

[133] Article 5(B)(2)(a) covers any '[i]ntra-basin use', 1995 Mekong Agreement, *supra* note 123.

Agreement.[134] As with some of the other principles reflected in the 1997 Convention, it would appear that the principle of prior notification of planned projects with possible adverse transboundary effects now forms part of customary international law.[135] The provisions of the 1997 Convention on planned measures could therefore inform the implementation of the prior notification and consultation principles, possibly assisting in the interpretation and application of the corresponding provisions of the 1995 Mekong Agreement.

The Mekong situation is perhaps a harbinger of developments concerning international watercourses generally, including those in Asia, as ever-increasing numbers of people compete for finite quantities of Earth's most precious natural resource. It is to be hoped that international law can help to stabilise relations between States in this field as the competition for shared freshwater resources intensifies.

134 1997 Convention, *supra* note 30, Arts 11–19.
135 See generally *Pulp Mills*, *supra* note 3.

5. Environmental impact assessment in preventing transboundary river pollution under international law: an analysis

Lan Hua

I. INTRODUCTION

Fresh water is an important resource for riparian countries, not only for the daily life of the public, but also for agricultural irrigation and hydroelectric power production. Clean, fresh water is also essential to the protection of the ecosystem and conservation of the river basins. With economic development, rivers have been increasingly utilised and exploited, creating crises and debate among the riparian countries regarding issues of pollution risk and environmental protection. With 261 of the world's river basins shared by more than one country,[1] the management of transboundary waters, especially pollution prevention, is a critical issue in building and maintaining friendly relationships among the riparian countries.

There is no doubt that riparian States are entitled to use the waters of a transboundary river within their territory, as confirmed by the Convention on the Law of Non-Navigational Uses of International Watercourses (1997 Convention on Watercourses).[2] This right is also supported by the United Nations (UN) Resolution of Permanent Sovereignty over Natural Resources.[3] However, a State is not allowed to knowingly use its territory

---

[1] Peter H Gleick, *The World's Water 2000–2001: The Biennial Report On Freshwater Resource* (Washington, DC: Island Press, 2000) at 28.

[2] See Convention on the Law of the Non-navigational Uses of International Watercourses, adopted 21 May 1997, 36 ILM 700 (entered into force 17 August 2014) (1997 Convention on Watercourses), Art 5.

[3] *Permanent Sovereignty over Natural Resources*, GA Res 1803 (XVII), UNGAOR, 17th Sess, Supp No 17, UN Doc A/5217 (1962) at 5.

for acts contrary to the rights of other States as established in the *Corfu Channel* case,[4] where it was held that a riparian country has the obligation of due diligence to prevent the risk of significant harm when using its water resources. Whether in regard to domestic rivers or transboundary waters, preventing and controlling pollution is preferable to implementing remedies after the pollution has occurred, as confirmed by the general commentaries on the Draft Articles on Prevention of Transboundary Harm from Hazardous Activities (Draft Articles on Prevention).[5] With regard to transboundary waters, effective pollution prevention and control requires political willingness and cooperation, and good faith implementation of specific legal frameworks at the domestic level. Among the legal frameworks established by States' practice, the Environmental Impact Assessment (EIA) mechanism is increasingly important and necessary in regulating the use of transboundary rivers. Many legal documents, including multilateral agreements relating to environmental protection, have incorporated the EIA principle, even though the provisions concerning EIA therein are quite general and lacking in details of EIA scope and content. A review of the dispute settlement provisions included in these legal documents is helpful in understanding the effectiveness of the EIA and its application to the utilisation and governance of transboundary rivers.

This chapter examines the legal roles and application of the EIA principle from the perspective of international law by analysing the practice of international dispute settlement relating to transboundary rivers, with the objective of developing suggestions for pollution control measures along the Lancang-Mekong River by the riparian countries in the context of the rapid development of the hydropower industry. This chapter consists of four parts: the first part is an introduction to the significance and necessity of EIA for the pollution control of transboundary rivers; the second part provides a textual analysis of the EIA mechanism in multilateral and bilateral agreements on transboundary damage or water use in order to determine the extent of EIA obligations for riparian countries; the third part focuses on the application of EIA to international disputes in international tribunals relating to the use of

---

[4] *Corfu Channel Case (UK v Albania)*, Merits, [1949] ICJ Rep 4 at 22.

[5] 'International Liability for Injurious Consequences Arising out of Acts Not Prohibited by International Law (Prevention of Transboundary Harm from Hazardous Activities)' in 'Report of the International Law Commission to the General Assembly covering the work of its fifty-third session, with commentaries, 2001' (UN Doc A/56/10), Ch V, in *Yearbook of the International Law Commission 2001*, vol II, Part Two (UN, 2001) (Draft Articles on Prevention).

transboundary waters; and the fourth part examines the possibility of incorporating an EIA mechanism into the legal frameworks governing the economic development and exploitation of the Lancang-Mekong River Basin, with the aim of preventing and/or minimising transboundary pollution and damage, followed by a short reflection and remarks.

II. LEGAL STATUS OF EIA IN TRANSBOUNDARY RIVERS GOVERNANCE

EIA is a process of evaluating the likely beneficial and adverse environmental impact of a proposed project of development, taking into account interrelated socio-economic, cultural and human health factors. As an environmental protection tool, EIA aims to predict environmental impact at an early stage in project planning and design, find ways and means to reduce adverse impact, shape projects to suit the local environment and present the predictions and options to decision-makers prior to the approval of the given project.[6]

A. Recognition of EIA in International Environmental Agreements

EIA was first introduced by domestic law in the United States (US) with the enactment of the National Environmental Policy Act of 1969 and was intended as a tool to reform the decision-making process by changing how policymakers approached environmental issues.[7] Although EIA was originally established at the national level, with the growing levels of environmental awareness, it has been used increasingly around the world. For example, in 1985 the European Economic Community (now the European Union (EU)) adopted its EIA directive, the Assessment of the Effects of Certain Public and Private Projects on the Environment.[8] Over time, more and more countries have incorporated EIA into their domestic laws.[9]

---

6 UN Environmental Programme (UNEP), *Resource and Guidance Manual for Environmental Impact Assessments* (UNEP Regional Office for West Asia, 2008) at 5.

7 *National Environmental Policy Act of 1969*, 42 USC § 4321 (1970).

8 EC *Council Directive 85/337/EEC of 27 June 1985 on the assessment of the effects of certain public and private projects on the environment*, [1985] OJ L 175/40.

9 Marceil Yeater and Lal Kurukulasuriya, 'Environmental Impact Assessment Legislation in Developing Countries' in Sun Lin and Lal Kurukulasuriya

At the international level, many environmental protection documents require risk assessment for activities that are likely to have a significant adverse impact on the environment. Principle 17 of the Rio Declaration provides that, '[EIA], as a national instrument, shall be undertaken for proposed activities that are likely to have a significant adverse impact on the environment and are subject to a decision of a competent national authority'.[10] Furthermore, the UNEP resolutions make clear the necessity of EIA in the transboundary context.[11]

Many multilateral and regional agreements gradually confirm EIA as a rule with legal binding force for the activities with adverse effects. Multilateral agreements with EIA rules include, but are not limited to, the following treaties:[12] the Convention on the Regulation of Antarctic Mineral Resource Activities (Article 4),[13] the Protocol on Environmental Protection to the Antarctic Treaty (Article 8),[14] the Convention on Biological Diversity (Article 14),[15] and the Convention on the Transboundary Effects of Industrial Accidents (Article 4).[16] The Convention on Environmental Impact Assessment in a Transboundary Context (1991

(eds), *UNEP's New Way Forward: Environmental Law and Sustainable Development* (UNEP Earthprint, 1995) 257–76 at 259; UN Economic Commission for Europe (UNECE), Environmental Series 5, *Application of Environmental Impact Assessment Principles to Policies, Plans and Programs*, UN Doc ECE/ENVWA/27 (UN, 1992) at 43.

[10] Rio Declaration on Environment and Development, in 'Report of the United Nations Conference on Environment and Development', 12 August 1992, UN Doc A/CONF.151/26 (Vol I), 31 ILM 874 (1992 Rio Declaration), Principle 17.

[11] *Environmental Law Guidelines and Principles on Shared Natural Resources*, UNEP, Principle 4, (approved by *Cooperation in the field of the environment concerning natural resources shared by two or more States,* GA Res 34/186, (1979); *Goals and Principles of Environmental Impact Assessment*, UNEP, Annex III, UN Doc UNEP/GC.14/17 (1987); *Adopting the Goals and Principles of the EIA*, Decision 14/25, UNEP, reprinted in *Report of the Governing Council on the Work of its Fourteenth Session*, UNGAOR, 42nd Sess, Supp No 25, Annex I at 77, UN Doc A/42/23 (1987).

[12] See Draft Articles on Prevention, *supra* note 5 at n 900.

[13] Convention on the Regulation of Antarctic Mineral Resource Activities, adopted 2 June 1988, 27 ILM 859 (not yet entered into force), Art 4.

[14] Protocol on Environmental Protection to the Antarctic Treaty, adopted 4 October 1991, 30 ILM 1455 (entered into force 14 January 1998), Art 8.

[15] Convention on Biological Diversity, adopted 5 June 1992, 1760 UNTS 79, 31 ILM 818 (entered into force 29 December 1993), Art 14.

[16] Convention on the Transboundary Effects of Industrial Accidents, 17 March 1992 (entered into force 19 April 2000), Art 4.

Espoo Convention) is notable because its primary aim is 'to prevent, reduce and control *significant* adverse transboundary environmental impact from proposed activities'.[17] It requires States to carry out an environmental impact assessment for all activities listed in Appendix I, which are 'likely to cause significant adverse transboundary impact' in another State.[18]

Regional agreements also emphasise EIA for environmental protection.[19] Such treaties include the Regional Convention for the Conservation of the Environment of the Red Sea and Gulf of Aden,[20] the ASEAN Agreement on the Conservation of Nature and Natural Resources (Article 14),[21] and the Convention for the Protection of the Natural Resources and Environment of the South Pacific Region (1986 Noumea Convention).[22] The 1986 Noumea Convention contains an explicit obligation in Article 16 to conduct EIA before embarking upon any major project that might affect the marine environment.[23]

B. Lack of EIA Content and Scope in International Environmental Agreements

Although the commentary to Article 7 of the Draft Articles on Prevention states as a general rule that 'any decision in respect of authorization of an

---

17 Convention on Environmental Impact Assessment in a Transboundary Context, adopted 25 February 1991, 1989 UNTS 309, 30 ILM 802 (entered into force 10 September 1997) (1991 Espoo Convention), Art 2.

18 Appendix I includes, among other things, oil refineries, coal gasification plants, thermal power stations (above 300MW heat), large diameter oil and gas pipelines, trading ports, waste disposal installations for incineration, chemical treatment or landfill of toxic and dangerous wastes, offshore hydrocarbon production and major storage facilities for petroleum, petrochemical and chemical products. *Ibid.*

19 See Draft Articles on Prevention, *supra* note 5 at n 900.

20 Regional Convention for the Conservation of the Environment of the Red Sea and Gulf of Aden, 14 February 1982 (entered into force 20 August 1985) accessed 7 January 2015 at http://www.unep.ch/regionalseas/main/persga/convtext.html.

21 ASEAN Agreement on the Conservation of Nature and Natural Resources, adopted 9 July 1985 (entered into force 16 November 1997) accessed 7 January 2015 at http://agreement.asean.org/media/download/20140119153602.pdf.

22 Convention for the Protection of the Natural Resources and Environment of the South Pacific Region, 24 November 1986, 26 ILM 38 (entered into force 22 August 1990).

23 *Ibid,* Art 16(2).

activity or project shall be based on an assessment of the possible transboundary harm caused by that activity, including any environmental impact assessment',[24] not all activities that are likely to result in environmental damage or harm are required to carry out an EIA. In addition, most existing international conventions actually do not specify the content of such assessments. For example, with regard to the prevention of damage to the maritime environment, although the United Nations Convention on the Law of the Sea (UNCLOS) does not specify the content of such assessment, Article 206 sets out the States' obligations when such EIA is deemed necessary, which reads as follows:

> When States have reasonable grounds for believing that planned activities under their jurisdiction or control may cause substantial pollution of or significant and harmful changes to the marine environment, they shall, as far as practicable, assess the potential effects of such activities on the marine environment and shall communicate reports of the results of such assessments in the manner provided in article 205.[25]

As argued by Ireland in the *MOX Plant* case, there are two basic requirements of EIA in the maritime field.[26] First, there must be '*reasonable grounds* for believing that planned activities may cause harmful changes to the maritime environment'.[27] Second, any assessment must be carried out '*as far as practicable*'.[28] Nevertheless, the required contents of an EIA are still ambiguous and difficult to monitor.

The basic criteria for the necessity of EIA are prescribed in international treaties; however, there is a lack of the specific contents and elements of an EIA in the transboundary context, which makes the application of EIA in practice difficult. The 1991 Espoo Convention serves as a worthy exception for it lays out the details of such assessment, including the basic obligations and procedures (such as notification and consultation) required for the EIA.[29] Nevertheless, it is worth bearing in mind that according to Article 38 of the International

---

24 Draft Articles on Prevention, *supra* note 5, Commentary to Art 7, at 157.

25 United Nations Convention on the Law of the Sea, adopted 10 December 1982, 1833 UNTS 3 (entered into force 16 November 1994) (UNCLOS), Art 206.

26 *The MOX Plant (Ireland v United Kingdom),* 'Memorial of Ireland' (26 July 2002), Ch 7, accessed 7 January 2015 at http://www.pca-cpa.org/showpage.asp?pag_id=1148.

27 *Ibid* at para 7.7.

28 *Ibid* at para 7.12.

29 1991 Espoo Convention, *supra* note 17, Arts 3 and 5.

Court of Justice (ICJ) Statute and the 1969 Vienna Convention on the Law of Treaties, treaty rules have legal binding force only with respect to contracting parties, unless the relevant rules are a codification of customary international law.[30] Therefore, if the riparian countries are not parties to the 1991 Espoo Convention, the Convention cannot be enforced against them, even when there is a project with transboundary harm or risk. Given the fact that the 1991 Espoo Convention has only 45 parties,[31] it cannot be said that its rules have gained wide acceptance.

Other international instruments also provide important guidance about the standard and content of an EIA in a project with transboundary risk, such as the Governing Council of UNEP, which adopted by consensus in 1987 the resolution of Goals and Principles of Environmental Impact Assessment in which the Governing Council recommended that the General Assembly endorse the Goals and Principles of Environmental Impact Assessment.[32] Although these documents illustrate important progress in the acceptance and use of the EIA mechanism, they are weak in terms of applicability to States, especially in the event of a dispute. According to Article 38(1)(d) of the ICJ Statute, judicial decisions and the teachings of the most highly qualified publicists of the various nations are only subsidiary means for determining the rules of law.

C. EIA in Transboundary Water Governance

At the international level, the documents regarding transboundary water governance include primarily (in chronological order): the Helsinki Rules on the Use of the Waters of International Rivers adopted by the International Law Association (1967 Helsinki Rules),[33] the 1997 Convention

---

30 Statute of the International Court of Justice, adopted 26 June 1945, 33 UNTS 993, Art 38(1)(a) (entered into force 24 October 1945) (ICJ Statute); Vienna Convention on the Law of Treaties, adopted 23 May 1969, 1155 UNTS 331 (entered into force 27 January 1980), Art 26.

31 See the status of the 1991 Espoo Convention, accessed 7 January 2015 at https://treaties.un.org/Pages/ViewDetails.aspx?src=TREATY&mtdsg_no=XXVII-4&chapter=27&lang=en.

32 *Report of the Governing Council on the work of its 14th session, Decision 14/25. Environmental impact assessment* (17 June 1987) accessed 7 January 2015 at http://www.unep.org/Documents.multilingual/Default.asp?DocumentID=100&ArticleID=1658&l=en.

33 Committee on the Uses of the Waters of International Rivers, *Helsinki Rules on the Use of the Waters of International Rivers adopted by the International Law Association at the Fifty-Second Conference* (Helsinki: International Law Association, 1967) (1967 Helsinki Rules).

Watercourses,[34] the Berlin Rules on Water Resources adopted by the International Law Association (2004 Berlin Rules),[35] and the Draft Articles on the Law of Transboundary Aquifers by the International Law Commission (ILC), which were adopted by the United Nations General Assembly (UNGA) as Resolution A/RES/63/124 in the 63rd session of the UNGA (2008 Draft Articles on Aquifers).[36]

Both the 1967 Helsinki Rules and the 2004 Berlin Rules are international guidelines regulating how transboundary rivers and their connected groundwaters may be used. The 1967 Helsinki Rules did not include an EIA requirement, because at that time the concept of EIA had just emerged in the US as a novel tool in environmental protection and had not been adopted widely outside the US. However, the 1967 Helsinki Rules recognised the obligation of preventing pollution and harm in the territory of a co-basin State,[37] which in the present context can be interpreted as requiring States to conduct EIA.[38] Chapter VI of the 2004 Berlin Rules, adopted to replace the 1967 Helsinki Rules and supplement the 1997 Convention on Watercourses, included a section on EIA, which provided not only the obligation of EIA, but also the scope, content and procedure of EIA, making the EIA a useful and practical mechanism.[39]

The 1997 Convention on Watercourses was drafted by the ILC and adopted by the UN in 1997 as a global legal framework establishing basic standards and rules for cooperation between watercourse States on the use, management and protection of waters across international boundaries. This Convention did not include the obligation of EIA, but its Article 12 provides the procedural requirement of notification for significant adverse effects.[40] According to the commentaries of the ILC on the

---

34 See 1997 Convention on Watercourses, *supra* note 2.

35 Water Resources Law Committee, *Berlin Conference (2004) – Fourth Report* (Berlin: International Law Association, 2004) (2004 Berlin Rules).

36 'Shared Natural Resources' in 'Report of the International Law Commission to the General Assembly covering the work of its sixtieth session' (UN Doc A/63/10), Ch 4, in *Yearbook of the International Law Commission 2008*, vol II, Part 2 (UN, 2008) (Draft Articles on Aquifers); *The Law of Transboundary Aquifers,* GA Res 63/124, UNGAOR, 63rd Sess, UN Doc A/RES/63/124 (2008).

37 1967 Helsinki Rules, *supra* note 33, Art X(1).

38 2004 Berlin Rules, *supra* note 35, Ch VI.

39 *Ibid* at 3.

40 1997 Convention on Watercourses, *supra* note 2, Art 12.

1997 Convention on Watercourses, Article 12 can be interpreted as an EIA requirement, albeit as a procedural one rather than as a substantive one.[41]

The 2008 Draft Articles on Aquifers, which have been supplemented by the Resolution of the Law of Transboundary Aquifers in 2011, followed the model of the 1997 Convention on Watercourses, by providing a set of rules on EIA, with Article 12 setting out the obligation of prevention and control of pollution, and Article 15 setting out the relevant procedures of EIA, such as notification and negotiation.[42]

The above four documents constitute the EIA rules relating to the utilisation of transboundary waters. Specifically, the 2004 Berlin Rules and the 2008 Draft Articles on Aquifers provide the procedural rules and criteria of EIA in transboundary waters. However, one should be aware of the deficiencies in these rules' legal effect. Among these documents, only the rules set out in the 1997 Convention on Watercourses are hard law – binding its contracting parties but having no binding force on non-party members except those incorporated by customary rules.[43] Whereas the rules of EIA in the 1967 Helsinki Rules, the 2004 Berlin Rules and the 2008 Draft Articles on Aquifers, being soft laws, are not binding even though highly authoritative institutions have adopted them. Thus, in practice, especially in dispute settlement, they are applicable only when verified as customary international rules.

In short, it is an established legal obligation of all countries to perform an EIA when a project or activity may pose significant harm or risk to transboundary waters. However, in transboundary waters governance, it is difficult to determine the existence of widely accepted legal rules or criteria regarding the content or scope of review of an EIA.

---

41 'Report of the International Law Commission to the General Assembly covering the work of its forty-sixth session, containing Draft articles on the law of the non-navigational uses of international watercourses and commentaries thereto and resolution on transboundary confined groundwater' (UN Doc A/RES/49/52) in *Yearbook of the International Law Commission 1994*, vol II, Part Two (UN, 1994) at 111.

42 Draft Articles on Aquifers, *supra* note 36; *The Law of Transboundary Aquifers*, GA Res 66/104, UNGAOR, 66th Sess, UN Doc A/RES/66/104 (2011).

43 1997 Convention on Watercourses, *supra* note 2. See Alistair Rieu-Clarke and Flavia Rocha Loures, 'Still Not in Force: Should States Support the 1997 UN Watercourses Convention?' (2009) 18(2) *Review of European Community & International Environmental Law* 185 at 195.

III. EIA IN TRANSBOUNDARY RIVERS DISPUTES

Analysis of the application of EIA in international disputes relating to the use of transboundary waters is helpful to understand the effectiveness and utility of the EIA mechanism as an environmental prevention tool.

A. Scope of the Disputes

There are many crises or conflicts between or among the riparian countries regarding the governance and utilisation of transboundary waters. However, compared with the large number of rivers and the numerous agreements on transboundary rivers, the disputes presented in international tribunals are rare. Before the establishment of the UN, cases came before the Permanent Court of International Justice (PCIJ) relating to the allocation of water and utilisation of shared waters, such as in the *River Oder* case,[44] *Diversion of Water from the Meuse Case*[45] and the *Oscar Chinn Case (Britain v Belgium)*.[46] However, EIA for transboundary harm was not a legal concept of that period and was not discussed in the courts' analysis.

This chapter thus focuses on the following four most relevant disputes, with three that have come before the ICJ, being the *Gabčíkovo-Nagymaros* case,[47] the *Pulp Mills* case,[48] the *San Juan River* case[49] and one that has come before an arbitral tribunal, being the *Kishenganga*

---

[44] *Territorial Jurisdiction of International Commission of River Oder (UK v Poland)*, Order of 15 August 1929, PCIJ (Ser A) No 23 at Annex 4.

[45] *Diversion of Water from the Meuse Case (Netherlands v Belgium)* (1937), PCIJ (Ser A/B) No 70.

[46] *Oscar Chinn Case (Britain v Belgium)* (1937), PCIJ (Ser A/B) No 63.

[47] *Case concerning Gabčíkovo-Nagymaros Project (Hungary/Slovakia)*, [1997] ICJ Rep 7 (*Gabčíkovo-Nagymaros*).

[48] *Case concerning Pulp Mills on the River Uruguay (Argentina v Uruguay)*, [2010] ICJ Rep 14 (*Pulp Mills*).

[49] *Construction of a Road in Costa Rica along the San Juan River (Nicaragua v Costa Rica)*, 'Application Instituting Proceedings Submitted by Nicaragua' (22 December 2011), accessed 7 January 2015 at http://www.icj-cij.org/docket/files/152/16917.pdf; *Certain Activities Carried Out by Nicaragua in the Border Area (Costa Rica v Nicaragua)*, 'Application Instituting Proceedings Submitted by Costa Rica' (18 November 2010) accessed 7 January 2015 at http://www.icj-cij.org/docket/files/150/16279.pdf. Proceedings joined: *Construction of a Road in Costa Rica along the San Juan River (Nicaragua v Costa Rica)*, 'Joinder of Proceedings' (17 April 2013) accessed 7 January 2015 at http://www.icj-cij.org/docket/files/152/17354.pdf.

Hydroelectric Project case.[50] Other cases relating to transboundary waters that have come before the ICJ, such as the *Dispute regarding Navigational and Related Rights (Costa Rica v Nicaragua)*[51] and the *Case Concerning the Frontier Dispute (Republic of Benin v Republic of Niger)*,[52] are not included because they focus on the delineation of rivers as borders or other territorial concerns or commercial rights, rather than on water as a consumable resource or a quantity to be managed, or an ecosystem that needs to be preserved or improved.

Since international courts and tribunals, especially the ICJ, play an important role in providing guidance on legal issues, the following is a chronological analysis of relevant cases related to the application of EIA.

B. The *Gabčíkovo-Nagymaros* Case between Hungary and Slovakia

1. Background and main issues of the dispute

The dispute arose primarily from the construction of dams that involved the implementation and interpretation of a bilateral treaty signed in 1977 between Hungary and Czechoslovakia (1977 Treaty).[53] The 1977 Treaty concerned the construction of a 'system of locks' on the Danube River to be operated jointly by both parties (Gabčíkovo-Nagymaros Project). Construction began in 1978.[54] The Hungarian government suspended works on its part of the Gabčíkovo-Nagymaros Project in 1989, and terminated the 1977 Treaty in 1992 in the face of growing domestic ecological concern, pressure and criticism.[55] Hungary argued that the ecological risks of the Gabčíkovo-Nagymaros Project, including reduction in water flows, damage to water quality, and the consequential loss

---

[50] *Indus Waters Kishenganga Arbitration (Pakistan v India)*, Final Award of 20 December 2013 (*Kishenganga*, Final Award) accessed 7 January 2015 at http://www.pca-cpa.org/showpage.asp?pag_id=1392 (*Indus Waters Kishenganga Arbitration*).

[51] *Dispute Regarding Navigational and Related Rights (Costa Rica v Nicaragua)*, [2009] ICJ Rep 213.

[52] *Case concerning the Frontier Dispute (Republic of Benin v Republic Niger)*, [2005] ICJ Rep 90.

[53] In 1993, the Slovak Republic became the Successor State to the Treaty after the division of Czechoslovakia. Treaty between the Hungarian People's Republic and the Czechoslovak Socialist Republic concerning the Construction and Operation of the Gabčíkovo-Nagymaros System of Locks, 16 September 1977, 1109 UNTS 236.

[54] *Gabčíkovo-Nagymaros, supra* note 47 at paras 17–18.

[55] *Ibid* at para 25.

of 'fluvial fauna and flora', were unacceptable.[56] Czechoslovakia subsequently initiated a 'unilateral diversion of the Danube' on its territory, known as Variant C, as an alternative to the original Gabčíkovo-Nagymaros Project, which resulted in a major reduction in the flow of the Danube River downstream into Hungary.

2. Application of EIA

The issue of EIA in the *Gabčíkovo-Nagymaros* case is not a focused one and the Court gives it little detailed explanation or discussion. In its judgment, the Court addressed the issue of EIA indirectly, rather than expressly, by noting that 'the awareness of the vulnerability of the environment and the recognition that environmental risks have to be assessed on a continuous basis'.[57] In the years since the 1977 Treaty's conclusion, such awareness and recognition have increased significantly.

However, Judge and Vice President Weeramantry in his separate opinion addressed the principle of EIA, clarifying its significance and legal status. Judge Weeramantry made two points in respect of EIA: first, he declared that it is an obligation under customary international law to carry out a proper EIA; and second, he stressed the importance of continuous EIA for a project as long as it continues in operation.[58] Continuous EIA means it is not a mere procedural requirement to be conducted before the start of a project that poses transboundary risk, but a substantive and ongoing assessment for the duration of the project's operation in order to prevent or reduce any unexpected environmental consequences.[59]

The *Gabčíkovo-Nagymaros* case was the first contentious case at the ICJ that raised direct questions of international environmental law and the utilisation of transboundary waters. The discussion of EIA in this case has had a positive impact on the development of international environmental law. In particular, the two key points of Judge Weeramantry's opinion on EIA have been cited frequently by leading academics and legal practitioners and have had an impact on framing the conduct and conversation of international bodies and academic commentary.[60]

---

56 *Ibid* at para 35.

57 *Ibid* at para 112.

58 *Gabčíkovo-Nagymaros, supra* note 47, Separate opinion of Vice-President Weeramantry at 111.

59 *Ibid.*

60 Many articles and books on EIA have cited Judge Weeramantry's points, e.g., Neil Craik, *The International Law of Environmental Impact Assessment:*

However, it is worth noting that Judge Weeramantry's discussion of the significance and legal status of EIA is set out in a separate opinion and does not constitute part of the majority opinion.

C. *Pulp Mills* Case between Argentina and Uruguay

The *Pulp Mills* case is not merely a case relating to a river treaty, but is also a case decided recently by the ICJ on the issue of EIA, especially with regard to the scope and content of EIA in the transboundary context.

1. Background of the dispute

The *Pulp Mills* case is a dispute regarding the interpretation and implementation of a bilateral agreement signed by Argentina and Uruguay in 1975, entitled the Statute of the River Uruguay (1975 Statute).[61] Argentina claimed that Uruguay breached its obligations under international law and the 1975 Statute by building pulp mills near one of Uruguay's transboundary rivers, including the procedural obligation of providing prior notification and the substantial obligation of preventing environmental damage.

The Court ruled in favour of Argentina's argument that Uruguay had violated the procedural requirement of the 1975 Statute,[62] but declined to rule on Argentina's argument regarding Uruguay's breach of substantive obligations arising under the 1975 Statute.[63]

2. Analysis on status and content of EIA in a transboundary context

The ICJ addressed the relationship between the need for an EIA, where the planned activity caused transboundary harm to a shared resource, and the obligations of the parties to prevent pollution and preserve the aquatic environment under Article 41(a) and (b) of the 1975 Statute.[64]

Both parties had no dispute on the legal status and necessity of EIA in a transboundary context, but disagreed on the scope and content of the

---

Process, Substance and Integration (Cambridge: Cambridge University Press, 2010) at 115; Erika Preiss, 'The International Obligation to Conduct an Environmental Impact Assessment: the ICJ Case Concerning the Gabčíkovo-Nagymaros Project' (1999) 7 *New York University Environmental Law Journal* 307; Ilias Plakokefalos, 'Prevention Obligations in International Environmental Law' (Research Paper, Amsterdam Law School, 2013) (unpublished) at 14.

[61] Statute of the River Uruguay (Argentina-Uruguay), adopted 26 February 1975, 1295 UNTS 340 (entered into force 18 September 1976).

[62] *Pulp Mills, supra* note 48 at para 267.

[63] *Ibid* at para 265.

[64] *Ibid* at paras 203–206.

EIA.[65] Argentina maintained that Uruguay failed to ensure 'full environmental assessments, prior to its decision to authorize the construction …' and that Uruguay failed to take account of all potential impacts from the mill.[66] Uruguay agreed that there existed the necessity to conduct an EIA on the Orion (Botnia) mill in accordance with international practice when the project poses potentially harmful transboundary effects to people, property and the environment of other States, but argued that international law did not impose any condition or requirement on the content of such assessment.[67] In sum, the content and scope of EIA is broader and more complete in the argument of Argentina than that of Uruguay.

As to the content of EIA, the Court maintained a conservative attitude. The ICJ declared that 'it is for each State to determine in its domestic legislation or in the authorization process for the project, the specific content of the environmental impact assessment required in each case, having regard to the nature and magnitude of the proposed development and its likely adverse impact on the environment as well as to the need to exercise due diligence in conducting such an assessment'.[68]

D. *San Juan River* Case between Nicaragua and Costa Rica

1. Background of the dispute

This is the latest case relating to EIA, instituted by Nicaragua on 22 December 2011, which concerned a road constructed by Costa Rica that runs parallel to the San Juan River.[69] Although the road is located solely on Costa Rican territory, Nicaragua contended that its construction resulted in harmful environmental effects on Nicaraguan territory, including the silting of the San Juan River, erosion of the river banks and harmful effects on the surrounding wetlands ecosystem.[70]

2. Application of EIA

As to the claim of EIA, both parties disputed the necessity of EIA. Nicaragua claimed that Costa Rica had not given Nicaragua information

---

65 *Ibid* at para 205.
66 *Ibid* at para 203.
67 *Ibid.*
68 *Ibid* at para 205.
69 *Construction of a Road in Costa Rica along the San Juan River (Nicaragua v Costa Rica)*, 'Application Instituting Proceedings Submitted by Nicaragua', *supra* note 49 at para 14.
70 *Ibid* at paras 4–10.

about the construction; therefore, they asked for an EIA and requested the Court to declare that Costa Rica had to present Nicaragua with an EIA that included the details of the construction.[71] Costa Rica rejected the obligation to prepare and provide Nicaragua with an EIA, arguing that it was unnecessary because the road was 50 metres away from the river and would not cause environmental problems for the San Juan River.[72] The Court, in its order of provisional measures, was of the view that, as the construction of the road only contributed 1 to 2 per cent of the total sediment load in the San Juan River, this was 'too small a proportion to have a significant impact on the river in the immediate future'.[73]

The Court reconfirmed the understanding of EIA in the *Pulp Mills* case and stated that the need for an EIA would depend on the 'issue of significant harm'.[74] According to the *Pulp Mills* case, the practice of EIA 'has gained so much acceptance ... it may now be considered a requirement under international law to undertake an EIA where the proposed industrial activity may have a significant adverse impact in a transboundary context, in particular, on a shared resource'.[75] Importantly, the Court in the *San Juan River* case clarified that the standard for mandating an EIA is lower than for finding an actionable damage ('may have a significant adverse impact' for the former, and 'significant harm' for the latter).[76]

E. The *Kishenganga Hydroelectric Project* Case between Pakistan and India

1. Background of the dispute

The *Kishenganga Hydroelectric Project* case is a dispute instituted by Pakistan against India in May 2010 under the Indus Waters Treaty 1960 (Indus Waters Treaty), involving the Kishenganga Hydroelectric Project

[71] *Ibid* at paras 49–52.

[72] *Ibid* at para 14.

[73] *Construction of a Road in Costa Rica along the San Juan River (Nicaragua v Costa Rica), Certain Activities Carried Out by Nicaragua in the Border Area (Costa Rica v Nicaragua)*, 'Request Presented by Nicaragua for the Indication of Provisional Measures' (Order of 13 December 2013) at para 34, accessed 7 January 2015 at http://www.icj-cij.org/docket/files/152/17838.pdf (*San Juan River*).

[74] *Ibid* at para 19.

[75] *Pulp Mills*, *supra* note 48 at para 204.

[76] *San Juan River*, *supra* note 73 at paras 19–21.

(KHEP) located on the Kishenganga/Neelum River.[77] The Indus Waters Treaty regulates the use by the two parties of the waters from the Indus system of rivers, but does not include a water sharing provision.[78]

The KHEP is designed to generate power by diverting water from a dam site on an area of higher elevation to an area of lower elevation through a system of tunnels.[79] Pakistan challenged, in particular, the permissibility and impact of the planned diversion by the KHEP of the waters and the effect that this diversion would have on Pakistan's own Neelum-Jhelum Hydroelectric Project (NJHEP).[80] The Court of Arbitration rendered its Final Award on 20 December 2013.[81]

2. Analysis of EIA in the dispute

The Court of Arbitration recognised that the Indus Water Treaty should be interpreted and applied in light of the 'customary international principles for the protection of the environment in force today'.[82] As to the EIA, the Court of Arbitration invoked the analysis of the ICJ in the *Pulp Mills* case, recognising EIA in the prevention of transboundary harm as 'contemporary customary international law', which requires States to take 'environmental protection into consideration when planning and developing Projects that may cause injury to a bordering State'.[83]

With respect to the environmental impact of the KHEP, both parties recognised the necessity of an EIA, but differed with regard to the content and scope of the environmental assessments. Pakistan had provided the Court of Arbitration with 'a holistic assessment of the interaction of a range of environmental indicators and attempted to capture the complexity of interactions within the river ecosystem'.[84] In

---

77 *Kishenganga,* Final Reward, *supra* note 50.

78 The Indus Waters Treaty 1960 (India-Pakistan), signed 19 September 1960, 419 UNTS 126.

79 *Indus Waters Kishenganga Arbitration (Pakistan v India)*, Partial Award of 18 February 2013 at para 127, accessed 7 January 2015 at http://www.pca-cpa.org/showpage.asp?pag_id=1392 (*Kishenganga,* Partial Award).

80 Permanent Court of Arbitration, Press Release, 'Court of Arbitration Renders its Final Award in the First Arbitration under the Indus Waters Treaty 1960' (21 December 2013) at 1, accessed 7 January 2015 at http://www.pca-cpa.org/showpage.asp?pag_id=1392.

81 *Kishenganga,* Final Award, *supra* note 50.

82 *Kishenganga,* Partial Award, *supra* note 79 at para 452.

83 *Ibid* at para 449.

84 *Kishenganga,* Final Award, *supra* note 50 at 97.

contrast, India had submitted a simpler assessment, drawing its conclusions from more limited data regarding the habitat for fish species.[85] While noting that there is no single 'correct' approach to environmental assessments, the Court of Arbitration expressed the view that Pakistan's in-depth assessment was more appropriate for a project of the magnitude of the KHEP, and emphasised that Pakistan's approach in these proceedings did not match its historical practices.[86] The Court of Arbitration therefore encouraged both parties to continue their in-depth and expansive attention to environmental concerns at all their hydroelectric projects.[87]

F. Remarks on the Application of EIA Requirements

The above dispute settlement practices in international tribunals illustrate the evolution of the recognition of the legal status of EIA and how it has been applied in the settlement of disputes relating to the utilisation of transboundary rivers. Three conclusions can be drawn from the review of the cases with regard to EIA. First, EIA is recognised as a legal obligation in customary international law, which means that even in the absence of a binding treaty rule providing the requirement of EIA, the operating State of a project that poses significant transboundary risk still has the obligation to conduct EIA. Second, the nature of the obligation to conduct an EIA in the transboundary context is not only a procedural requirement, but also a substantive one. Third, the specific content and scope of EIA is not well established under international law. Neither the ICJ in the *Pulp Mills* case nor the ILC in its Draft Articles on Prevention attempted to provide for an international minimum standard for the content and scope of EIA.[88] Presently, under the requirement of due diligence and good faith, the scope and content of an EIA is left to the discretion of the State itself. Given that it is probable that the scope and content of an EIA for a specific project or activity will differ considerably amongst contesting States, this will lead to inconsistent outcomes.

Furthermore, a review of the ICJ dispute settlement practice indicates that such recognition and application of the EIA obligation runs parallel to the development of EIA in treaty practice, which reflects the dynamic effects of treaty rules and customary rules in transboundary rivers

---

85 *Ibid* at para 98.
86 *Ibid* at paras 99–101.
87 *Ibid* at para 101.
88 See Draft Articles on Prevention, *supra* note 5 at 158; see also Ilias Plakokefalos, *supra* note 60 at 15.

governance. However, it is worth noting that the application and evolution of the EIA obligation is not as advanced as the efforts and resolutions of international institutions, no matter whether it be an intergovernmental institution, such as the ILC, or a non-governmental organisation, such as the International Law Association.

IV. EIA BASED ON HYDROPOWER PROJECTS BUILDING IN THE LANCANG-MEKONG BASIN

A. Background of Hydropower Projects Building in the Lancang-Mekong River Basin

As developed and developing countries attempt to cope with climate change and increasing pressures to become more energy efficient and meet various emission reduction targets, more and more hydropower dams are being planned or are under construction. Although dams can play important roles in flood moderation, irrigation and the production of clean energy through hydropower, they still meet a series of worries, environmental concerns and controversies from the public, especially with regard to the pollution and damage to the river basin ecosystem. In India, people in the north-east state of Assam have been fighting for several years against a proposed gargantuan network of dams across the upper reaches of its rivers in the Brahmaputra Valley.[89] In China, the cascade of proposed hydroelectric projects on one of Asia's longest undammed rivers, the Nu River, has caused great controversy in south-west China.[90] In Bhutan, the Kurichu dam project was also challenged by the public because of environmental concerns.[91]

---

[89] Tanmoy Sharma, 'Mega dams: campaigning against the plans of the Indian government' *Open India* (20 January 2012) accessed 7 January 2015 at https://www.opendemocracy.net/openindia/tanmoy-sharma/mega-dams-campaigning-against-plans-of-indian-government.

[90] Kristen McDonald, 'Hydropower on the Nu: one river, many perspectives' *China Dialogue* (19 March 2008) accessed 7 January 2015 at https://www.chinadialogue.net/article/1816-Hydropower-on-the-Nu-one-river-many-perspectives-.

[91] Anjana Mahanta, 'The Case of Kurichu in the Indo-Bhutan Context: Transboundary Hydropower Projects and Downstream Flooding' (Paper submitted for the Sustainable Mountain Development Summit, Nagaland, India 25–27 September 2013) accessed 7 January 2015 at http://www.indiawaterportal.org/sites/indiawaterportal.org/files/kurichu.pdf.

Such is the case with many of the world's transboundary rivers, including the Mekong River Basin. In Myanmar, the Myitsone hydroelectric project in Kachin state was planned and under construction, until President Thein Sein suspended its construction on 30 September 2011 because of the risk of permanent harm to the Irrawaddy River's ecosystem.[92] The Laotian government announced the construction of the Xayaburi dam, the first mainstream dam on the lower Mekong River, despite widespread protest.[93] In fact, according to the data published by the governments of the Lancang-Mekong River Basin and other relevant reports, there are a number of dams that have been constructed or have been proposed or are under construction throughout the Lancang-Mekong River Basin.[94] Major hydropower projects have been proposed in Thailand, Laos, Cambodia and Vietnam. For example, Laos has constructed new dams, such as Houay Ho, Nam Leuk and Theun-Hinboun, while other dams, such as Xayaburi, Pak Lay, Ban Koum are under construction.[95] Vietnam has constructed a number of new dams as well, including Buon Tua Sra, Plei Krong and Yali Falls.

B. Remarks on Promotion of Environmental Cooperation and Pollution Control in the Mekong River Basin

There are two relevant sets of rules with regard to pollution control in the Mekong River Basin. The first is the institutional and legal framework for environmental cooperation and promotion in the area of the Mekong River Basin, which is administered through the Mekong River Commission (MRC). The second is the established general rule of international law that requires EIA when a project or activity poses significant transboundary environmental impact, as concluded by the ICJ in the *Pulp Mills* case. Between them, a new and specific regional agreement on the utilisation of the Mekong River among the riparian countries may offer a

---

92 Rachel Harvey, 'Burma dam: Work halted on divisive Myitsone project' *BBC* (30 September 2011) accessed 7 January 2015 at http://www.bbc.co.uk/news/world-asia-pacific-15121801?.

93 'Laos forges ahead with controversial Mekong dam' *China Dialogue* (18 October 2012) accessed 7 January 2015 at https://www.chinadialogue.net/blog/5222–Laos-forges-ahead-with-controversial-Mekong-dam-/en.

94 Peter King, Jeremy Bird and Lawrence Haas, *The Current Status of Environmental Criteria for Hydropower Development in the Mekong Region: A Literature Compilation* (WWF-Living Mekong Programme, 2007).

95 Lao People's Democratic Republic, Ministry of Energy and Mines, Department of Energy Business, *Powering Progress*: *Power Projects*, accessed 7 January 2015 at http://www.poweringprogress.org/new/power-projects.

better approach for addressing the river's transboundary pollution issues while keeping a balance of the different or even competing interests of the concerned countries.

The MRC is the intergovernmental body whose mission is 'to promote and co-ordinate sustainable management and development of water and related resources for the countries' mutual benefit and the people's well-being by implementing strategic programs and activities and providing scientific information and policy advice'.[96] Members of the MRC, which include Cambodia, Laos, Thailand and Vietnam, concluded an agreement on the Cooperation for the Sustainable Development of the Mekong River Basin in 1995.[97] The parties agreed to cooperate in all fields of sustainable development, utilisation, management and conservation of the water and related resources of the Mekong River Basin.[98] Although there is no express EIA procedure or requirement, Articles 2 and 3 indicate that EIA is a legal obligation of the parties when proposed projects pose significant transboundary risk.[99] Such mechanisms have helped to promote environmental cooperation among the parties and prevention of pollution in the transboundary or shared waters through the adoption of relevant documents. For example, the 2010 MRC Hua Hin Declaration and the MRC guidelines helped to facilitate the incorporation of EIA at the national level among MRC members.[100]

However, such arrangements by the MRC have two inherent weaknesses. The first is the MRC's influence. Given that the MRC includes only four downstream countries of the six riparian countries, with the two notable exceptions being the two upper basin countries, China and Myanmar, the EIA requirements in the resolutions or proposals of the

---

96 Mekong River Commission, *Mission of the Mekong River Commission*, accessed 7 January 2015 at http://www.mrcmekong.org/about-mrc/vision-and-mission/.

97 Mekong River Commission, *Agreement on the Cooperation for the Sustainable Development of the Mekong River Basin*, adopted 5 April 1995, 34 ILM 864 (not yet in force).

98 *Ibid* at Art 1.

99 *Ibid.*

100 Mekong River Commission, *MRC Hua Hin Declaration: Meeting the Needs, Keeping the Balance: Towards Sustainable Development of the Mekong River Basin*, adopted 5 April 2010, accessed 7 January 2015 at http://www.mrcmekong.org/assets/Publications/Events/MRCSummit2010/MRC-Hua-Hin-Declaration-05-Apr-10.pdf; Mekong River Commission, *Guidelines on Implementation of Procedures for Water Use Monitoring*, adopted 5 April 2006, accessed 7 January 2015 at http://www.mrcmekong.org/assets/Uploads/Tech-Guidelines-PWUM.pdf.

MRC may be disregarded or rejected by non-members. Although channels exist for the exchange of information between the MRC and non-members, non-members are not legally bound by the MRC's decisions. That being said, matters that come before the MRC may not consider or include the concerns or interests of the upper basin countries, as the MRC is a group of downstream countries.

The second weakness is the nature and strength of the MRC's documents. Although it may be easier to adopt soft laws such as guidelines or proposals, the legal effect and enforceability of these provisions is limited. Relying upon an informal legal system alone to legitimate and limit claims to use shared water resources is inherently unstable.

In the context of the rapid development of hydropower projects that may cause transboundary harm to the water and ecosystem, it is necessary and also a better choice to conclude a regional agreement regarding environmental protection and pollution prevention. Under a regional framework, not only can the criteria and measures be set out in greater detail, such as the content, scope and procedures regarding EIA, but also the balance of different or even competing interests between upper and lower riparian countries can be maintained in order to encourage broad participation in the development of equitable and reasonable controls in the utilisation of the transboundary waters. As to EIA, the 1991 Espoo Convention regime may serve as an appropriate model by which the projects needing EIA can be listed, preventative measures can be detailed, and the criteria for other projects can be clarified.

Under international law, there is a positive obligation for all riparian countries to conduct EIA with good faith when a proposed and ongoing project or activity poses significant transboundary impact. However, for the Langcang-Mekong River Basin, in order to deal with the problem of competing uses of the transboundary waters, it may be a better approach for all riparian States to reach a regional agreement. This approach is supported by the Draft Articles on Prevention.[101] Furthermore, the ILC also urges States to take national and international action to implement the principles contained in its Draft Articles.[102]

---

101 Draft Articles on Prevention, *supra* note 5.

102 'Report of the International Law Commission to the General Assembly covering the work of its fifty-eighth session, 2006' (UN Doc A/61/10) at 105.

6. State responsibility and transboundary marine pollution

Robert Beckman

I. INTRODUCTION

Pollution of the marine environment is caused by human activities conducted in and on the oceans. This chapter examines the responsibilities owed by States to ensure that damage to the marine environment caused by such activities is prevented, reduced and controlled. It examines potential remedies available to States that have been affected by transboundary marine pollution, and the ways in which liability for damage can be attributed to States who fail to give effect to their duties and obligations to protect and preserve the marine environment from activities conducted under their jurisdiction and control.

The primary source of international law that governs the oceans is the 1982 United Nations Convention on the Law of the Sea (UNCLOS).[1] UNCLOS is a framework agreement setting out the rights and obligations of States with respect to all uses of the oceans. It is now almost universally accepted and many of its provisions, including those on protection and preservation of the marine environment, are generally regarded as binding on all States under customary international law.[2]

UNCLOS is important to marine pollution because Part XII on protection and preservation of the marine environment is an integral part of the Convention. Since no 'reservations' to UNCLOS are permissible,[3]

1 United Nations Convention on the Law of the Sea, adopted 10 December 1982, 1833 UNTS 397 (entered into force 16 November 1994) (UNCLOS).

2 Robin R Churchill and A Vaughan Lowe, *The Law of the Sea,* 3rd edn (New York: Juris Publishing, 1999) at 24. As of October 2014, there were 166 parties. For the status of UNCLOS, see https://treaties.un.org/pages/ViewDetailsIII.aspx?&src=TREATY&mtdsg_no=XXI~6&chapter=21&Temp=mtdsg3&lang=en (accessed 8 January 2015).

3 UNCLOS, *supra* note 1, Art 309.

States cannot become parties to UNCLOS without accepting its detailed provisions on the prevention, reduction and control of marine pollution.

UNCLOS is not a stand-alone convention, but a framework convention. UNCLOS recognises the competence of specialised international organisations to proscribe rules to protect the marine environment. Many of the provisions of UNCLOS refer to laws, regulations, rules, standards and recommended practices of the 'competent international organization'.[4] The International Maritime Organization (IMO) is recognised as the competent international organisation with respect to the safety of maritime navigation and the protection of the marine environment from ship-source pollution.[5] Many of the provisions of UNCLOS incorporate by reference the conventions and codes of the IMO on the protection and preservation of the marine environment. Because the general provisions in UNCLOS automatically incorporate the rules, standards, procedures and practices set out in IMO instruments that are being constantly updated by the IMO, the provisions in UNCLOS on prevention, reduction and control of pollution from ships remain current and relevant.

Although UNCLOS is the primary source of law that governs all uses of the oceans, its provisions on responsibility and liability for pollution of the marine environment must be read together with the International Law Commission's 2001 Articles on Responsibility of States for Internationally Wrongful Acts (Draft Articles on State Responsibility).[6] Several of the provisions in the Draft Articles on State Responsibility have been cited by international courts and arbitral tribunals as evidence of general international law, and they have therefore become the starting point for analysing the general international law on the responsibility of States.[7]

Like the provisions of the 1969 Vienna Convention on the Law of Treaties, the Draft Articles on State Responsibility are 'secondary rules'. The ILC Articles set out the general conditions under international law

---

[4] *Ibid*, Arts 22(3)(a), 41(4), 60(3), 197, 201, 204(1), 207(4), 208(5), 210(4), 211–214, 216–218, 222, 251, 262, 271, 297(1)(c).

[5] Secretariat of the International Maritime Organization (IMO), *Implications of the United Nations Convention on the Law of the Sea for the International Maritime Organization*, LEG/MISC.7 (19 January 2012) (*Implications of UNCLOS for the IMO*) at 7–8.

[6] 'State Responsibility' in 'Report of the International Law Commission covering the work of its fifty-third session, with commentaries, 2001' (UN Doc A/56/10), Ch IV, in *Yearbook of the International Law Commission 2001*, vol II, Part Two (UN, 2001) (Draft Articles on State Responsibility).

[7] James R Crawford, *Brownlie's Principles of Public International Law*, 8th edn (Oxford: Oxford University Press, 2012) at 540.

for a State to be considered responsible for wrongful actions or omissions in breach of its primary obligations, and the legal consequences that flow therefrom. The 'primary rules' of international law are the obligations imposed on States in treaties and in rules of customary international law governing particular areas of law. For example, the primary rules governing the law of the sea are set out in UNCLOS. Secondary rules in the 1969 Vienna Convention and in the Draft Articles on State Responsibility set out general rules on matters such as the legal consequences which result if a State Party breaches the primary rules set out in UNCLOS.

Article 235 of UNCLOS sets out the general principles concerning responsibility and liability with respect to protection and preservation of the marine environment. Article 235(1) provides that States are *responsible* for the fulfilment of their international obligations concerning the protection and preservation of the marine environment, and that they shall be *liable* in accordance with international law. These provisions should be read in light of the Draft Articles on State Responsibility and the decisions of international courts and tribunals.

A State is 'responsible' under international law if it fails to fulfil its international obligations to prevent, reduce and control pollution of the marine environment. If the responsibility of a State is engaged, it is under an obligation to cease continuing an act that is contrary to its obligations, and to offer appropriate assurances and guarantees of non-repetition, if circumstances so require.[8] The responsible State is also under an obligation to make full 'reparation' for the injury *caused by* the internationally wrongful act.[9] In other words, a State is liable for damage resulting from pollution of the marine environment by ships or persons subject to its jurisdiction and control only if a causal link is established between the failure of the State to fulfil its obligations and the damage to the marine environment. The existence of a causal link between the State's failure and the damage is a necessary element in establishing liability and cannot be presumed.

The rules on responsibility and liability have been clarified by the 2011 Advisory Opinion of the Seabed Disputes Chamber of the International Tribunal for the Law of the Sea (ITLOS) concerning *Responsibilities and Obligations of States Sponsoring Persons and Entities with respect to*

---

[8] Draft Articles on State Responsibility, *supra* note 6, Art 30.
[9] *Ibid*, Art 31(1).

Activities in the Area (*Advisory Opinion on Seabed Activities*).[10] Although the Advisory Opinion was on specific questions relating to the responsibilities and obligations of States sponsoring entities to conduct exploration and exploitation activities of the deep seabed beyond areas of national jurisdiction, much of the Chamber's discussion and analysis is relevant to other activities conducted by persons under the jurisdiction and control of a State that may cause pollution of the marine environment.

II. GENERAL OBLIGATIONS CONCERNING POLLUTION OF THE MARINE ENVIRONMENT

A. Due Diligence Obligations under Article 194 of UNCLOS

Article 194(1) imposes a general obligation on States to take all measures consistent with UNCLOS that are necessary to prevent, reduce and control pollution of the marine environment *from any source*, using for this purpose the best practicable means at their disposal and in accordance with their capabilities. Thus, the obligation to take measures is qualified by language concerning common but differentiated responsibilities as set out in Principle 7 of the 1992 Rio Declaration on the Environment and Development (1992 Rio Declaration).[11]

Article 194(2) imposes an obligation on States regarding transboundary pollution from activities under their jurisdiction and control, that is, activities by ships flying their flag, by entities engaged in seabed activities subject to their jurisdiction, and so on. It provides that:

> States shall take all measures necessary to ensure that activities under their jurisdiction or control are so conducted as not to cause damage by pollution to other States and their environment, and that pollution arising from incidents or activities under their jurisdiction or control does not spread beyond the areas where they exercise sovereign rights in accordance with this Convention.

---

[10] *Responsibilities and Obligations of States Sponsoring Persons and Entities with respect to Activities in the Area,* Advisory Opinion, 1 February 2011, ITLOS Rep 2011, 10 (*Advisory Opinion on Seabed Activities*).

[11] Rio Declaration on Environment and Development, in 'Report of the United Nations Conference on Environment and Development', 12 August 1992, UN Doc A/CONF.151/26 (Vol I), 31 ILM 874 (1992 Rio Declaration).

Article 194(3) provides that the measures in Part XII of UNCLOS on protection and preservation of the marine environment shall deal with all sources of pollution of the marine environment, including pollution from hazardous and noxious substances, pollution from vessels, and pollution from installations and devices used in exploration or exploitation of the natural resources of the seabed and subsoil.

Article 194(5) provides that the measures taken in accordance with Part XII shall include those necessary to protect and preserve rare or fragile ecosystems as well as the habitat of depleted, threatened or endangered species and other forms of marine life. In other words, it requires States to take measures not only to prevent pollution, but to protect and preserve marine biological diversity.

These provisions in UNCLOS establishing a 'responsibility to ensure' must be read in light of the reasoning in the *Advisory Opinion on Seabed Activities* in which the Seabed Disputes Chamber defines the 'responsibility to ensure' as one of 'due diligence'.[12] The provisions in UNCLOS establishing a responsibility to ensure[13] set out obligations that States Parties must fulfil by exercising their power over entities under their jurisdiction and control, such as ships flying their flag or companies subject to their national jurisdiction. Such 'due diligence' obligations may be characterised as obligations 'of conduct' and not 'of result'.[14]

These due diligence obligations require a State to take measures within its legal system, including the adoption of laws and regulations as well as administrative measures to ensure that such laws and regulations are enforced. In paragraph 197 of its judgment in the *Pulp Mills* case, the International Court of Justice (ICJ) described such due diligence obligations in a treaty as follows:

> It is an obligation which entails not only the adoption of appropriate rules and measures, but also a certain level of vigilance in their enforcement and the exercise of administrative control applicable to public and private operators, such as the monitoring of activities undertaken by such operators ...[15]

---

12 *Advisory Opinion on Seabed Activities*, *supra* note 10 at paras 107–116.

13 UNCLOS, *supra* note 1, Arts 194(2), 210(3), 217(1),(2), 235(2).

14 *Advisory Opinion on Seabed Activities*, *supra* note 10 at para 110.

15 *Case concerning Pulp Mills on the River Uruguay (Argentina v Uruguay)*, [2010] ICJ Rep 14 (*Pulp Mills*) at para 197.

B. Environmental Impact Assessment

Articles 204 to 206 of UNCLOS impose general obligations on States to monitor the risks or effects of pollution of the marine environment. Article 204 provides that States should keep under surveillance the effects of activities that they engage in or permit. Article 206 provides that when States have reasonable grounds for believing that planned activities under their jurisdiction or control may cause substantial pollution of or significant and harmful changes to the marine environment, they shall, as far as practicable, assess the potential effects of such activities on the marine environment.

The rather vague language of Article 206 has been clarified (and developed) by both the ICJ and ITLOS. In paragraph 204 of its judgment in the *Pulp Mills* case, the Court stated:

> It may now be considered a requirement under general international law to undertake an environmental impact assessment where there is a risk that the proposed industrial activity may have a significant adverse impact in a transboundary context, in particular, on a shared resource. Moreover, due diligence, and the duty of vigilance and prevention that it implies, would not be considered to have been exercised, if a party planning works liable to affect the regime of the river or quality of its waters did not undertake an environmental impact assessment on the potential effects of such works.

In paragraph 145 of its 2011 *Advisory Opinion on Seabed Activities*, the ITLOS Seabed Disputes Chamber affirmed that the obligation to conduct an environmental impact assessment (EIA) is a general obligation under both UNCLOS and customary international law.[16] In paragraph 146 it implied that the obligation to conduct an EIA is required under Article 206 of UNCLOS:

> As regards the Convention, article 206 states the following:
>
> > 'When States have reasonable grounds for believing that planned activities under their jurisdiction or control may cause substantial pollution of or significant and harmful changes to the marine environment, they shall, as far as practicable, assess the potential effects of such activities on the marine environment and shall communicate reports of the results of such assessments in the manner provided in article 205.'[17]

---

[16] *Advisory Opinion on Seabed Activities*, *supra* note 10 at para 145.

[17] UNCLOS, *supra* note 1, Art 206 (in which Article 205 refers to an obligation to publish reports).

C. Duty to Cooperate on Activities that May Cause Transboundary Pollution

The decisions of ITLOS have clarified the principles of general international law that apply to planned activities that may cause transboundary marine pollution of the marine environment. ITLOS addressed this issue in the course of its decisions in two cases on provisional measures. In the *MOX Plant* case, the Tribunal stated in paragraph 82 of its Order for Provisional Measures that the 'duty to cooperate' is a fundamental principle in the prevention of pollution of the marine environment under Part XII of the Convention and general international law.[18] And the Tribunal stated:

> Ireland and the United Kingdom shall cooperate and shall, for this purpose, enter into consultations forthwith in order to:
>
> (a) exchange further information with regard to possible consequences for the Irish Sea arising out of the commissioning of the MOX plant;
> (b) monitor risks or the effects of the operation of the MOX plant for the Irish Sea;
> (c) devise, as appropriate, measures to prevent pollution of the marine environment which might result from the operation of the MOX plant.[19]

In the *Case concerning Land Reclamation by Singapore in and around the Straits of Johor (Malaysia v Singapore)*, the Tribunal repeated its statement that the duty to cooperate is a fundamental principle in the prevention of pollution of the marine environment under Part XII of the Convention and general international law.[20] It further stated that the record of this case shows that there was insufficient cooperation between the parties up to the submission of the Statement of Claim on 4 July 2003.[21] The Tribunal then stated that 'given the possible implications of land reclamation on the marine environment, prudence and caution

---

[18] *MOX Plant case (Ireland v United Kingdom),* Provisional Measures, Order of 3 December 2001, ITLOS Case No 10, 41 ILM 405 (2002) (*MOX Plant*) at para 82.

[19] *Ibid* at para 89.

[20] *Case concerning Land Reclamation by Singapore in and around the Straits of Johor (Malaysia v Singapore),* Provisional Measures, Order of 8 October 2003, ITLOS Case No 12 (*Land Reclamation*) at para 92.

[21] *Ibid* at para 97.

require that Malaysia and Singapore establish mechanisms for exchanging information and assessing the risks or effects of land reclamation works and devising ways to deal with them in the areas concerned'.[22]

The duty to cooperate was also stressed in the Joint Declaration of Judges *Ad Hoc* Hossain and Oxman, but they framed the 'fundamental principle' as a 'due regard obligation' rather than a duty to cooperate. They stated that:

> Our decisions to join in supporting the unanimous Order of the Tribunal are informed by a fundamental principle on which the Convention is built. The right of a State to use marine areas and natural resources subject to its sovereignty or jurisdiction is broad but not unlimited. It is qualified by the *duty to have due regard* to the rights of other States and to the protection and preservation of the marine environment.[23]

Although there is no express language in UNCLOS providing for a duty to cooperate or a due regard obligation concerning planned activities under the jurisdiction and control of one State that may pollute the marine environment of another State, ITLOS has in effect read this fundamental principle into Part XII of UNCLOS. The duty to cooperate set out in these cases seems to include a duty to give timely notice to the potentially affected State of the planned activities, to provide it with relevant information about the planned activities and to consult with it in good faith. It also implies an obligation to exchange information and to negotiate in good faith in order to balance the competing interests of the two States. Thus, it can be argued that ITLOS has interpreted the provisions in UNCLOS in light of principles of international environmental law that developed subsequent to UNCLOS by reading Principle 19 of the 1992 Rio Declaration into Part XII of UNCLOS. Principle 19 reads as follows:

> States shall provide prior and timely notification and relevant information to potentially affected States on activities that may have a significant adverse transboundary environmental effect and shall consult with those States at an early stage and in good faith.

---

22 *Ibid* at para 99.

23 *Ibid*, Joint Declaration of Judges *Ad Hoc* Hossain and Oxman (emphasis added), accessed 8 January 2015 at https://www.itlos.org/fileadmin/itlos/documents/cases/case_no_12/C12.decl.Hossain-Oxman.E.pdf.

D. Precautionary Approach and Precautionary Principle

The precautionary approach is set out in Principle 15 of the 1992 Rio Declaration as follows:

> In order to protect the environment, the precautionary approach shall be widely applied by States according to their capabilities. Where there are threats of serious or irreversible damage, lack of full scientific certainty shall not be used as a reason for postponing cost-effective measures to prevent environmental degradation.[24]

ITLOS was asked to consider the legal status of the precautionary approach in both the *MOX Plant* and *Land Reclamation* cases. It did not expressly mention the precautionary approach in either case. Rather, the Tribunal stated that 'prudence and caution' required the two States to take cooperative measures in order to fulfil their 'duty to cooperate', which the Tribunal stated was a fundamental principle of general international law.[25] In the *MOX Plant* case, the Tribunal stated that 'prudence and caution require that Ireland and the United Kingdom cooperate in exchanging information concerning risks or effects of the operation of the MOX plant and in devising ways to deal with them, as appropriate'.[26] In the *Land Reclamation* case, the Tribunal stated that 'given the possible implications of land reclamation on the marine environment, *prudence and caution* require that Malaysia and Singapore establish mechanisms for exchanging information and assessing the risks or effects of land reclamation works and devising ways to deal with them in the areas concerned'.[27]

It is sometimes asserted that the precautionary approach has become a principle of general international law[28] and that the effect of the principle is to reverse the burden of proof concerning the possible impact of a given activity.[29] However, the wording of Principle 15 does not seem to support the position that a State intending to undertake a given activity

---

24 1992 Rio Declaration, *supra* note 11, Principle 15.

25 *MOX Plant*, *supra* note 18 at para 82; *Land Reclamation*, *supra* note 20 at para 92.

26 *MOX Plant*, *ibid* at para 84.

27 *Land Reclamation*, *supra* note 20 at para 99.

28 Patricia Birnie, Alan Boyle and Catherine Redgwell, *International Law and the Environment,* 3rd edn (Oxford: Oxford University Press, 2009) at 27–8.

29 Philippe Sands, *Principles of International Environmental Law,* 2nd edn (Cambridge: Cambridge University Press, 2003) at 273.

has the burden of proving that the activity will not harm the environment.[30] The comment of the ICJ in the *Pulp Mills* case seems to support this conclusion, as the Court stated in paragraph 164:

> Regarding the arguments put forward by Argentina on the reversal of the burden of proof and on the existence, vis-à-vis each Party, of an equal onus to prove under the 1975 Statute, the Court considers that while a precautionary approach may be relevant to the interpretation and application of the provisions of the Statute, it does not follow that it operates as a reversal of the burden of proof.[31]

In its *Advisory Opinion on Seabed Activities,* ITLOS suggests that it agrees with the ICJ on the precautionary approach and hints that it may be willing to read it into UNCLOS:

> The Chamber observes that the precautionary approach has been incorporated into a growing number of international treaties and other instruments, many of which reflect the formulation of Principle 15 of the Rio Declaration. In the view of the Chamber, this has initiated a trend towards making this approach part of customary international law. This trend is clearly reinforced by the inclusion of the precautionary approach in the Regulations and in the 'standard clause' contained in Annex 4, section 5.1, of the Sulphides Regulations. So does the following statement in paragraph 164 of the ICJ Judgment in *Pulp Mills on the River Uruguay* that 'a precautionary approach may be relevant in the interpretation and application of the provisions of the Statute' (i.e., the environmental bilateral treaty whose interpretation was the main bone of contention between the parties). This statement may be read in light of article 31, paragraph 3(c), of the Vienna Convention, according to which the interpretation of a treaty should take into account not only the context but 'any relevant rules of international law applicable in the relations between the parties'.[32]

E. Notification in Case of Emergencies

Article 198 of UNCLOS provides that when a State becomes aware of cases in which the marine environment is in imminent danger of being damaged or has been damaged by pollution, it shall immediately notify other States it deems likely to be affected by such damage.[33] This article is very similar to Principle 18 of the 1992 Rio Declaration, which provides that 'States shall immediately notify other States of any natural

---

30 1992 Rio Declaration, *supra* note 11, Principle 15.
31 *Pulp Mills*, *supra* note 15 at para 164.
32 *Advisory Opinion on Seabed Activities*, *supra* note 10 at para 135.
33 UNCLOS, *supra* note 1, Art 198.

disasters or other emergencies that are likely to produce sudden harmful effects on the environment of those States'.[34]

F. Recourse for Compensation under National Legal Systems

Article 235(2) of UNCLOS recognises that individuals or companies may be liable for damage caused by pollution of the marine environment. It imposes an obligation on States Parties to ensure that prompt and adequate compensation or other relief is available under their national legal system for damage caused by individuals or companies who are subject to their jurisdiction.[35]

III. SPECIFIC OBLIGATIONS CONCERNING POLLUTION OF THE MARINE ENVIRONMENT

According to the reasoning of ITLOS in its 2011 *Advisory Opinion on Seabed Activities*, the obligations of States are not limited to the due diligence obligation and other general obligations.[36] Under UNCLOS and related instruments, States Parties also have obligations with which they have to comply independently of their obligation to ensure that activities under their jurisdiction and control are so conducted as not to cause damage by pollution to other States and their environment. These obligations may be characterised as 'direct obligations'.

The obligations on States Parties to adopt laws and regulations and take other measures to prevent, reduce and control pollution of the marine environment from specific activities under their jurisdiction and control is very specific and detailed in many cases. This is because many of the obligations in UNCLOS incorporate by reference the detailed provisions in the conventions and codes adopted by the International Maritime Organization (IMO) or other 'competent international organizations'.[37]

---

34 1992 Rio Declaration, *supra* note 11, Principle 18.

35 UNCLOS, *supra* note 1, Art 235(2).

36 *Advisory Opinion on Seabed Activities, supra* note 10 at para 121.

37 UNCLOS, *supra* note 1, Arts 197, 201, 207(4), 208(5), 210(4), 211(1)–(2) and (5)–(6), 212(3), 213–14, 216(1), 217(1) and (4), 218(1), 220(7), 222.

A. Obligations of Flag States to Prevent Pollution from Vessels

The most detailed provisions in UNCLOS regarding pollution of the marine environment are those governing ship-source pollution. The UNCLOS provisions strike a balance between the right of all States to have ships flying their flag exercise the right of navigation on the seas, and the interests of coastal States and the international community as a whole in preventing, reducing and controlling pollution of the marine environment by vessels. Whilst freedoms of the seas are enjoyed by States under UNCLOS, States may authorise companies and individuals to exercise these freedoms by registering their vessels with the State and flying its flag. These vessels remain under the jurisdiction and control of the flag State.

Article 94 imposes an obligation on States to take such measures for ships flying their flag as are necessary to ensure safety at sea, and such measures shall include those necessary to ensure that the master, officers and, to the extent appropriate, the crew are fully conversant with and required to observe the applicable international regulations concerning the safety of life at sea, the prevention of collisions, and the prevention, reduction and control of marine pollution. In addition, Article 211(2) provides that States have an obligation to adopt laws and regulations governing pollution from ships flying their flag that 'at least have the same effect' as the *generally accepted international rules and standards* adopted by the IMO. These provisions in effect mean that flag States must take steps to ensure that ships flying their flag comply with the very detailed rules and standards in the International Convention for the Prevention of Pollution of the Sea, known as MARPOL 73/78,[38] as amended, as well as the IMO codes concerning the carriage of dangerous goods.[39]

---

[38] International Convention for the Prevention of Pollution from Ships (MARPOL), as amended by the Protocol relating to the 1973 International Convention on the Prevention of Pollution from Ships, adopted 17 February 1978, 1340 UNTS 61 (entered into force 2 October 1983) (MARPOL 73/78).

[39] It is not entirely clear which of the annexes to MARPOL would be included as 'generally accepted international rules and standards'. Parties to MARPOL are automatically bound by Annexes I and II, but the other four annexes must be ratified separately. As of October 2014, there were 152 parties to MARPOL 73/78 and Annexes I & II on oil pollution, 138 parties to Annex III on harmful substances in packaged form, 131 parties to Annex IV on sewage, 144 parties to Annex V on garbage and 75 parties to Annex VI on air pollution. Since only about half of the parties to MARPOL are parties to the Annex VI, the argument can be made that it does not fall within the phrase 'generally accepted

Therefore, even if a flag State is not a party to MARPOL 73/78, it must take measures to ensure that its ships observe MARPOL 73/78 (as amended) and it must adopt laws and regulations on the prevention of pollution from ships that are 'at least as effective' as those in MARPOL 73/78 and the IMO codes on the carriage of dangerous goods.[40] Furthermore, Article 217(1) of UNCLOS imposes specific obligations on flag States with respect to enforcement, including the following:

> States shall ensure compliance by vessels flying their flag or of their registry with applicable international rules and standards, established through the competent international organization or general diplomatic conference, and with their laws and regulations adopted in accordance with this Convention for the prevention, reduction and control of pollution of the marine environment from vessels and shall accordingly adopt laws and regulations and take other measures necessary for their implementation. Flag States shall provide for the effective enforcement of such rules, standards, laws and regulations, irrespective of where a violation occurs.[41]

B. Obligation on Flag States to Prevent Pollution from Dumping

Dumping is defined in Article 1 of UNCLOS as 'any deliberate disposal of wastes or other matter from vessels, aircraft, platforms or other man-made structures at sea'.[42] Dumping does not include the placement of matter on the seabed for a purpose other than dumping, nor does it include the disposal of wastes governed from ships or platforms that is regulated by the provisions in MARPOL 73/78.[43]

Article 210 provides that States shall adopt laws and regulations and adopt other measures to prevent, reduce and control pollution of the marine environment by dumping. It also provides that national laws, regulations and measures shall be 'no less effective' in preventing, reducing and controlling such pollution than the 'global rules and

---

international rules and standards'. To complicate matters further, some of the IMO codes and guidelines that were initially adopted in non-binding resolutions have been subsequently incorporated into IMO conventions and made legally binding. For example, the International Code for the Construction and Equipment of Ships Carrying Dangerous Chemicals in Bulk (IBC Code) has been incorporated into MARPOL. On this very complicated issue, see *Implications of UNCLOS for the IMO*, *supra* note 5.

40 UNCLOS, *supra* note 1, Art 211(2).

41 *Ibid*, Art 217(1).

42 *Ibid*, Art 1(5)(a).

43 *Ibid*, Art 1(5)(b).

standards'.[44] The 'global rules and standards' are those set out in the 1972 Convention on the Prevention of Marine Pollution by Dumping of Wastes and Other Matter (1972 Convention).[45] It is not clear, however, whether the global rules and standards also include the 1996 Protocol to the 1972 Convention. The reason for this doubt is that 87 States are parties to the 1972 Convention, but only 44 of the 170 members of the IMO are parties to the 1996 Protocol.[46] On the other hand, the reference in this article is to 'global rules and standards', not to 'generally accepted international rules and standards'.[47] The study of the IMO Secretariat on the relationship between UNCLOS and the IMO Conventions does not take a clear position on this issue, but it does point out that there are no fundamental inconsistencies between UNCLOS and the 1972 Convention.[48]

The enforcement of laws on dumping is the responsibility of coastal States and flag States. Article 216 provides that the laws and regulations adopted in accordance with the Convention and applicable rules and standards shall be enforced by the flag State with respect to vessels flying its flag. In effect this means that a State Party to UNCLOS must adopt and enforce laws and regulations on dumping that are 'no less effective' than those in the 1972 Convention even if it is not a party to the 1972 Convention.

C. Pollution from Seabed Activities Subject to National Jurisdiction

Seabed activities subject to national jurisdiction are activities on the seabed and ocean floor in the EEZ or on the continental shelf, including construction and use of artificial islands, installations and structures, the exploration and exploitation of oil and gas resources, dredging and mining.

Article 208 of UNCLOS provides that coastal States shall adopt laws and regulations and take other measures to prevent, reduce and control

---

[44] *Ibid,* Art 210(6).

[45] Convention on the Prevention of Marine Pollution by Dumping of Wastes and Other Matter, adopted 13 November 1972, 1046 UNTS 120 (entered into force 30 August 1975); 1996 Protocol to the 1972 Convention on the Prevention of Marine Pollution by Dumping of Wastes and Other Matter, adopted 7 November 1996, 36 ILM 1 (entered into force 24 March 2006).

[46] For the status of these treaties, see http://www.imo.org/About/Conventions/StatusOfConventions/Documents/Status%20-%202014.pdf (accessed 8 January 2015).

[47] UNCLOS, *supra* note 1, Art 210(4).

[48] *Implications of UNCLOS for the IMO*, *supra* note 5 at 67.

pollution of the marine environment from such activities. It also provides that such laws, regulations and measures 'shall be no less effective than international rules, standards and recommended practices and procedures'.[49] Article 208 also provides that States shall establish global and regional rules, standards and recommended practices and procedures to prevent, reduce and control pollution of the marine environment from such activities.[50] However, no such global rules, standards and recommended practices and procedures have been established.[51] Therefore, there are no international norms by which one can judge the national laws and regulations. The *Montara Incident*[52] in the Timor Sea and the *Deep Water Horizon Incident*[53] in the Gulf of Mexico have triggered calls in some quarters for the international community to take steps to establish international rules, standards and recommended practices and procedures for deep water drilling. In addition, the fact that some Pacific island States are beginning to licence companies to engage in mining activities on their continental shelf has raised the issue of whether the international community should also establish international rules, standards and recommended practices and procedures to govern mining activities within national jurisdiction, especially if such activities might

---

49 UNCLOS, *supra* note 1, Art 208(3).

50 *Ibid*, Art 208(5).

51 However, some IMO conventions establish rules and standards for certain matters relating to seabed activities. For example, the 1990 International Convention on Oil Pollution Preparedness, Response and Cooperation (OPRC) applies to oil pollution from offshore installations, and the 1972 London Convention applies to the disposal of platforms on the seabed.

52 'On 21 August 2009 the Montara wellhead platform drill rig owned by PTTEP Australasia suffered a well head accident, resulting in the uncontrolled discharge of oil and gas. The discharge of oil and gas was stopped on 3 November 2009.' Australian Government, Department of the Environment, 'Montara Oil Spill', accessed 8 January 2015 at http://www.environment.gov.au/topics/marine/marine-pollution/montara-oil-spill.

53 'On the evening of 20 April 2010, a gas release and subsequent explosion occurred on the Deepwater Horizon oil rig working on the Macondo exploration well for BP in the Gulf of Mexico … Eleven people died as a result of the accident and others were injured … The fire burned for 36 hours before the rig sank, and hydrocarbons leaked into the Gulf of Mexico before the well was closed and sealed.' BP, 'Deepwater Horizon accident and response', accessed 8 January 2015 at http://www.bp.com/en/global/corporate/gulf-of-mexico-restoration/deepwater-horizon-accident-and-response.html.

pose a risk to the marine environment in maritime zones of other States or the marine environment of the high seas.[54]

D. Pollution from or through the Atmosphere

Article 212 provides that States shall adopt laws and regulations to prevent and reduce pollution of the marine environment from or through the atmosphere. There are three categories of pollution of the marine environment from or through the atmosphere.

The first category concerns pollution of the marine environment from or through the atmosphere from ships exercising their right to navigate on the oceans. Flag States have the responsibility to regulate this and to follow the rules, standards and recommended practices and procedures on air pollution from ships that are established by the IMO under Annex VI of MARPOL.[55] This issue is also being discussed at the IMO in the context of the negotiations on climate change.[56]

The second category concerns pollution of the marine environment from or through the atmosphere from aircraft exercising the right to fly over the oceans. The State of registry of the aircraft has the responsibility to regulate this, and to comply with the rules, standards and recommended practices and procedures adopted by the International Civil Aviation Organization (ICAO).[57]

---

[54] Tina Hunter and Madeline Taylor, 'Deep Sea Bed Mining in the South Pacific: a background paper' (Centre for International Minerals and Energy Law, 2013) accessed 8 January 2015 at http://www.law.uq.edu.au/documents/cimel/Deep-Sea-Bed-Mining-in-the-South-Pacific.pdf; Cristelle Maurin, 'Pacific Region Faces Seabed Mining Challenge', *Asia Times* (18 April 2013) accessed 8 January 2015 at http://atimes.com/atimes/China_Business/CBIZ-01-180413.html; Michelle Allsopp, Clare Miller, Rebecca Atkins, Steve Rocliffe, Imogen Tabor, David Santillo and Paul Johnston, Greenpeace Research Laboratories Technical Report, 'Review of the Current State of Development and the Potential for Environmental Impacts of Seabed Mining Operations' (March 2013) accessed 8 January 2015 at http://www.greenpeace.to/greenpeace/wp-content/uploads/2013/07/seabed-mining-tech-review-2013.pdf.

[55] UNCLOS, *supra* note 1, Art 217(1).

[56] Sophia Kopela, 'Climate Change and the International Maritime Organization: Another Breakthrough at the Marine Environment Protection Committee?' 17: 24 *ASIL Insights* (15 November 2013) accessed 8 January 2015 at http://www.asil.org/insights/volume/17/issue/24/climate-change-and-international-maritime-organization-another.

[57] UNCLOS, *supra* note 1, Art 212(1); see International Civil Aviation Organization: A United Nations Specialized Agency, accessed 8 January 2015 at http://www.icao.int/Pages/default.aspx.

The third category concerns pollution of the marine environment from or through the atmosphere from activities inside the territory of a State.[58] This type of pollution is in effect a form of land-based pollution, or pollution of the marine environment from land-based activities. Since the land-based activities that are the source of the pollution take place within the territory of a State, they are arguably better dealt with through other conventions rather than through a convention that governs ocean activities.

However, the general principles governing transboundary pollution would apply to pollution from or through the atmosphere from land-based activities in another State. Every State has an obligation to ensure that activities under its jurisdiction and control do not pollute the marine environment of another State or the marine environment of the high seas.[59] As the definition in Article 1 of UNCLOS provides, there is pollution of the marine environment if there are deleterious effects such as harm to living resources and marine life, hazards to human health, hindrance to marine activities, including fishing and other legitimate uses of the sea, impairment of quality for use of sea water and reduction of amenities.

E. Pollution from Land-based Sources

Article 207 of UNCLOS provides that States shall adopt laws and regulations to prevent, reduce and control pollution of the marine environment from land-based sources, including rivers, estuaries, pipelines and outfall structures, taking into account internationally agreed rules, standards and recommended practices and procedures.[60]

Article 207 also provides that States, acting especially through competent international organisations or diplomatic conference, shall endeavour to establish global and regional rules, standards and recommended practices and procedures to prevent, reduce and control pollution of the marine environment from land-based sources, taking into account characteristic regional features, the economic capacity of developing States and their need for economic development.[61]

The international community has been unable to establish global and regional rules, standards and recommended practices and procedures as

---

[58] Sands, *supra* note 29 at 323, 428; Birnie, Boyle and Redgwell, *supra* note 28 at 342–4.

[59] UNCLOS, *supra* note 1, Art 194(2).

[60] *Ibid*, Art 207(1).

[61] *Ibid,* Art 207(4).

called for in Article 207.[62] One of the major reasons for this is that most of the land-based sources of marine pollution are from activities that have historically been matters essentially within the domestic jurisdiction of States. They include the construction of sewer systems, the regulation of hazardous wastes in industry, the use of fertilisers and pesticides, and so on. Many States, especially developing countries, are reluctant to agree to any legally binding instruments that limit their development policies within their territory.

Nevertheless, the general principles on transboundary pollution would apply to pollution of the marine environment from land-based sources. Every State has an obligation to ensure that activities under its jurisdiction and control do not pollute the marine environment of another State or the marine environment of the high seas.[63] Therefore, if it can be established that activities within the territory of a State are causing pollution of the marine environment, that State would in principle be under an obligation of due diligence to take steps to prevent, reduce and control the pollution caused by the activity.

IV. RESPONSIBILITY OF STATES FOR POLLUTION OF THE MARINE ENVIRONMENT

A. Invocation of the Responsibility of a State

An important contribution of the Draft Articles on State Responsibility is that they set out when a State can invoke the responsibility of another State for breaches of its obligations under international law. These provisions are especially important when considering the obligations undertaken by States when they become parties to an international treaty that contains provisions obligating States to take measures to protect the marine environment.

The general principle that an 'injured State' can invoke the responsibility of another State is set out in Article 42 of the Draft Articles on State Responsibility, which reads as follows:

> A State is entitled as an injured State to invoke the responsibility of another State if the obligation breached is owed to:

---

[62] Myron H Nordquist (ed), *1982 United Nations Convention on the Law of the Sea: A Commentary*, vol IV (The Hague: Martinus Nijhoff Publishers, 1990) at 133.

[63] UNCLOS, *supra* note 1, Art 194(2).

a) that State individually; or
b) a group of States including that State, or the international community as a whole, and the breach of the obligation:
   i) specially affects that State; or
   ii) is of such a character as radically to change the position of all the other States to which the obligation is owed with respect to the further performance of the obligation.

The obligations of a State under UNCLOS to adopt and effectively enforce laws and regulations to prevent, reduce and control pollution of the marine environment are obligations owed to a group of States, that is, to all States Parties to the Convention. If a breach of those obligations threatens to cause significant pollution of the marine environment in a maritime zone subject to the jurisdiction of a coastal State, that coastal State would be specially affected pursuant to Article 42(b)(i). Therefore, it would have the status of an injured State under Article 42 of the ILC Articles. The requirement is that a State is an injured State only if it is 'specially affected', that is, it must be affected by the breach in a way that distinguishes it from the generality of other States to which the obligation is owed.[64]

Article 48 of the Draft Articles on State Responsibility sets out the circumstances under which a State *other than an injured State* can invoke the responsibility of a State. The relevant language reads as follows:

1. Any State other than an injured State is entitled to invoke the responsibility of another State in accordance with paragraph 2 if:
   a. The obligation breached is owed to a group of States including that State, and is established for the protection of a collective interest of the group.

The official commentary to Article 48 states that obligations coming within the scope of this Article must be 'collective obligations', that is, they must apply between a group of States and they must have been established in some collective interest, such as protection and preservation of the environment.[65] A State entitled to invoke responsibility under Article 48(1) is 'acting not in its individual capacity by reason of having suffered injury, but in its capacity as a member of the group of States to which the obligation is owed'.[66]

---

[64] Draft Articles on State Responsibility, *supra* note 6, Commentary to Art 42, at para (1), 117.
[65] *Ibid*, Commentary to Art 48, at para (6), 126.
[66] *Ibid,* Commentary to Art 48, at para (1), 126.

Article 48 would enable any State Party to UNCLOS to invoke the responsibility of a State if it breaches its obligations to adopt and effectively enforce laws and regulations to protect the marine environment, as these obligations are established for the collective interest of all States Parties. The State invoking the responsibility of the flag State would be acting in its capacity as a member of the group of States to which the obligation is owed, that is, as one of the States Parties to UNCLOS.

The invocation of the responsibility of a State is understood to require the taking of measures of a relatively formal character. It is not sufficient to criticise a State for breaching its international obligations, to call on a State to observe its international obligations, or to make a formal protest. To invoke the responsibility of another State, a State must raise or present a claim against the State or commence proceedings against the State before an international court or tribunal.[67] Article 43 of the Draft Articles on State Responsibility set out the procedure for notice of a claim by an injured State which invokes the responsibility of another State. It must give notice of its claim to that State and specify: (a) the conduct that the responsible State should take in order to cease the wrongful act, if it is continuing; and (b) what form reparation should take.[68]

B. Consequences of an Internationally Wrongful Act

If there is an internationally wrongful act, the Draft Articles on State Responsibility establish that two consequences follow for the responsible State. First, it must cease the unlawful conduct, and offer assurances and guarantees of non-repetition; second, it is under an obligation to make full 'reparation' for the injury caused by the internationally wrongful act.[69]

If an injured State has the right to invoke the responsibility of another State, it can seek remedies such as assurances and guarantees of non-repetition or an order that the conduct of the State is a breach of its international obligations without having to prove that the internationally wrongful act caused injury to it. However, if an injured State is seeking reparation for injury or damage caused to it, it must show that the injury was caused by the internationally wrongful act.[70]

---

[67] *Ibid,* Commentary to Art 42, at para (2), 116.
[68] *Ibid,* Art 43(2).
[69] *Ibid*, Arts 30 and 31.
[70] *Ibid,* Commentary to Art 31, at para (9), 92.

C. Importance of the Dispute Settlement Regime in Part XV of UNCLOS

In most cases of environmental harm it is not possible to invoke the responsibility of another State by the commencement of proceedings against the State before an international court or tribunal. This is because the jurisdiction of international courts and tribunals is based on the principle of consent. No international court or arbitral tribunal has jurisdiction to hear the dispute unless both States consent.

UNCLOS is in many respects an exception to this rule because the general principle in Part XV on the Settlement of Disputes is that disputes concerning the interpretation or application of any provision in UNCLOS are subject to the system of *compulsory procedures entailing binding decisions*.[71] When States become parties to UNCLOS, they consent in advance to the system of compulsory binding dispute settlement in Part XV of the Convention.

The 'default' rule in UNCLOS is that if there is a dispute between two States concerning the interpretation or application of any provision in the Convention, it is subject to the system of compulsory procedures entailing binding decisions in Section 2 of Part XV. States are obligated to first exchange views to try to resolve the dispute by following the procedures set out in Section 1 of Part XV.[72] However, where no settlement has been reached by recourse to Section 1, the dispute may be unilaterally submitted *at the request of any party to the dispute* to the court or tribunal having jurisdiction under Section 2.[73]

Which court or tribunal has jurisdiction to hear a dispute depends on whether the parties to the dispute have exercised their right to select a procedure for resolving disputes to which they are parties.[74] Under Article 287, a State is free to choose, by means of a written declaration, one or more of four procedures for the settlement of disputes concerning the interpretation or application of the Convention: (1) adjudication before the ICJ; (2) adjudication before ITLOS; (3) arbitration under Annex VII of UNCLOS; or (4) special arbitration under Annex VIII of

---

[71] Tommy Koh and S Jayakumar, 'Negotiating Process of the Third United Nations Conference on the Law of the Sea', in Myron H Nordquist (ed), *United Nations Convention on the Law of the Sea 1982: A Commentary*, Vol I (The Hague: Martinus Nijhoff Publishers, 1985) 29–134.

[72] UNCLOS, *supra* note 1, Art 283.

[73] *Ibid*, Art 286.

[74] *Ibid*, Art 288(1).

UNCLOS.[75] A State may indicate its choice of procedure when signing, ratifying or acceding to UNCLOS, or at any time thereafter.[76] If two States party to a dispute have elected the same procedure, the dispute may only be referred to that procedure, unless the parties otherwise agree.[77] If the States party to the dispute have not elected the same procedure, or if one of them has not made a choice of procedure, the dispute may be submitted only to arbitration under Annex VII, unless the parties otherwise agree.[78]

Another very important feature of UNCLOS with respect to disputes concerning the pollution of the marine environment is that a State party to a dispute, which is referred to dispute settlement under Section 2 of Part XV, may also request provisional measures *to prevent serious harm to the marine environment*.[79] The only prerequisite is that the court or tribunal to which the dispute has been duly submitted must first determine that *prima facie* it has jurisdiction under Part XV or Part XI.[80] Such provisional measures are legally binding.[81] Even if a dispute is being referred to an arbitral tribunal, a State Party may request provisional measures from ITLOS pending the establishment of the arbitral tribunal if ITLOS 'considers that *prima facie* the tribunal which is to be constituted would have jurisdiction and that the urgency of the situation so requires'.[82]

D. Potential Cases on the Responsibility of States for Pollution of the Marine Environment

Following from the above analysis, there are several types of disputes concerning the interpretation and application of the provisions on the protection of the marine environment in UNCLOS that are subject to the dispute settlement procedures in UNCLOS. They would include the following:

1. A dispute on whether a flag State is in breach of its obligations under Articles 94, 194, 211 and 217 to prevent, reduce and control

---

75 *Ibid*, Art 287(1).
76 *Ibid*.
77 *Ibid*, Art 287(4).
78 *Ibid*, Art 287(5).
79 *Ibid*, Art 290(1).
80 *Ibid*.
81 *Ibid*, Art 290(6).
82 *Ibid*, Art 290(5).

pollution of the marine environment by ships flying its flag when such breaches cause or threaten serious pollution of the marine environment of the State instituting proceedings.

2. A dispute on whether a coastal State is in breach of its obligations under Articles 194, 208, 214, 235, 60 and 80 to prevent, reduce and control pollution from the marine environment from seabed activities under its jurisdiction when such breaches cause or threaten serious pollution of the marine environment of the State instituting proceedings.
3. A dispute on whether a flag State or a coastal State is in breach of its obligations under Articles 194, 206, 210 and 216 to prevent, reduce and control pollution of the marine environment by dumping when such breaches cause or threaten serious pollution of the marine environment of the State instituting proceedings.
4. A dispute on whether a coastal State is in breach of its obligations under Articles 194, 204, 206, 207 and 213 to prevent pollution of the marine environment from land-based activities under its jurisdiction and control, as well as its duty to cooperate and its obligation to conduct an environmental impact assessment, when activities cause or threaten serious pollution of the marine environment of the State instituting proceedings.

The relief sought in a case would depend upon the particular facts, but it could include a request for an order to cease the activities, an assurance of cessation of the activities and guarantees of non-repetition, and an order that the conduct of the State is a breach of its international obligations. If the State in breach of its obligations has failed to adopt laws and regulations as required in UNCLOS or has failed to take administrative measures to ensure effective compliance with such laws and regulations, the remedy sought might be an order for the State concerned to take such measures by a particular date. If the case involves a failure by a State to fulfil its 'duty to cooperate' and its obligation to conduct an EIA, the State instituting proceedings could seek an order requiring the offending State to cooperate with it by sharing information on the activities it intends to carry out, by entering into consultations with it on how to conduct the EIA and how to minimise the potential impact of the activities on the marine environment of the affected State. Finally, if the State instituting proceedings can prove that the breaches have caused damage to the marine environment in its maritime zones, it could also seek an order for the payment reparations, including compensation for clean up and restoration.

If the threat to its marine environment is serious and imminent, the State instituting procedures could also request an order for provisional measures under Article 290 to prevent serious harm to the environment. As explained earlier, even if the dispute would go to an arbitral tribunal under Annex VII, provisional measures can be sought from ITLOS pending the constitution of the Annex VII arbitral tribunal.

V. CONCLUSIONS

The general principles governing transboundary pollution of the marine environment are set out in UNCLOS and in the principles of general international law. The provisions in UNCLOS are being interpreted and applied by international courts and arbitral tribunals in light of evolving principles of international environmental law. This is a very positive development as it will ensure that UNCLOS remains a living document that can be adapted to meet the increased concerns about the fate of the oceans.

The general principles governing transboundary pollution of the marine environment are the same as those governing transboundary air pollution, transboundary pollution of shared freshwater resources and other forms of transboundary pollution. However, the fact that UNCLOS is a universally accepted global convention governing the oceans makes transboundary marine pollution different in three important respects.

First, the principles and rules governing States are more clear and certain because they are set out in UNCLOS. States have a general due diligence obligation under Article 194 of UNCLOS to ensure that activities under their jurisdiction and control are so conducted as to not cause damage by pollution to other States and their environment. In addition, UNCLOS contains provisions establishing direct obligations on States to prevent, reduce and control pollution of the marine environment from specific activities subject to their jurisdiction and control. This is especially so with respect to ship-source pollution because the provisions in UNCLOS incorporate by reference the detailed provisions contained in other conventions and instruments.

Second, the dispute settlement regime in Part XV of UNCLOS gives injured States Parties the right to institute proceedings against other States Parties who have breached their obligations under UNCLOS to prevent, reduce and control pollution of the marine environment. When they become parties to UNCLOS, States give their consent in advance to the procedures in Section 2 of Part XV. As a result, if a dispute arises concerning the interpretation or application of any provision in the

Convention, and it cannot be resolved by negotiation and consultation between the parties, either party to the dispute may unilaterally institute proceedings against the other State.

Third, the rules on the responsibility of States for internationally wrongful acts and the dispute settlement mechanisms in UNCLOS give injured States the right to seek remedies to prevent transboundary pollution of the marine environment. Such remedies may include a request for a State to conduct an EIA or to fulfil its duty to cooperate when it plans to engage in activities within its jurisdiction and control which may cause significant pollution of the marine environment of another State. This process is aided by the fact that the dispute settlement procedures in UNCLOS expressly provide that the State which commences proceedings has the right to request *provisional measures* to prevent serious harm to the marine environment.

Although UNCLOS provides States with a mechanism to hold States *responsible* if they fail to fulfil their obligations to prevent, reduce and control pollution of the marine environment, and to seek remedies to prevent and reduce such pollution, it does not provide an easy mechanism for holding States *liable* for damage caused to the marine environment. States are entitled to reparations from another State for damage to their marine environment only if the internationally wrongful act caused pollution of the marine environment. In other words, a State can be *responsible* for breaching its international obligations, but it may not be *liable* to pay reparations for damages unless it can be established that there is a causal link between the breach of the international obligation and the damage to the marine environment. However, this is not as serious a defect as it might appear. This is because remedies that force States to comply with their obligations to prevent, reduce and control pollution are far more important than remedies that require States to pay compensation after pollution has occurred.

In conclusion, the compulsory procedures entailing binding decisions in Part XV of UNCLOS provide scope for ensuring universal adherence to the provisions in UNCLOS concerning the prevention, reduction and control of pollution of the marine environment. However, the existence of remedies is not sufficient. What is needed is a group of environmentally conscious States who are willing to pay the financial and diplomatic costs involved in bringing proceedings against States that fail to fulfil their international obligations under UNCLOS to protect and preserve the marine environment.

7. Transboundary pollution from offshore activities: a study of the Montara offshore oil spill

Youna Lyons

I. INTRODUCTION

The 2009 Montara offshore oil spill and its consequences were anything but expected. Originating from an offshore well in the Timor Sea 140 nautical miles off the coast of Australia and 50 nautical miles from the exclusive economic zone (EEZ) boundary with Indonesia, it is a quintessential example of transboundary pollution from offshore activities and is still the subject of unresolved and heated debates between Australia and Indonesia.

This chapter first briefly examines the circumstances of the Montara blowout and spill, the extent of its transboundary effects and the conclusions from the investigations carried out by Australia. The remainder of the chapter reviews the administrative and legal mechanisms and steps taken by the government of Australia and PTT Exploration and Production Australasia (PTTEP-AA), the oil operator, and their compliance with international law. It distinguishes three successive phases: first, the oil spill preparedness; second, the oil spill response; and third, the mechanisms for compensation. For each phase, the factual, administrative and legal circumstances of the Montara spill are contrasted with the applicable international legal framework.

This chapter does not discuss remedies that may be sought by public and private entities before the judicial courts of Australia and Indonesia or other agreed dispute resolution mechanisms. It focuses on the respective rights and duties of the State of Australia and Republic of Indonesia in the context of a transboundary oil spill from offshore activities.[1]

[1] A large and varied number of lawsuits have been commenced by private entities on the basis of loss suffered from the spill, before national courts of

II. THE MONTARA INCIDENT: FACTS

A. Incident: The Uncontrolled Hydrocarbon Release

On 21 August 2009, a platform well located in the Timor Sea was subject to an uncontrolled release of hydrocarbons, which lasted 74 days.[2] The platform, West Atlas, was a jack-up rig owned by the Norwegian-Bermudan Seadrill and operated by PTT Exploration and Production Australasia (PTTEP-AA), an Australian subsidiary of PTTEP, itself a subsidiary of Petroleum Authority of Thailand (PTT), a Thai National Oil Company. It was located halfway between the main coast of Australia and the coast of Pulau Roti, around 135 nautical miles (approximately 250 kilometres) offshore from both. The EEZ boundary between Australia and Indonesia was 50 nautical miles away from the well.

Difficulties were encountered in plugging the well and stopping the spill, so another mobile rig was brought from Pulau Batam (just south of Singapore) (the West Triton rig), to drill a hole into the leaking oil well and pump mud in to alleviate pressure and stop the oil flow. Contracted by PTTEP-AA on 23 August 2009, West Triton arrived at the well site on 11 September 2009 and through a trial and correction process, proceeded to drill 2.6 kilometres into the seabed to intercept the casing. These efforts were unsuccessful until 1 November 2009. About 3,400 barrels of heavy weight mud were used to stop the leak (after an undetermined quantity of lighter mud had unsuccessfully been injected) and the release of hydrocarbon stopped on 3 November 2009.[3] It was the third largest oil spill in Australia and the worst of its kind in Australia's offshore petroleum industry.[4]

Australia and Indonesia. They include, among others, private parties (such as marine resource users) suing PTTEP-AA, as well as Australia's administrative agencies responsible for combating the spill and clean-up employees of PTTEP-AA suing their employer and Australia's administrative agencies suing PTTEP-AA.

[2] Australia Maritime Safety Authority, 'Response to the Montara Wellhead Platform Incident: Report of the Incident Analysis Team' (March 2010) accessed 9 January 2015 at http://www.amsa.gov.au/forms-and-publications/publications/montara_iat_report.pdf.

[3] Montara Commission of Inquiry, 'Report of the Montara Commission of Inquiry' (17 June 2010) accessed 9 January 2015 at http://www.industry.gov.au/AboutUs/CorporatePublications/MontaraInquiryResponse/Documents/Montara-Report.pdf at 263.

[4] *Ibid* at 38.

Estimations of the amount of oil release vary between 2,000 barrels per day (equivalent to 318,000 litres), according to the Australian Department of Resources, Energy and Tourism and 400 barrels per day according to PTTEP-AA.[5] Similarly, the extent of the ocean surface impacted by the oil spill and dispersants used to break up the oil varies across sources. Calculations based on satellite imagery suggest that 90,000 square kilometres of ocean space would have been affected. However, the penetration within the water column seems to have not been monitored and was thus not recorded.[6]

PTTEP-AA handed the management of response operations to the Australia Maritime Safety Authority (AMSA). Equipment necessary to combat the spill was brought from Singapore, Geelong and Darwin in Australia. The task force deployed was composed of personnel from the relevant commonwealth bodies and different States and Territories of Australia, as well as New Zealand (under a Memorandum of Understanding signed between them for that purpose), Oil Spill Response Ltd in Singapore and private third parties. Dispersant spraying started on 23 August 2009 and continued until 1 November 2009. In total, 184,000 litres of dispersants were reported to have been sprayed using six types of dispersants, which had been pre-approved under Australia's National Plan for combating pollution of the sea by oil and other noxious substances (National Plan).[7]

Under the National Plan, an incident analysis team was established by AMSA to assess the adequacy of the response to the spill, provide

---

[5] *Ibid* at 26. The Report of the Montara Commission of Inquiry points out that the methods available to measure the daily flow of the oil were not used, which prevented any fully reliable assessment. PTTEP Australia, 'Montara Environmental Monitoring Program' (August 2012) at 9, accessed 9 January 2015 at http://www.ppr-digital.com.au/docs/brochures_pttep.pdf. PTTEP-AA declared that the spill was 300,000 barrels, which is equivalent to 4,750 tonnes. Based on the extreme figure proposed by the Department of Resources, Energy and Tourism, the total amount spilled could have been more than 10,000 tonnes. By contrast, the Gulf of Mexico oil spill was more than 700,000 tonnes, and the Amoco Cadiz was around 225,000 tonnes. World Ocean Review, 'Marine Resources – Opportunities and Risks' (2014) at 42, accessed 9 January 2015 at http://worldoceanreview.com/wp-content/downloads/wor3/WOR3_english.pdf.

[6] Report of the Montara Commission of Inquiry, *supra* note 3 at 303.

[7] Types of dispersants used include: Slickgone NS, Slickgone LTSW, Ardrox 6120, Tergo R40, Corexit 9500, Corexit 9527 and Tergo 40. See AMSA, 'Major Oil Spills in Australia – Montara Well Head Platform', accessed 9 January 2015 at https://www.amsa.gov.au/environment/major-historical-incidents/Montara_Wellhead/.

strategic recommendations for improvement and ultimately improve Australia's future major incident responses. This first report was followed by the creation of the Montara Commission of Inquiry (Commission), which was charged with carrying out in-depth investigations into the cause of the incidents and scrutinising the quality of the response to the spill. The Report of the Montara Commission of Inquiry was made publicly available together with a draft Government response on 24 November 2010.[8]

B. Sea Area Concerned

The Timor Sea and its living and non-living resources have been the subject of long and complex negotiations between Australia and Indonesia, which have led to a number of maritime boundary treaties being concluded. Initially these were to govern parts of the seabed only,[9] then subsequently, there was the 1997 Treaty establishing an EEZ Boundary and Certain Seabed Boundaries (1997 EEZ Treaty).[10] Australia and Indonesia have also concluded special arrangements on fisheries to allow access rights for Indonesian traditional fishers into a negotiated area in the Australian EEZ (which includes the water surrounding Ashmore Reef and Cartier Island),[11] an arrangement for joint development of hydrocarbons in another part of the Timor Sea and more recently the 2002

---

[8] Report of the Montara Commission of Inquiry, *supra* note 3. On 25 May 2011, the Australian government released its final response to the report, accessed 9 January 2015 at http://ris.finance.gov.au/files/2011/06/Montara_RIS.pdf (Government's response).

[9] Agreement between the Government of the Commonwealth of Australia and the Government of the Republic of Indonesia Establishing Certain Seabed Boundaries, concluded 18 May 1971, 974 UNTS 308 (entered into force 8 November 1983); Agreement between the Government of the Commonwealth of Australia and the Government of the Republic of Indonesia establishing certain sea-bed boundaries in the area of the Timor and Arafura seas, supplementary to the Agreement of 18 May 1971, concluded 9 October 1972, 974 UNTS 320 (entered into force 8 November 1983).

[10] Treaty between the Government of Australia and the Government of the Republic of Indonesia Establishing an Exclusive Economic Zone Boundary and Certain Seabed Boundaries, signed 14 March 1997, 1997 ATNIF 4 (1997 EEZ Treaty).

[11] Memorandum of Understanding between the Government of Australia and the Government of the Republic of Indonesia regarding the Operations of Indonesian Traditional Fishermen in Areas of the Australia Exclusive Fishing Zone and Continental Shelf, signed 7 November 1974, accessed 9 January 2015 at http://press.anu.edu.au/apem/boats/mobile_devices/apb.html.

Timor Sea Treaty between the Government of East Timor and the Government of Australia.[12]

This complex patchwork of agreements results in different areas of the seabed and the water column in the Timor Sea being subject to unique combinations of rights and obligations between Australia and Indonesia. Notable examples are the areas of overlapping jurisdiction where Australia has sovereign rights over the seabed resources, but Indonesia has sovereign rights over the resources of the water column above the seabed. Another example is the area of the Australian EEZ where Indonesian fishermen have rights of access and freedom to fish within prescribed limits. The Montara oil spill appears to have penetrated these two areas as well as areas where Indonesia has jurisdiction over both the seabed and the water column.

The oil spill started from a well on the continental shelf of Australia and first spread in the water column of Australia's EEZ. In Australia, attention was concentrated on the potential spread of oil to the sensitive and protected habitat and resources of Ashmore Reef, Cartier Island and the North West coast of Australia. However, satellite imagery on 1 September 2009 showed that patches of weathered oil had crossed the boundary between the EEZ of Australia and Indonesia and expanded beyond.[13] The first area of water column affected by the oil spill beyond Australia's EEZ boundary was located above a part of the seabed under the jurisdiction of Australia pursuant to the 1997 EEZ Treaty. Later satellite imagery showed that sheens of oil could also be seen beyond this area of overlapping jurisdiction on the Indonesian side of the seabed boundary. Other parts appear to have also penetrated the eastern part of the areas where Indonesian fishermen have access rights.

C. The Montara Commission of Inquiry

The Report of the Montara Commission of Inquiry presents 100 findings and 105 recommendations,[14] stating that PTTEP-AA did not observe 'sensible oilfield practices' at the Montara Oilfield and that major shortcomings in the company procedures were widespread and systemic

---

[12] Timor Sea Treaty between the Government of East Timor and the Government of Australia, signed 20 May 2002, 2003 ATS 13 (entered into force 2 April 2003).

[13] Australian Government, Media Release, 'Indonesia: Montara Oil Spill' (2 November 2009) accessed 9 January 2015 at http://www.indonesia.embassy.gov.au/jakt/MR09_086.html.

[14] See Report of the Montara Commission of Inquiry, *supra* note 3.

and led to the blowout.[15] The primary cause of the spill seems to have been a failure of the primary and secondary well control barriers and cemented casings, which did not comply with PTTEP-AA's own Well Construction Standards or other sensible oilfield practice.[16]

The well operations had been suspended for four months in 2009 when the West Atlas Rig was deployed elsewhere. Following the return of the drilling rig to the H1 Montara well, the only secondary barrier that had been installed was removed. The blowout occurred 15 hours after the untested and unprotected primary barrier had been left exposed.[17]

The Report notes that this blowout 'was not the reflection of one unfortunate accident, or of bad luck. What happened … was an accident waiting to happen; the company's systems and processes were so deficient and its key personnel so lacking in basic competence, that the Blowout can properly be said to have been an event waiting to occur'.[18]

However, the Commission also found that if the regulator[19] had been more diligent, it could have prevented these substandard practices.[20] More particularly, the Commission considered that the Northern Territory Department of Resources made a major error when it approved the drilling programme, which involved leaving the H1 Well with only one barrier for 36 hours.[21] The Commission acknowledges the move from a prescriptive to more 'objective' regulatory system (also called a 'risk based' regulatory system), where the onus to maintain safety to minimise the risk of a major accident event is placed on operators and where operators are granted flexibility on how to best manage hazards and manage risk.[22] However, it emphasises that the regulator must ensure that a company's procedures meet the statutory standard of good oilfield practice. The Commission contrasts the Montara circumstances with the system in place in the state of Victoria where approval for licences is embedded in a monitoring, inspection, audit and compliance regime.[23]

Australia's Commonwealth government responded to the deficiencies identified by the Commission with a set of measures, one of which is the establishment of a single national regulator for offshore petroleum

---

15 *Ibid* at 6.
16 See *ibid*, 49–160.
17 *Ibid* at 8.
18 *Ibid* at 11.
19 In this instance, the Northern Territory Department of Resources.
20 Report of the Montara Commission of Inquiry, *supra* note 3 at 16.
21 *Ibid* at 213.
22 See *ibid*, 177–8.
23 *Ibid* at 15.

activities, the National Offshore Petroleum Safety and Environmental Management Authority (NOPSEMA) and the institutional re-organisation of other authorities involved.[24]

This chapter does not discuss the national issues raised by the Report with respect to the deficiencies in oil spill preparedness and response in Australia, nor does it investigate the details of PTTEP-AA's responsibility in the blowout. The focus is the transboundary aspects of the spill, the applicable international legal framework and the extent to which Australia and Indonesia complied with international law in the three stages of preparedness, response and compensation.

III. THE INTERNATIONAL LEGAL FRAMEWORK APPLICABLE BETWEEN AUSTRALIA AND INDONESIA TO TRANSBOUNDARY OIL POLLUTION FROM OFFSHORE ACTIVITIES

A. Provisions in the 1982 United Nations Convention on the Law of the Sea

The 1982 United Nations Convention on the Law of the Sea (UNCLOS)[25] sets out the general framework for all marine and maritime activities, including jurisdiction over maritime areas and specific provisions regarding pollution from offshore activities and transboundary pollution.[26] With respect to pollution from offshore activities, Article 194 provides that 'States shall take all measures necessary to prevent, reduce and control pollution of the marine environment from any source', including measures 'designed to minimize to the fullest possible extent', 'pollution from installations and devices used in exploration or exploitation of the natural resources of the seabed and subsoil, in particular *measures for preventing accidents* and *dealing with emergencies*'.[27] Although UNCLOS highlights the importance of preparedness (or preventative measures) and contingency planning in the context of offshore installations and devices used in exploration or exploitation of the natural resources of the seabed and subsoil, it does not detail the specific steps

[24] Government's response, *supra* note 8 at 7.

[25] United Nations Convention on the Law of the Sea, adopted 10 December 1982, 1833 UNTS 3 (entered into force 16 November 1994) (UNCLOS).

[26] *Ibid*, Art 194(c).

[27] Emphasis added.

that States must take in this context. Article 208, however, calls on States to 'adopt laws and regulations to prevent, reduce and control pollution of the marine environment arising from or in connection with seabed activities subject to their jurisdiction'.[28] It further provides that States 'shall establish global and regional rules, standards and recommended practices and procedures' to that effect.[29] However, international attempts to negotiate a global instrument have been unsuccessful.[30]

Nevertheless, the 'obligation to ensure' that activities are conducted so as not to cause damage to the marine environment and that measures are taken to prevent accidents and deal with emergencies has been interpreted by the International Court of Justice (ICJ)[31] as an obligation of due diligence. The Advisory Opinion of the Seabed Dispute Chamber of the International Tribunal for the Law of the Sea (ITLOS) on *Responsibilities and Obligations of States Sponsoring Persons and Entities with Respect to Activities in the Area* (*Advisory Opinion on Seabed Activities*) continued to develop the notion of due diligence.[32] It states that the obligation 'to ensure' is 'not an obligation to achieve' but rather 'an obligation to deploy adequate means, to exercise best possible efforts, to do the utmost to obtain this result'.[33] In the Montara case, 'the result' would be the objective of Article 194 of UNCLOS, which is to 'prevent, reduce and control pollution'. The *Advisory Opinion on Seabed Activities* further reasons that 'due diligence' is a 'variable concept' that 'may change over time as measures considered sufficiently diligent at a certain moment may become not diligent enough in light, for instance, of new scientific or technological knowledge'.[34] Finally, it indicates that this

---

28 UNCLOS, *supra* note 25, Art 208(1).

29 *Ibid*, Art 208(5).

30 See Robert Beckman's chapter in this book (Chapter 6). See also J Ashley Roach, 'International Standards for Offshore Drilling' in Myron H Nordquist, John Norton Moore, Aldo Chircop, and Ronán Long (eds), *The Regulation of Continental Shelf Development: Rethinking International Standards* (The Hague: Martinus Nijhoff Publishers, 2013) 105–50.

31 *Case concerning Pulp Mills on the River Uruguay (Argentina v Uruguay)*, [2010] ICJ Rep 14 (*Pulp Mills*). The judgment states (at para 223): 'The Court observes that the obligation to prevent pollution and protect and preserve the aquatic environment of the River Uruguay [...] and the exercise of due diligence implied in it, entail a careful consideration of the technology to be used.'

32 *Responsibilities and Obligations of States Sponsoring Persons and Entities with respect to Activities in the Area,* Advisory Opinion, 1 February 2011, ITLOS Rep 2011, 10 (*Advisory Opinion on Seabed Activities*) at para 117.

33 *Ibid* at para 110.

34 *Ibid* at para 117.

obligation of due diligence includes the obligation to apply a precautionary approach and best environmental practices.[35]

Other provisions in UNCLOS that are applicable in the context of transboundary pollution, irrespective of whether it occurred from offshore activities, include:

- notification of imminent or actual damage;[36]
- cooperating on activities that may cause transboundary pollution and jointly developing and promoting contingency plans for responding to pollution incidents;[37]
- monitoring of the risks or effects of pollution;[38]
- publication of the reports presenting the result of the monitoring studies;[39] and
- assessing potential effects of activities.[40]

---

[35] The *Advisory Opinion on Seabed Activities* adds that these two obligations may also apply as direct obligations of States rather than just through the obligation of due diligence. *Ibid* at para 122.

[36] Article 198 of UNCLOS (*supra* note 25) provides that '[w]hen a State becomes aware of cases in which the marine environment is in imminent danger of being damaged or has been damaged by pollution, it shall immediately notify other States it deems likely to be affected by such damage, as well as the competent international organizations'.

[37] Article 199 of UNCLOS (*supra* note 25) provides that '[s]tates shall cooperate, to the extent possible, in eliminating the effects of pollution and preventing or minimizing the damage. To this end, States shall jointly develop and promote contingency plans for responding to pollution incidents in the marine environment'.

[38] Article 204 of UNCLOS (*supra* note 25) provides that States shall endeavour 'to observe, measure, evaluate and analyse, by recognized scientific methods, the risks or effects of pollution of the marine environment. In particular, States shall keep under surveillance the effects of any activities which they permit or in which they engage in order to determine whether these activities are likely to pollute the marine environment'.

[39] Article 205 of UNCLOS (*supra* note 25) further provides that 'States shall publish reports of the results obtained pursuant to article 204 …'.

[40] Article 206 of UNCLOS (*supra* note 25) provides that '[w]hen States have reasonable grounds for believing that planned activities under their jurisdiction or control may cause substantial pollution of or significant and harmful changes to the marine environment, they shall, as far as practical, assess the potential effects of such activities on the marine environment and shall communicate reports of the results of such assessments in the manner provided in article 205'.

B. Application of the Convention on Oil Pollution Preparedness, Response and Cooperation

On a global level, the 1990 Convention on Oil Pollution Preparedness, Response and Cooperation (OPRC)[41] fleshes out the legal duties necessary to ensure adequate preparedness, contingency planning and response to a spill. At a regional level, various agreements have been adopted for regional seas such as the Protocol concerning Cooperation in Preventing Pollution from Ships and, in Cases of Emergency, Combating Pollution of the Mediterranean Sea[42] (also applicable to offshore installations) and the Protocol for the Protection of the Mediterranean Sea against Pollution Resulting from Exploration and Exploitation of the Continental Shelf and the Seabed and its Subsoil.[43] These agreements also include specific requirements for adequate equipment and devices to be available as well as contingency planning and notification requirements. However, no such treaty has been signed between Indonesia and Australia. The following question thus arises: What is the applicable legal framework with respect to preparedness, response and compensation in the event of a transboundary oil spill from offshore activities in Indonesia and Australia? First, the obligations contained in UNCLOS apply to Indonesia and Australia. Second, although Indonesia has not ratified OPRC and Australia has, the standards set out in OPRC may be given effect between these two States by application of UNCLOS.

As Indonesia is not a party to OPRC, it cannot invoke OPRC directly against Australia, even though Australia is expected to have implemented OPRC as a party to the convention. Nevertheless, according to UNCLOS (a convention that has been adopted by both Australia and Indonesia), coastal States shall 'adopt laws and regulations and take other measures to prevent, reduce and control pollution of the seabed connected with seabed activities'.[44] Furthermore, 'these regulations and measures shall be no less effective than international rules, standards and recommended

---

41 1990 Convention on Oil Pollution Preparedness, Response and Cooperation, adopted 30 November 1990, 1891 UNTS 51 (entered into force 13 May 1995) (OPRC).

42 Protocol concerning Cooperation in Preventing Pollution from Ships and, in Cases of Emergency, Combating Pollution of the Mediterranean Sea, opened for signature 25 January 2002 (entered into force 17 March 2004).

43 Protocol for the Protection of the Mediterranean Sea against Pollution Resulting from Exploration and Exploitation of the Continental Shelf and the Seabed and its Subsoil, adopted 14 October 1994 (not yet entered into force).

44 UNCLOS, *supra* note 25, Art 208.

practices and procedures'.[45] Consequently, to the extent that OPRC qualifies as the 'international rules, standards and recommended practices and procedures', national measures must be at least as effective.

Based on the observation that 107 countries have ratified OPRC (as of 28 July 2014), including most of the main offshore oil and gas producing States,[46] it seems sound to consider that OPRC qualifies as the 'international rules, standards and recommended practices and procedures' referred to in Article 208 of UNCLOS.[47]

The analysis below is thus primarily based on UNCLOS and OPRC, although similar arguments and conclusions could be made on the basis of UNCLOS and customary international law without referring to OPRC.

IV. OIL SPILL PREPAREDNESS

A. Obligation of Preparedness

Successful oil spill response depends heavily on the quality of the preparedness and contingency planning. In this context, OPRC provides for the following specific obligations on the parties:[48]

- undertake individually or jointly, to take all appropriate measures [according to the Convention] to prepare for and respond to an oil pollution incident;
- require that operators of offshore units have oil pollution emergency plans (coordinated with the national system in place and approved by the national authority in charge);
- establish a national system for responding promptly and effectively to oil pollution incidents, including a national contingency plan for preparedness and response; and
- establish (either unilaterally, or through bilateral or multilateral cooperation) a minimum level of pre-positioned oil spill combating

---

45 *Ibid.*

46 See IMO, Status of multilateral Conventions and instruments in respect of which the International Maritime Organization or its Secretary-General performs depositary or other functions, accessed 9 January 2015 at http://www.imo.org/About/Conventions/StatusOfConventions/Documents/Status%20-%202014.pdf.

47 For a discussion on international rules and standards and their application by reference, see Robert Beckman's chapter in this book (Chapter 6).

48 OPRC applies to oil spills from fixed as well as floating offshore installations. See OPRC, *supra* note 41, Art 2(4).

> equipment, commensurate with the risk involved, programmes for its use, programmes of exercises and training, detailed plans and communication capabilities as well as coordination arrangements.[49]

In the context of a transboundary spill, these obligations must be interpreted together with the obligation to cooperate.

B. General Duty to Cooperate Before a Spill

With respect to contingency planning, OPRC acknowledges the importance of mutual assistance and international cooperation, including the exchange of information respecting the capabilities of States to respond to oil incidents and the preparation of oil pollution contingency plans. OPRC also expresses awareness of the need to promote international cooperation and to enhance existing national, regional and global capabilities concerning oil pollution preparedness and response, taking into account the special needs of developing countries. In line with these objectives, parties commit to endeavour to conclude bilateral or multilateral agreements for oil pollution preparedness and response.[50] However, there is no specific requirement provided on the scope or content of such agreements. Specific provisions are limited to each State's own jurisdiction despite the underlying objectives expressed in the preamble that emphasise the importance of cooperation.

Outside OPRC, the duty to cooperate on activities that may cause transboundary pollution has also been recognised by international courts and tribunals. It is deduced from the general obligations of cooperation and due diligence in the protection of the marine environment. In the *Land Reclamation* case between Malaysia and Singapore, ITLOS stated that 'given the possible implications of land reclamation on the marine environment, prudence and caution require that Malaysia and Singapore establish mechanisms for exchanging information and assessing the risks or effects of land reclamation works and devising ways to deal with them in the areas concerned'.[51]

Indeed, the importance of cooperation at the early stage of contingency planning was recognised by Indonesia and Australia as it has been

---

49 *Ibid*, Art 6.
50 *Ibid*, Art 10.
51 *Case concerning Land Reclamation by Singapore in and around the Straits of Johor (Malaysia v Singapore)*, Provisional Measures, Order of 8 October 2003, ITLOS Case No 12, at para 99.

throughout the ASEAN[52] and the rest of the world under the auspices of the International Maritime Organization (IMO)[53] through numerous programmes designed to improve contingency planning.

When considered in the context of pollution from offshore activities in an area where such pollution would be likely to become transboundary, a sensible interpretation of these principles would be that neighbouring States should:

- inform each other of activities under their jurisdiction which entail a serious risk of transboundary pollution;
- agree on a contingency plan, should pollution occur; and
- cooperate prior to the start of an activity which carries a transboundary risk of pollution in order to establish a baseline of the marine environment (within the respective jurisdiction of each State) before commencing the activity.

These 'sensible' steps presuppose, however, that the States concerned agree that a specific activity under their jurisdiction or control carries a sufficient risk of transboundary pollution. Although this may generally not correspond to current State practice, there are strong arguments to support the view that exploration and production of blocks overlapping or contiguous with maritime boundaries should trigger such steps.

This said, whether the drilling for oil a few miles or further from a maritime boundary would be considered as carrying a sufficiently serious risk of transboundary pollution so as to trigger the application of these obligations is far more debatable.

C. Bilateral Agreements between Indonesia and Australia

Countries in the region do recognise the need for bilateral and regional contingency planning in regional seas, enclosed and semi-enclosed seas

---

[52] The ASEAN-OSRAP (Oil Spill Response Action Plan) and the ASEAN-OSPAR (Oil Spill Preparedness and Response) were respectively launched in 1993 and 1994. Michael G Faure and James Hu (eds), *Prevention and Compensation of Marine Pollution Damage: Recent Developments in Europe, China and the US* (Alphen aan den Rijn: Kluwer Law International, 2006) at 111. For recent development of ASEAN-OSRAP, see http://www.ipieca.org/system/files/uploads/GISEA_Newsletter-vol2.pdf (accessed 9 January 2015).

[53] See IMO/IPIECA Global Initiative, accessed 9 January 2015 at http://www.ipieca.org/focus-area/oil-spill-preparedness; IMO's Integrated Technical Co-operation Programme, accessed 9 January 2015 at http://www.imo.org/OurWork/Environment/PollutionResponse/Pages/Default.aspx.

as well as in some straits such as the Straits of Malacca and Singapore, either with or without support from the IMO. In the Asia-Pacific Region, the transportation working group of the Asia Pacific Economic Cooperation (APEC) has taken on the mission of verifying the status of contingency planning in the region and promoting cooperation in this respect.[54] However, cooperation still seems to be mostly organised within bilateral agreements or multilateral agreements and for specific geographical areas such as the Straits of Malacca and Singapore or the Sulawesi Sea. Institutionalised, organised and systematic sharing of domestic legislation, lists of legal dispersants and the quantity and location of equipment available to combat spills are not yet available in the APEC region as it is for instance in the Northwest Pacific Region within the Northwest Pacific Action Plan for Marine Environmental Emergency Preparedness and Response Regional Activity Centre (NOWPAP-MERRAC).[55]

Indonesia and Australia signed a Memorandum of Understanding on Oil Pollution Preparedness and Response on 3 October 1996 (1996 MoU). However, the 1996 MoU does not appear to have been implemented and it is of note that the Montara spill did not seem to have triggered reliance on its terms by either Indonesia or Australia. Nevertheless, both countries have provided the International Owners Pollution Federation (ITOPF) with necessary information to comply with the IMO's spill reporting procedure. Information in the ITOPF country profiles for both Australia and Indonesia include the government authority charged with overall responsibility for dealing with oil spills from ships, institutional response arrangements and policy on oil spill response.[56] It is noted that this information has been compiled for oil pollution from shipping rather than offshore activities and may therefore be partly inadequate in circumstances of an oil spill from offshore activities. Nevertheless, it is a very useful starting point from which to build consultation mechanisms.

---

[54] 'APEC Concept Paper for the Development of Broader Regional Cooperation among APEC Economies on Oil Spills Preparedness and Response Arrangements' (28th APEC Transportation Working Group Meeting, Vancouver, 28 September 2006) accessed 9 January 2015 at http://www.apec-tptwg.org.cn/new/Archives/tpt-wg28/Maritime/2006_TPT-WG-28_MEG_017.doc.

[55] Northwest Pacific Action Plan for Marine Environmental Emergency Preparedness and Response Regional Activity Centre, accessed 9 January 2015 at http://merrac.nowpap.org/plan/connector/1/data/plan/basic/Glist/1//.

[56] See for instance Indonesia's profile, accessed 9 January 2015 at http://www.itopf.com/knowledge-resources/countries-regions/countries/indonesia/.

The 1997 EEZ Treaty also contains provisions for transboundary pollution in the areas of overlapping jurisdiction. However, they mostly reiterate the terms of UNCLOS, especially with respect to the obligation of States to take effective measures as may be necessary to prevent, reduce and control pollution of the marine environment.[57] The 1997 EEZ Treaty also reiterates the obligation of the two States to cooperate with each other in relation to the exercise of their respective rights and jurisdiction.[58]

D. Steps Taken for Preparedness and Cooperation Prior to the Montara Transboundary Spill

With respect to the oil spill preparedness of West Atlas, the drilling ship, it seems that although she may have had a written emergency plan, no adequate oil combating equipment was on board.[59] In addition, although it is clear that Indonesia and Australia had domestic policies in place for oil pollution preparedness and response, it is unclear whether the bilateral agreements and arrangements between Indonesia and Australia were sufficient to uphold their obligation of preparedness under UNCLOS and OPRC or their duty to cooperate. The Report of the Montara Commission of Inquiry does not mention that Indonesia and Australia have cooperated to develop joint contingency planning or other means of joint preparation in case of a transboundary spill. Rather, it seems that they were not prepared for a transboundary spill. Although it is difficult to characterise a clear breach of obligations under OPRC, whether the contingency plans in place included adequate preparation to face a transboundary spill that could extend into Indonesia's EEZ is more questionable.

As discussed in the next section of this chapter, the response to the spill suggests that there were no workable mechanisms in place for consultation and cooperation between Australia and Indonesia that could be used in an emergency response situation, such as occurred in the Montara oil spill. An example of this is the claim presented by Indonesia on the poor level of information from Australia and the fact that it did

---

57 1997 EEZ Treaty, *supra* note 10, Art 7(j).

58 *Ibid*, Art 7(n).

59 This is inferred from incident reports that all emphasise the immediate transfer of responsibility for the response to the spill to AMSA and the inability of Atlas to adequately respond to the spill. See Report of the Montara Commission of Inquiry, *supra* note 3, 22–24, 279–85. However, it is noted that the Commission does not appear to have investigated the oil spill response material available on West Atlas nor its oil spill contingency plan.

not get answers on, among others, the type(s) of dispersant used and the location where they were used.[60] This suggests that all necessary processes for cooperation were not in place. The fact that the stockpiles used to respond to the spill had to be brought from Singapore, Geelong (Victoria, Australia) and Darwin (Northern Territory, Australia)[61] and that the drilling rig used to drill the relief well also had to sail from Pulau Batam (Indonesia, next to Singapore), which took over two weeks, also suggest a potential lack of preparedness and cooperation.

The completion of an environmental baseline prior to undertaking activities that carry a risk of transboundary pollution is worthy of particular attention. The lack of a before-spill baseline with which to compare the results of the monitoring of the impacts of a spill has been highlighted as a major issue, following both the Montara and Gulf of Mexico spills. Environmental Impact Assessments (EIA) required from oil operators by the national legislation of the State with jurisdiction and control over oil mining activities appear to generally not include the constitution or contribution to the constitution of such a baseline. By contrast, the model adopted by the International Seabed Authority for activities in areas beyond national jurisdiction is interesting in that any exploration activity requires that a prior environmental baseline study be done and shared with the scientific community.[62]

V. OIL SPILL RESPONSE

This section focuses on the obligations of States in response to an oil spill under OPRC and the 1997 EEZ Treaty and contrasts them with the response to the Montara spill. It also discusses two major issues which arose from the oil spill response to the Montara incident: the first is the dispersant used in Indonesia's waters and the contention of damage from

---

[60] Christine Mason, 'Submission by the West Timor Care Foundation, Kupang (West Timor), Republic of Indonesia, to the Draft Government Response to the Report of the Montara Commission of Inquiry' (West Timor Care Foundation, 16 February 2011) at 7, 12, accessed 9 January 2015 at http://www.industry.gov.au/AboutUs/CorporatePublications/MontaraInquiryResponse/Documents/11_West_Timor_Care_Foundation.pdf (Submission by the West Timor Care Foundation).

[61] Report of the Montara Commission of Inquiry, *supra* note 3 at 280.

[62] International Seabed Authority, 'Regulations on Prospecting and Exploration for Polymetallic Nodules in the Area' (Regulations on Polymetallic Nodules) Reg 31, Art 4.

dispersant, and the second is the lack of monitoring of the transboundary part of the spill.

A. Rules under OPRC

As already mentioned, OPRC recalls in its preamble the importance of mutual assistance and international cooperation not only for contingency planning but also to exchange reports of incidents of significance that may affect the marine environment of a coastline and related interests of States. It also recalls the 'polluter pays' principle as a general principle of international law and further provides for the following specific obligations in case of an oil spill:

- Parties must adopt oil pollution reporting procedures, in accordance with the requirements developed by the IMO;
- 'Polluting State' must assess the nature, extent and possible consequences of the oil pollution incident, transfer this assessment to all other interested States, as well as details of any action it has taken or intends to take until the actions have been taken; and
- Parties have the obligation to cooperate and provide advisory services, technical support and equipment for the purpose of responding to an oil pollution incident.

B. Rules under the 1997 EEZ Treaty

In its article on areas of overlapping jurisdiction, the 1997 EEZ Treaty emphasises the importance of the marine environment and provides that:

- the Parties shall take effective measures as may be necessary to prevent, reduce and control pollution of the marine environment; and
- each Party shall be liable in accordance with international law for pollution of the marine environment caused by activities under its jurisdiction.

In effect, the 1997 EEZ Treaty mostly reiterates the language of UNCLOS and general rules of international law. It does not provide for additional obligations to those applicable under UNCLOS and OPRC.

C. Institutional Organisation of Oil Spill Response and Notification

Response operations were guided by AMSA under the National Plan. As the Combat Agency, AMSA was charged with the operational responsibility of oil spill response. The Northern Territory Department of Resources was the Statutory Agency under the National Plan, charged with the oversight of the response, institution of prosecutions and the recovery of clean-up costs.

As the Combat Agency, AMSA established an Incident Coordination Group, which met every day from 24 August to 9 November 2009. It brought together representatives from the Department of the Environment, Water, Heritage and the Arts (Commonwealth) (DEWHA), the Department of Resources, Energy and Tourism (Commonwealth), PTTEP-AA, the Australian Fisheries Management Authority (AFMA), the Department of Prime Minister and Cabinet, Australian Marine Oil Spill Centre (AMOSC) and the Department of Foreign Affairs and Trade. It must be noted that this coordination effort, as it is described in the government reports published in Australia after the spill, did not include representatives from Indonesia.[63]

However, there is no doubt that Indonesia was officially informed that sheens of oil were reaching Indonesia's EEZ on 1 September 2009. It is unclear whether Australia notified Indonesia prior to this (officially or unofficially) of the risks associated with this development. However, given the importance and location of the Montara oil spill, it would be surprising if no exchange took place during 21 August 2009 and 1 September 2009. It would have been natural for Indonesia to ask for information on the risk of the spill reaching waters under its jurisdiction.

A media release from the Australian Embassy in Jakarta also indicates that Indonesian officials visited Darwin from 30 September 2009 to 2 October 2009 to observe first-hand Australia's response to the spill. Furthermore, the Australian Minister for Environment met the Indonesian Minister for Environment on 28 October 2009.[64]

There is therefore no doubt that Australia met its international obligation of notification of imminent or actual damage to the environment under Article 198 of UNCLOS. What is open for discussion, however, is

---

63 Department of Resources, Energy and Tourism, 'Final Government Response to the Report of the Montara Commission of Inquiry' (2011) accessed 9 January 2015 at http://www.industry.gov.au/AboutUs/CorporatePublications/MontaraInquiryResponse/Documents/FinalMontaraCommissionInquiryReport.pdf.

64 Australian Government, Media Release, *supra* note 13.

the timing and extent of the consultation and the cooperation in the design of the response operations for the part of the spill that is located on the Indonesian side of the EEZ boundary.

D. Dispersant Used in Indonesia's Waters and Contention of Damage from Dispersant

This section discusses the application of the obligations to report, monitor the effects, consult and cooperate in the context of the application of dispersant by Australia to combat the Montara oil spill.

The Commission reports that dispersant operations commenced two days after the blowout and continued until the spill was stopped; 184,000 litres of dispersant types that were pre-approved by Australia were deployed, based on available stocks and taking into account the logistics of transporting the quantities needed to support the response operations.[65] Reports suggest that dispersant was first applied within Australia's EEZ and later in Indonesia's EEZ as judged necessary by the Australian response team.[66]

The Report of the Montara Commission of Inquiry, the press release from the Australian Embassy in Jakarta and PTTEP-AA's declarations emphasise that the response team made the decision to use dispersant as one of the three priority treatment options on the basis that it was the best method to protect sensitive areas along Ashmore Reef, Cartier Island and potentially the Western Australian coast.[67] None of the arguments developed in the reports seem to take into account the impact of this choice on Indonesia's resources. Furthermore, it is noted that the massive use of dispersants in open ocean waters has been questioned by many in Australia due to the potential impact on sub-surface organisms such as fish and their larvae, coral spawns and benthic communities. The Net Benefit Analysis (NEBA) conducted by AMSA showed that even in the event of coral spawning during dispersant application, the application of dispersant would prevent greater damage to particularly sensitive habitats around Ashmore Reef and Cartier Island.

Dispersants are expected to break-up oil slicks into small oil droplets and suspend them into the upper five metres of the water column. This process improves the dissolution of the broken oil droplets and subsequently their biodegradation by microorganisms that occur naturally in the marine environment. Dispersant is considered to be particularly

---

[65] Report of the Montara Commission of Inquiry, *supra* note 3 at 282–5.
[66] *Ibid* at 283.
[67] *Ibid* at 281–2.

useful in situations where a large quantity of oil threatens sensitive habitats in shallow or intertidal areas. However, dispersant is also a pollutant. The decision to deploy dispersant and the determination of the quantity to be used should therefore be made with caution.

The Commission considered that AMSA's initial decision to choose dispersant as the primary response solution was consistent with information available at the time. However, it found that water sampling did not include sub-surface waters affected by dispersants, despite the known effect of dispersant to sink hydrocarbons into the water column. The Report states that water sampling needed to include the vertical and horizontal distribution of the oil and dispersant to be adequate for environmental assessment purposes. The information should have been available throughout the period of treatment of the spill and should have informed decisions with respect to response strategy and dispersant use.

This emphasis on the importance of ongoing monitoring of the effect of dispersant before applying more is consistently noted by topic experts.[68] Long-term effects of dispersants are also understudied and are a cause for concern among the scientific community, for instance with respect to the endocrine disruptor components of dispersants.[69] The Dispersant Guidelines adopted by the IMO in 1995[70] have been under review following the Montara and Gulf of Mexico oil spills and a new draft is still in progress. The recommended level of caution is likely to be increased due to the lessons learned from both incidents and subsequent research. Pollution risk from dispersants is also the reason why compensation for damages from oil spills from shipping accidents, which fall under the 1972 International Convention on Civil Liability for Oil Pollution Damage, can be refused in circumstances where the use of dispersant is judged excessive.

Of note is the fact that the Commission did not question whether the decision taken by AMSA was appropriate for Indonesia's marine environment. Its objective was to assess the adequacy of the response for Australia's marine environment. Further, another consequence of dispersant use is that it renders the oil, which is still present in the water

---

[68] See e.g., Merv Fingas (ed), *Oil Spill Science and Technology* (Burlington, MA: Elsevier Inc, 2011).

[69] This is not a suggestion that the dispersants used by Australia contained such endocrine disruptors as the author did not find proof of this. It is merely a presentation of the issues.

[70] International Maritime Organization/United Nations Environment Programme, 'IMO/UNEP Guidelines on Oil Spill Dispersant Application Including Environmental Considerations' (1995).

column albeit at an expected lower concentration, less visible or invisible (where the use of dispersant is fully successful) on satellite imagery because it only catches surface oil. If, as it seems, dispersant has been largely applied on the spill that was crossing into Indonesia's side of the EEZ boundary, satellite imagery cannot be a fully reliable means of determining the extent of the oil spill on Indonesia's side of the boundary. This further emphasises the importance of water sampling for Indonesia.

Discussions and criticism on the use of dispersants in the Montara spill have been echoed and expanded following the Gulf of Mexico oil spill. The degree of caution exercised by AMSA on the basis of information available at the time might be satisfactory, but these events have changed the rules and led to an increase in the scrutiny expected by oil responders. This will soon be reflected in the new Dispersant Guidelines in progress at the IMO.

In conclusion, although there is no evidence to suggest that the use of dispersant by Australia was inappropriate, the lack of monitoring of dispersant (identified in the Report) invites further questions as to whether the obligation to adopt the precautionary approach and to conduct an environmental impact assessment have been fully satisfied. There is also no evidence that Indonesia was consulted with respect to the response strategy employed on their side of the boundary. Given the potential effects of dispersants, the obligation of cooperation and consultation required Australia to consult Indonesia before deciding on the application of dispersants and to organise jointly the monitoring of the effect. Finally, in the event of another spill, respect of the new IMO standards on dispersants would require higher scrutiny.

E. Monitoring of the Spill and of Environmental Impacts and Related Cost Issues

The Commission highlighted two additional gaps in the documentation of the spill and of its consequences. First, there has been no accurate estimate of the amount of oil released on a daily basis.[71] Although means could and should have been deployed, it was not done. Second, the terms of the scientific monitoring programme were only agreed with PTTEP-AA on 9 October 2009. The national authority in charge was not in a position to start the monitoring prior to acceptance by PTTEP-AA due to the lack of funding provisions and available resources and

[71] Report of the Montara Commission of Inquiry, *supra* note 3 at 26.

equipment under the National Plan. This situation was judged unacceptable by the Commission as it highlighted a major gap in the application of environmental legislation to Commonwealth waters. One of its findings is that despite ongoing monitoring since October, it is now unlikely that the full extent of the environmental damage from the Montara oil spill will ever be established.[72]

One of the explanations provided by the Report on the late start of monitoring of environmental impacts from the spill (unlike the operational monitoring of the oil spill itself which started straight away) was based on an organisational deficiency of the Australian response system and a gap in the National Plan.[73] As a result, there was no legal basis on which DEWHA could require PTTEP-AA to undertake and/or cover the costs of monitoring. Unfortunately, agreement on the terms of the scientific (environmental) monitoring was only reached six weeks into the spill. The Report states that this prolonged delay in undertaking scientific monitoring was '*unacceptable*' and restricted the scope for assessment of the environmental damage from the spill.[74]

As an example, sampling of fish from the commercial fishery had still not occurred by mid-December, seven weeks after the uncontrolled release of oil and gas had been stopped.[75] This compromises the reliability of the assessment especially for swimming fauna, which would have by then simply escaped the area.

The Report of the Montara Commission of Inquiry also points out that the ultimate cause of the delayed scientific monitoring response might be that Australian law does not require ecological damages to be assessed.[76] This provided arguments for PTTEP-AA to try to reduce the scope of the monitoring of habitats or species under a qualified risk from pollution.

However, this discussion on the limitations and delay in the monitoring of the extent of the spill and its effect concerns primarily Australia's EEZ and territorial sea. Although to some extent the geographical reach of the surface oil has been studied by Australia's entities in charge of oil spill response, the monitoring of impact carried out later by PTTEP-AA did not concern Indonesia's EEZ and territorial sea. The extent of the spill in the water column (vertical rather than horizontal reach) has not been monitored either.

---

72 *Ibid.*
73 *Ibid* at 290.
74 *Ibid* at 291.
75 *Ibid.*
76 *Ibid.*

Based on the above information, it seems that the specific obligation (i) to supply to the affected State detailed information relating to the spill, its likely consequences and actions taken; and (ii) of cooperation and provision of advisory services may not have been fully met. It also seems that the standards of monitoring and reporting of impacts followed by Australia for the transboundary part of the spill did not comply with current standards of due diligence set by international law as recognised in the *Pulp Mills* case and in the ITLOS Advisory Opinion on *Responsibilities and Obligations of States Sponsoring Persons and Entities with Respect to Activities in the Area,*[77] at least not initially. Indonesia argues that 'the Australian Government failed to "respond" in any meaningful way besides meeting with the Government of the Republic of Indonesia to update them on the situation'.[78] A specific point made is the alleged failure by the Government of Australia to 'facilitate expeditious transboundary mobility of personnel, material and equipment in the case of an emergency' as provided in the 1996 MoU.[79] By contrast, recent reports refer to joint efforts between Australia and Indonesia to assess the effect of the spill and the prejudice suffered.[80]

VI. COMPENSATION

A. Current Deadlocks Between Indonesia and PTTEP-AA and Australia

The Government of Indonesia claims US$2.4 billion in compensation for the damage suffered from the Montara oil spill.[81] Provincial legislative speaker Ibrahim Agustinus Medah says that he has witnessed first-hand the suffering of various local fishing communities, as well as seaweed farms and pearl farms in the three districts of Rote Ndao, Sabu Raijua and Kupang. 'We lost at least 6,000 oysters by September 2009, just a few weeks after the oil spill began,' he said. 'Each oyster costs an

---

[77] *Advisory Opinion on Seabed Activities*, *supra* note 32.

[78] Submission by the West Timor Care Foundation, *supra* note 60.

[79] *Ibid.*

[80] 'Indonesia and Australia Join Forces For Montara Oil Spill', *The Jakarta Globe* (16 July 2010) accessed 9 January 2015 at http://www.thejakartaglobe.com/archive/indonesia-and-australia-join-forces-for-montara-oil-spill/.

[81] PTTEP, 'Montara Lessons Learned' (July 2012) at 22, accessed 9 January 2015 at http://www.pttep.com/inc/download.aspx?file=../download/DocumentSSHE_ReportFile_24804.MONTARA%20Lessons%20Learned.pdf.

average Rp 1 million [$110], so the loss to local farmers was Rp 6 billion. The seaweed stocks, too, were lost during the same period … These days you get fish being sold for up to Rp 50,000 each, when before they would only cost between Rp 20,000 and Rp 25,000 each … The reduced supply is driving the price up, which affects everyone, not just the fishermen.'[82]

The West Timor Care Foundation also prepared two reports, one preliminary one addressed to the Commission (albeit late, as it is dated 11 May 2010)[83] and a more detailed one on 16 February 2011 in response to the Draft Government Response to the Report of the Montara Commission of Inquiry.[84] These reports indicate that Indonesia has undertaken research and is in a position to prove the damage of the Montara spill on seaweed farmers as well as water contamination around the West Timorese region by the oil spill. The latter is based on water samplings showing contamination with an oil that matches with a 95 per cent confidence the characteristics of the Montara crude oil. One of the reports also discusses the use of two types of dispersants (Corexit EC9500 and EC9527A), which have been forbidden in the United Kingdom since 1998, and indicates that Indonesia was not informed of the quantities of dispersant used in their water nor the type of dispersant, despite these questions being posed to AMSA. However, the *Jakarta Globe* also reports that some fishermen say the reduced catches and disruptions to seaweed and oyster farming may have been caused by inclement weather rather than the oil slick.[85]

PTTEP-AA's position in response is that to date no verifiable, valid, scientific evidence has been presented to them in support of these claims.[86] It bases its position on the various reports and the monitoring programme by the Australian government. However, as previously discussed, this justification is weak given that none of these reports had the

---

[82] Eras Poke, 'East Nusa Tenggara Fishermen Fear for Future after Oil Spill', *The Jakarta Globe* (28 June 2000) accessed 9 January 2015 at http://www.thejakartaglobe.com/archive/east-nusa-tenggara-fishermen-fear-for-future-after-oil-spill/.

[83] West Timor Care Foundation, Submissions to the Commission of Inquiry into the Uncontrolled Release of Oil and Gas from the Montara Wellhead Platform (11 May 2010).

[84] Submission by the West Timor Care Foundation, *supra* note 60.

[85] Poke, *supra* note 82.

[86] PTTEP Australasia, Fact Sheet, 'Government of Indonesia Compensation Claim' at 1 (copy on file with author).

objective of investigating the extent of the spill in Indonesian waters or the impact of the spill on Indonesian resources.

Nonetheless, it remains that the spill started far away from the coast of Indonesia and, under Australian law, it seems that Indonesian victims have the onus of proving their losses in order to justify any awards of compensatory damage. However, Indonesia appears not to have presented any such documented claims.

Australia's response to Indonesia's claim of US$2.4 billion[87] appears similar to that of PTTEP-AA. Public communications suggest that discussions with Indonesia are tripartite, involving representatives of both PTTEP-AA and of the Australian government. The principle of State responsibility has not been discussed; the only issue seems to be the determination of the loss suffered and the causality link between the alleged loss and the oil spill.

While the need to prove the damage suffered and the causality link are standard and expected, three less usual elements require further attention:

- the lack of proof may result from the delay in starting the scientific monitoring, due to the arguments presented by PTTEP-AA before they agreed monitoring terms and the lack of cost recovery provisions in Australian law; it seems unfair that PTTEP-AA and the Australian government could now benefit from their own negligence;
- the other reason for the lack of proof is the lack of general resources in Indonesia to proceed with their own monitoring and assessment; and
- finally, Indonesia's reliance on marine resources is greater than that of Australia; further, the people who rely on these resources for subsistence are less likely to have the adaptation capability to move where fish stocks may have moved or move to another industry. This again makes for an unbalanced situation.

As has been mentioned earlier in this chapter, OPRC provides for an obligation to provide advisory services, technical support and equipment for the purpose of responding to an oil incident, upon the request of an affected party. Arguably, Australia provided this when it expanded the oil spill response operations into Indonesia although it is unclear whether Indonesia requested it.

---

87 See PTTEP, 'Montara Lessons Learned', *supra* note 81.

B. The Issue of Ecological Damage

The Report of the Commission of Inquiry highlights that the Australian Institute of Marine Science noted (in its submission to the Commission) that Australia does not have legislation requiring ecological damage to be assessed nor compensated for. This can be contrasted with the United States legislation that allows to some extent the recovery of damage to the environment because it focuses on damage to natural resources.[88]

However, the current international regime for compensation for damage from oil pollution from shipping[89] only allows compensation for economic loss suffered as a result of a spill, including clean-up costs and reinstatement expenses under strict restrictions.[90] Ecological damage, such as the loss of ecosystem services or long-term rehabilitation measures, cannot be recovered. Despite criticism of this and international attempts to extend the definition of recoverable damages, States Parties have so far refused to do so.[91] It must be pointed out that this analogy is theoretical as Indonesia has not ratified any of the relevant conventions on pollution damage from shipping and especially not the 1969 Civil Liability Convention.

By contrast, the standard contract adopted for deep seabed mining in the Area by the International Seabed Authority includes a comprehensive definition of the marine environment, such that liability for damage to the marine environment would include environmental damage per se. Marine environment is defined as including 'the physical, chemical, geological

---

[88] For a comprehensive discussion of compensation of damage to the marine environment, including the United States Oil Pollution Act 1990, see Louise Angélique de la Fayette, 'New Approaches for Addressing Damage to the Marine Environment' (2005) 20 *International Journal of Marine and Coastal Law* 167 at 184.

[89] International Convention on Civil Liability for Oil Pollution Damage, adopted 29 November 1969, 973 UNTS 3 (entered into force 19 June 1975; superseded by the 1992 Protocol to the 1969 International Convention on Civil Liability for Oil Pollution Damage, 1956 UNTS 255) (1969 Civil Liability Convention); International Convention on the Establishment of an International Fund for Compensation for Oil Pollution Damage, 18 December 1971, 11 ILM 284 (entered into force 16 October 1978; superseded by the 1992 Protocol to Amend the 1971 International Convention on the Establishment of an International Fund for Compensation for Oil Pollution Damage, 1953 UNTS 330).

[90] IPIECA and ITOPF, 'Oil Spill Compensation: A Guide to the International Conventions on Liability and Compensation for Oil Pollution damage' (2007) accessed 9 January 2015 at http://www.ipieca.org/publication/oil-spill-compensation.

[91] De la Fayette, *supra* note 88.

and biological components, conditions and factors which interact and determine the productivity, state, condition and quality of the marine ecosystem, the waters of the seas and oceans and the airspace above those waters, as well as the seabed and ocean floor and subsoil thereof'.[92] Maybe this could serve as a model for mining on the continental shelf.

VII. CONCLUSION

This study of the Montara oil spill shows first and foremost that Australia and Indonesia were taken by surprise and were unprepared to respond to (i) a transboundary spill and (ii) a spill from a remote offshore installation.

Although this oil spill case does not provide clear grounds to argue that either Australia or Indonesia has breached their international obligations, it provides a typical example of 'paper' cooperation that fails to meet its purpose. It is also a useful case study to improve the institutional, legal and practical resources in place for the next transboundary oil spill response, which may be bigger and involve resources located closer to shore. This case study also exemplifies the importance of consultation and cooperation prior to a spill, that is, the importance of careful, detailed and pragmatic contingency planning beforehand. Without these, fast response in consultation is not possible. Since the Montara incident, the 2011 Strategic Action Programme for the Arafura and Timor Seas has included regional response coordination arrangements.[93]

Lastly, the response to the Montara oil spill would have been helped by clear international obligations for pollution from offshore activities. National legislation is often proving weaker when there are no clear international rules creating obligations on States to enact strict regulations. Furthermore, the fact that there is no international compensation regime in place for transboundary spills from offshore activities makes the adoption and implementation of OPRC even more important, especially where offshore activities are close to maritime boundaries such as the South China Sea. The initiative started by Indonesia following the Montara incident may bring progress on this front because the IMO agreed to provide guidance for bilateral and multilateral agreements in

---

92 Regulations on Polymetallic Nodules, *supra* note 62, Reg 1(3)(c).

93 2001 Strategic Action Programme for the Arafura and Timor Seas, Arafura & Timor Seas Ecosystem Action Program, accessed 9 January 2015 at http://www.atsea-program.org/publication-1/atsea-strategic-action-programme/view.

the context of liability and compensation issues concerning transboundary pollution damage from offshore oil exploration and exploitation activities following the 97th session of its Legal Committee.[94]

[94] IMO Legal Committee, 101st Session, 'Any Other Business: (i) Liability and compensation issues connected with transboundary damage from offshore oil exploration and exploitation activities', LEG 101/11, 14 February 2014, accessed 9 January 2015 at https://imo.amsa.gov.au/public/2014/leg101.html.

8. Preventing transboundary nuclear pollution: a post-Fukushima legal perspective

Günther Handl*

I. INTRODUCTION

It has often been suggested that truly significant advances in the safety of complex technologies, industrial production processes, transportation or weapons systems follow only in the wake of major accidents. In this vein, regulatory initiatives following the accidents at the Three Mile Island[1] and Chernobyl[2] power plants seem to confirm that 'legislation by catastrophe' is a hallmark also of the nuclear power industry. Similarly, in the aftermath of Fukushima the question is not about whether the accident will turn out to be a catalyst for significant regulatory change,[3]

* The author wishes to thank Jan P Fladeboe and Andreas Molin for sharing relevant information, James Duggan for assisting with library resources, and Xu Xuelei for her research assistance.

[1] See e.g., John Kemeny, *Report of the President's Commission on the Accident at Three Mile Islands: The Need for Change: The Legacy of TMI* (President's Commission on the Accident at Three Mile Islands, 1979) at 61–80, whose recommendation targeted in particular safety-related organisation, procedure and practices to which both US industry and the government responded.

[2] See generally Johan Rautenbach, Wolfram Tonhauser and Anthony Wetherall, 'Overview of the International Legal Framework Governing the Safe and Peaceful Uses of Nuclear Energy – Some Practical Steps' in Joint Report by the Organisation for Economic Co-operation and Development (OECD) Nuclear Energy Agency and the International Atomic Energy Agency (IAEA), *International Nuclear Law in the Post-Chernobyl Period* (Paris: OECD, 2006) at 7; Günther Handl, 'Aprés Tschernobyl: Quelques refléxions sur le programme législative multilateral à l'ordre du jour' (1988) 92 *Revue Général de Droit International Public* 5.

[3] The likelihood of such change is already evident from the IAEA, *IAEA Action Plan on Nuclear Safety*, IAEA Doc GOV/2011/59-GC(55)/14 (2011). For

but rather its precise direction and international scope. What is evident, at any rate, is that the 2011 accident has triggered a massive review of nuclear power plant safety both at national and international levels.[4] In this sense, Fukushima persuasively demonstrates, yet again, that nuclear safety matters are inherently a matter of international concern.[5] This is

---

an overview of some national regulatory changes already put in place, see *Nuclear Safety: Countries' Regulatory Bodies Have Made Changes in Response to the Fukushima Daiichi Accident*, Report to the Chairman, Subcommittee on Transportation and Infrastructure, Committee on Environment and Public Works, US Senate, GAO-14-109 (March 2014) (US Nuclear Safety). Moreover, at the European level, on 8 July 2014, the Council of the European Union adopted a revised nuclear safety directive, *Council Directive 2014/87/Euratom of 8 July 2014 amending Directive 2009/71/Euratom establishing a Community framework for the nuclear safety of nuclear installations* [2014] OJ L 219/42 (European Council Directive).

[4] Thus, over the last couple of years the IAEA has hosted several international experts' meetings and published a series of reports on a variety of Fukushima-related safety topics. For details, see Fukushima Nuclear Accident, accessed 9 January 2015 at http://www.iaea.org/newscenter/focus/fukushima/. Moreover, Fukushima's 'lessons to be learned' were the specific subject of the Second Extraordinary Meeting of the Contracting Parties to the Convention on Nuclear Safety (27–31 August 2012); the Sixth Review Meeting of the Convention on Nuclear Safety (24 March–4 April 2014), as noted by André-Claude Lacoste, 'Statement of the President of the 6th RM of the CNS', presented on 24 March 2014; and of a conference co-sponsored by Japan and the IAEA, the Fukushima Ministerial Conference on Nuclear Safety (15–17 December 2012). See also The National Diet of Japan, *The Official Report of the Fukushima Nuclear Accident Independent Investigation Commission* (2012); OECD/NEA, *The Fukushima Daiichi Nuclear Power Plant Accident: OECD/NEA Nuclear Safety Response and Lessons Learned* (OECD, 2013) (OECD/NEA Nuclear Safety Response); EC, *Communication from the Commission to the Council and the European Parliament on the comprehensive risk and safety assessments ('stress tests') of nuclear power plants in the European Union and related activities*, COM(2012) 571 final (4 October 2012); US, National Research Council, *Lessons Learned from the Fukushima Nuclear Accident for Improving Safety and Security of U.S. Nuclear Plants* (Washington, DC: The National Academies Press, 2014).

[5] Ironically, Fukushima itself may not really qualify as an accident involving significant transboundary pollution. Although a very substantial amount of radioactive materials released from the power plant entered the sea, the strong diluting effect of ocean currents lowered ambient concentrations to levels below any public health concern in areas beyond Japan's territorial sea. See e.g., WHO, *Health Risk Assessment from the Nuclear Accident after the 2011 Great East Japan Earthquake and Tsunami, Based on a Preliminary Dose Estimation* (WHO, 2013) at 9 (concluding that 'no discernible increase in health risks from

obviously so because unscheduled events at nuclear power facilities may lead to transboundary radiological effects. Such events are also capable of causing significant transboundary psychological,[6] economic and political consequences,[7] as evidenced, most notably, by Fukushima's accelerating effect upon Germany's energy transformation or '*Energiewende*'.[8] In sum, in today's hyper-connected world a nuclear accident anywhere truly is an accident everywhere.

Although non-radiological transboundary repercussions are certainly part of what might be called States' 'common interest and common responsibility' to prevent future nuclear accidents,[9] they cannot readily be accommodated within the concept of *transboundary pollution* – the basic theme of this book – and therefore will not be discussed further. Instead, the focus here will be on actual or threatened transboundary physical effects associated with nuclear fuel cycle activities.[10] Among these

---

the Fukushima event is expected outside Japan'); Jonathan Amos, 'North American Scientists Track Incoming Fukushima Plume' *BBC News* (25 February 2014) accessed 9 January 2015 at http://www.bbc.com/news/science-environment-26329323.

[6] See e.g., Younghwan Kim, Minki Kim and Wonjoon Kim, 'Effect of Fukushima Nuclear Disaster on the Global Public Acceptance of Nuclear Energy' (2013) 61 *Energy Policy* 822.

[7] Note, e.g., the IAEA, *Declaration by the IAEA Ministerial Conference on Nuclear Safety in Vienna on 20 June 2011,* IAEA Doc INFIRC/821 (2011) (Declaration by the IAEA Ministerial Conference) at para 4, in which the Ministers emphasise that 'nuclear accidents may have transboundary effects and raise the concerns of the public about the safety of nuclear energy and the radiological effects on people and the environment'.

[8] The German government decided to cancel plans for an extension of the national use of nuclear power beyond 2022 and instead shut down seven reactors immediately. See e.g., Mycle Schneider and Anthony Froggatt, *World Nuclear Industry Status Report 2013* (Mycle Schneider Consulting, July 2013). Fukushima affected many other countries' nuclear energy policies. See e.g., Ed Crooks and Sylvia Pfeifer, 'Too Hot to Handle', *The Financial Times* (16 March 2011) 9, accessed 9 January 2015 at http://www.ft.com/cms/s/0/a45a7e88-5004-11e0-9ad1-00144feab49a.html#axzz39UdGj22I.

[9] See e.g., the Swiss Proposal to the Plenary Session on CNS Effectiveness, in *Final Summary Report*, 2nd Extraordinary Meeting of the Contracting Parties to the Convention of Nuclear Safety (27–31 August 2012) Doc CNS/ExM/2012/04/Rev.2 (CNS 2nd Extraordinary Meeting).

[10] In other words, 'transboundary nuclear pollution' as used here denotes radiation effects in the territory, upon people and resources of one State due to commercial nuclear activities within the jurisdiction or control of another State, i.e., the nuclear installation State, or as in the case of the maritime transport of radioactive materials, the flag State.

effects, only those caused by an unscheduled event or accident will be of special interest[11] as it is the risk of transboundary radiation effects associated with abnormal occurrences that represents the quintessential international legal and policy challenge posed by nuclear power operations. This challenge is clearly a topical one on account of Fukushima. However, it is likely to remain a concern also for the foreseeable future because of potential safety problems associated with the ageing of existing nuclear power plants worldwide[12] and the projected increase in global nuclear power generation.[13]

The notion of 'transboundary nuclear risk' that covers not only accident prevention but also accident management is a function of a complex, indeed dizzying, ensemble of variables. Any attempt to address the international legal aspects of transboundary nuclear risk management comprehensively would therefore be well beyond the scope of the present

---

[11] This is not to deny that transboundary radiological effects associated with normal operating conditions at nuclear fuel cycle facilities are of legitimate international concern. Such concern is not inspired by the discredited notion that the transboundary deposit of radionuclides by itself amounts to an actionable 'violation of sovereignty'. See e.g., *Case concerning Aerial Herbicide Spraying (Ecuador v Colombia)*, 'Counter-Memorial of the Republic of Colombia' (29 March 2010), [2010] ICJ Pleadings (vol 1) at paras 8.34–8.40 – but by the fundamental radiological protection principle according to which any increase in radiation exposure should, as far as reasonably possible, be avoided (ALARA – 'As Low As Reasonably Achievable' principle for radiation safety). However, from an international legal point of view, routine releases of radionuclides that lead to a measurable increase in transboundary environmental radioactivity represent 'run-of-the mill' transboundary pollution scenarios.

[12] The risk associated with ageing power plants is described as 'a continuing time-dependent degradation of material due to service conditions, including normal operation and transient conditions; this can affect the capability of engineered structures, systems and components (SSCs) to perform their required function … The ageing of nuclear power plants (NPPs) could affect safety and reduce safety margins if detection or corrective action does not take place before the degradation of key SSCs, or before the loss of functional capability occurs'. IAEA, *Nuclear Safety Review for the Year 2012,* Doc GC(56)/INF/2 (July 2012) at 31.

[13] See e.g., Eduardo Porter, 'Unavoidable Answer for the Problem of Climate Change', *The New York Times* (19 November 2013) accessed 9 January 2015 at http://www.nytimes.com/2013/11/20/business/economy/unavoidable-answer-to-problem-of-climate-change.html?pagewanted=all&_r=0; World Nuclear Organization, 'Emerging Nuclear Energy Countries' (February 2014) accessed 9 January 2015 at http://www.world-nuclear.org/info/Country-Profiles/Others/Emerging-Nuclear-Energy-Countries/. But see Schneider and Froggatt, *supra* note 8.

chapter. Rather, the focus of inquiry here is more limited, namely on the role of international law in setting the basic parameters of the process through which the realisation of transboundary nuclear risks might be prevented. Specifically, this chapter first reviews transboundary nuclear pollution prevention from the perspective of customary international law, then from the viewpoint of the international nuclear safety regime centred principally on the Convention on Nuclear Safety (CNS),[14] and the Joint Convention on the Safety of Spent Fuel Management and on the Safety of Radioactive Waste Management (Joint Convention).[15] Finally, it concludes with a review of two special situations in which one State's nuclear power or fuel cycle activities impinge directly upon the territorial sovereignty of other States, hence giving rise to enhanced legal claims for protection against transboundary risks of nuclear harm: (i) the 'border siting' of nuclear facilities whose emergency planning zones would necessarily extend into neighbouring States' territory; and (ii) the maritime transportation of radioactive materials through foreign States' territorial seas.

II. TRANSBOUNDARY NUCLEAR RISK MANAGEMENT: THE BASIC INTERNATIONAL LEGAL FRAMEWORK

A. Customary International Law

That customary international law obliges States to ensure that activities within their jurisdiction and control respect the environment of other States or of areas beyond national control is by now a commonplace, though fundamentally important notion.[16] Today, it is equally accepted

---

[14] IAEA Doc INFCIRC/449 (5 July 1994) (CNS).

[15] IAEA Doc INFCIRC/546 (24 December 1997) (Joint Convention). This chapter will not discuss aspects of accident management, such as transboundary emergency notification and assistance, transboundary liability or the processing of claims for compensation. For discussion of some of the accident management issues to emerge post-Fukushima, see IAEA, 'International Experts' Meeting on Severe Accident Management in the Light of the Accident at the Fukushima Daiichi Nuclear Power Plant' (17–20 March 2014) accessed 9 January 2015 at http://www-pub.iaea.org/iaeameetings/46832/International-Experts-Meeting-on-Severe-Accident-Management-in-the-Light-of-the-Accident-at-the-Fukushima-Daiichi-Nuclear-Power-Plant.

[16] See *Legality of the Threat or Use of Nuclear Weapons*, Advisory Opinion, [1996] ICJ Rep 226 at 241–2, para 29.

that this obligation requires a State not only to redress transboundary harmful effects *ex post facto*, but also to take measures to control and regulate in advance sources of potential significant transboundary harm. Indeed, this customary principle of prevention is a long-standing one and, as the International Law Commission (ILC) specifically acknowledges, was already highlighted in the arbitral award in the *Trail Smelter* case.[17] The latter's version of the 'no significant harm' formula[18] was generally understood to cover situations in which transboundary harm was merely threatened, provided the risk was significant in the sense of entailing a high probability or near certainty of non-trivial harmful consequences. On the other hand, the *Trail Smelter* tribunal's dictum did not seem to extend to situations in which a given transboundary risk – though of low probability – threatened extremely serious or potentially catastrophic consequences. However, writing at the time of the 1972 Stockholm Conference on the Human Environment, Frederic Kirgis addressed head-on the issue of the *Trail Smelter*'s normative reach of transboundary risk involving low probability/serious consequence scenarios. Building on the risk of harm as a composite of probability and consequence, Kirgis proposed modifying the *Trail Smelter* standard so that a 'potentially greater harm call[s] for abstention from conduct under a proportionately lesser showing that harm will occur'.[19]

This modified-burden-of-proof approach attracted instant support in academia.[20] However, its seemingly impeccable logic has not been

---

[17] See 'International Liability for Injurious Consequences Arising out of Acts Not Prohibited by International Law (Prevention of Transboundary Harm from Hazardous Activities)' in 'Report of the International Law Commission to the General Assembly covering the work of its fifty-third session, with commentaries, 2001' (UN Doc A/56/10), Ch V, in *Yearbook of the International Law Commission 2001*, vol II, Part Two (UN, 2001) (Draft Articles on Prevention) at 378.

[18] *Trail Smelter Case (United States of America, Canada)* (1938 and 1941) 3 RIAA 1905 (*Trail Smelter*) at 1965: '[U]nder the principles of international law … no State has the right to use or permit the use of its territory in such a manner as to cause injury … in or to the territory of another or the property or persons therein, when the case is of serious consequence and the injury is established by clear and convincing evidence.'

[19] Frederic Kirgis, 'Technological Challenge to the Shared Environment: United States Practice' (1972) 66 *American Journal of International Law* 290 at 294.

[20] See e.g., Luzius Wildhaber, 'Die Öldestillieranlage Sennwald und das Völkerrecht der grenzüberschreitenden Luftverschmutzungen' (1975) 31 *Annuaire Suisse de droit international* 97 at 119; Albrecht Randelzhofer and Bruno

validated by international practice bearing on transboundary nuclear risks.[21] This is true even of practice related to circumstances in which a transboundary nuclear risk appears especially accentuated on account of, say, a nuclear facility's location at or very close to an international boundary[22] or an internationally shared natural resource, such as a river or lake. There is no need to retrace in detail all the relevant evidence. Suffice it to say instead that treaty practice, diplomatic exchanges regarding nuclear risks in border areas or relevant international jurisprudence provide at best inconclusive evidence in support of the modified-burden-of-proof approach.[23] Equally significantly, in the final analysis, the ILC's Draft Articles on Prevention of Transboundary Harm from Hazardous Activities[24] do not consider anything less than a threat of certain transboundary harm – no matter how potentially catastrophic its nature or scope – as entailing an international obligation of prevention.

The ILC defines 'risk of significant transboundary harm' as including events of a high probability/significant consequences as well as low

---

Simma, 'Das Kernkraftwerk an der Grenze' in Dieter Blumenwitz and Albrecht Randelzhofer (eds), *Festschrift für Friedrich Berber* (Munich: Beck, 1973) 389 at 416–17; Günther Handl, 'An International Legal Perspective on the Conduct of Abnormally Dangerous Activities in Frontier Areas: The Case of Nuclear Power Plant Siting' (1978) 7 *Ecology Law Quarterly* 1 at 35–36.

[21] Günther Handl, *Grenzüberschreitendes nukleares Risiko und völkerrechtlicher Schutzanspruch* (Berlin: Duncker und Humblot, 1992) at 27–28.

[22] For further discussion of the issue of due diligence and border siting of nuclear power facilities, see *infra* Part IV.A.

[23] See generally Günther Handl, '*Trail Smelter* in Contemporary International Environmental Law: Its Relevance in the Nuclear Energy Context' in Rebecca Bratspies and Russell Miller (eds), *Transboundary Harm in International Law: Lessons from the Trail Smelter Arbitration* (Cambridge: Cambridge University Press, 2006) 125 at 133–7. See also *infra* Part IV.B. Indeed, in a somewhat analogous transboundary risk context in the *Aerial Herbicide Spraying* case, Ecuador recently merely suggested that Colombia had failed to act with due diligence or to take appropriate precautionary measures to prevent transboundary drifts of the herbicide, not that Colombia was required to abstain from conducting aerial spraying in the border region per se. See *Case concerning Aerial Herbicide Spraying (Ecuador v Colombia)*, 'Memorial of Ecuador' (28 April 2009) ICJ Pleadings (vol 1) at para 8.36: 'In alleging that Colombia has failed to exercise due diligence with respect to herbicide spraying, Ecuador is not arguing that spraying must not take place at all within Colombian territory. In the present case Ecuador's argument is precisely – and simply – that Colombia manifestly failed to take all the appropriate precautionary measures within its power to prevent transboundary drift from causing significant harm in Ecuador.'

[24] See Draft Articles on Prevention, *supra* note 17.

probability/catastrophic consequences nature.[25] But it does not espouse the modified-burden-of-proof approach's rationale. According to draft Article 3, a State must 'take all appropriate measures to prevent significant transboundary harm or at any rate to minimize the risk thereof'. In other words, whereas the former – certain harm – attracts an obligation of prevention, the latter – less than certain harm – attracts a mere obligation to minimise its risk of realisation. This differentiation between the source State's obligations is thus based on the probability of harm alone, rather than on the composite of probability and consequences of the event. In short, in this normative scheme, a 'mere' risk of significant transboundary harm does not attract an obligation of prevention, even when such a risk might threaten potentially catastrophic consequences.[26] Of course, in certain circumstances, neighbouring (source and affected) States may well agree ad hoc that the carrying on of a particular hazardous activity which, though not per se prohibited by international law, carries a significant risk of transboundary harm of a low probability/high consequence nature would be unreasonable, hence internationally impermissible. However, as a general rule, a source State will be deemed internationally permitted to engage in such transboundary risk creation as long as it exercises due diligence with the aim of minimising or preventing the risk from being realised.

In its *Pulp Mills* decision, the International Court of Justice (ICJ) characterised the nature of a State's customary legal obligation to prevent transboundary pollution as one of due diligence. The ICJ then went on to note that '[i]t is an obligation which entails not only the adoption of appropriate rules and measures, but also a certain level of vigilance in their enforcement and the exercise of administrative control applicable to public and private operators …'.[27] Such measures must, as the Seabed

---

[25] See *ibid*, Art 2(a).

[26] Admittedly, the ILC's commentary in the Draft Articles on Prevention at Art 3 acknowledges that the extent of a source State's due diligence obligation to minimise a transboundary risk of harm is directly proportional to the risk of harm threatened. See *ibid* at 394. However, it does not acknowledge also that, at the extreme end of the spectrum of threatened consequences, the obligation to minimise would – consistent with Kirgis' modified burden of proof approach – turn into one of prevention. For further details, see Handl, *supra* note 23 at 137–9.

[27] *Case concerning Pulp Mills on the River Uruguay (Argentina v Uruguay),* [2010] ICJ Rep 14 (*Pulp Mills*) at para 197. While the specific starting point of the Court's pronouncement was the bilateral treaty governing the matter, the Court's characterisation of the parties' due diligence obligation draws on customary international law, hence is of general import. See *ibid* at paras 64–65.

Dispute Chamber of the International Tribunal for the Law of the Sea (ITLOS) put it, be 'reasonably appropriate' for securing compliance by persons under the State's jurisdiction.[28] Other factors broadly indicative of a source State's due diligence include, of course, the degree of its cooperation with risk-exposed neighbouring States generally and its compliance with procedural obligations aimed at preventing or minimising transboundary harm, specifically.[29] Beyond these general parameters, it is the conventional nuclear safety regime itself that can provide a clearer picture as to the degree of due diligence a nuclear installation State may have to adopt in any given situation in order to comply with its customary international legal obligation regarding the prevention of transboundary nuclear pollution.

B. Transboundary Risk Management and the Nuclear Safety Conventions

Today the management of transboundary nuclear risks is – largely indirectly[30] – governed also by an elaborate treaty-based safety regime centred on the CNS and the Joint Convention (Conventions).[31] However, in the view of some, especially early commentators, the Conventions do not appreciably or sufficiently advance transboundary nuclear risk management beyond the parameters established under customary international law.[32] For example, Peter Cameron referring to the Joint

---

[28] *Responsibilities and Obligations of States Sponsoring Persons and Entities with respect to Activities in the Area,* Advisory Opinion, 1 February 2011, ITLOS Rep 2011, 10 at para 228.

[29] See Draft Articles on Prevention, *supra* note 17, Arts 4–9.

[30] One exception is the express reference to contracting parties' obligation to consult potentially affected neighbouring States about a proposed nuclear installation/facility site. See CNS, *supra* note 14, Art 17(iv); Joint Convention, *supra* note 15, Art 13(1)(iv). Moreover, Article 27 of the Joint Convention also addresses the transboundary movement of spent fuel and radioactive waste. For further details, see *infra* Part IV.

[31] Other conventions of note: IAEA, *Convention on Early Notification of a Nuclear Accident*, IAEA Doc INFCIRC/335 (18 November 1986); IAEA, *Convention on Assistance in Case of a Nuclear Accident or Radiological Emergency*, IAEA Doc INFCIRC/336 (18 November 1986). However, both conventions focus on accident management, rather than on the prevention of the realisation of the transboundary risk of harm, the focus of the present inquiry.

[32] See e.g., Menno Kamminga, 'The IAEA Convention on Nuclear Safety' (1995) 44 *International and Comparative Law Quarterly* 872; Simon Caroll, 'Transboundary Impacts of Nuclear Accidents: Are the Interests of Non-Nuclear States Adequately Addressed by International Nuclear Safety Instruments?'

Convention's safety requirements asserts that 'it is hard to see how anyone could find anything offensive with the "motherhood and apple pie" sentiments in these "requirements"'.[33] Worse, at times the conventional nuclear safety framework has come in for savage criticism.[34]

At first glance, some of this criticism might seem well justified. The CNS, the nature of whose normative provisions is also fairly representative of the Joint Convention, underwrites rather lofty and comprehensive safety objectives,[35] but fails to follow through with detailed and specific technical provisions that would flesh out in operationally meaningful fashion the broad normative principles it espouses. Or, it features safety-related provisions which, though normatively specific, are subject to very significant qualifications.[36] Finally, practically all the installation

---

(1996) 5 *Review of European Community and International Environmental Law* 205; Karen McMillan, 'Strengthening the International Legal Framework for Nuclear Energy' (2000–2001) 13 *Georgetown International Environmental Law Review* 983.

[33] Peter Cameron, 'Joint Convention on the Safety of Spent Fuel Management and on the Safety of Radioactive Waste Management' in Nathalie Horbach (ed), *Contemporary Developments in Nuclear Energy Law: Harmonizing Legislation in CEEC/NIS* (The Hague: Kluwer Law International, 1999) 117 at 127. He goes on to say that the Joint Convention's safety requirements 'are merely hortatory, encouraging Contracting Parties to take action but defining the obligation in the softest of soft terms …'. See also Louise de la Fayette, 'International Environmental Law and the Problems of Nuclear Safety' (1993) 5 *Journal of Environmental Law* 31 at 68; Jack Barkenbus and Charles Forsberg, 'Internationalizing Nuclear Safety: The Pursuit of Collective Responsibility' (1995) 20 *Annual Review of Energy and the Environment* 179 at 198.

[34] See, in particular, Katia Boustany, 'The Development of Nuclear Law-Making or the Art of Legal "Evasion"' (1998) 61 *Nuclear Law Bulletin* 39.

[35] These objectives as set forth in Article 1 of the CNS are to achieve and maintain a high level of nuclear safety worldwide through the enhancement of national measures and international cooperation including, where appropriate, safety-related technical cooperation; to establish and maintain effective defences in nuclear installations against potential radiological hazards in order to protect individuals, society and the environment from harmful effects of ionising radiation from such installations; and to prevent accidents with radiological consequences and to mitigate such consequences should they occur. CNS, *supra* note 14, Art 1.

[36] A good illustration of this phenomenon is Article 6 of the CNS, which calls for any nuclear installation whose safety cannot be upgraded to be shut down. However, its language is hortatory rather than mandatory; moreover, its normative significance is further compromised by the fact that it admits of balancing safety risks against countervailing considerations, such as a contemplated shutdown's 'social, environmental and economic impact' on the country.

State's conventional safety obligations,[37] including those listed in Article 6 (soft as they already are) are couched as obligations of conduct rather than of result,[38] thereby incorporating a standard of 'mere' due diligence. Disappointment over the substance of many of the Conventions' individual safety provisions is thus certainly understandable. However, to consider their normative weakness as indicative of the international safety regime's lack of robustness as a whole, as some critics have done, is to fail to recognise the nature and inherent dynamic of the larger regulatory context of nuclear safety: the Conventions' perceived deficiencies are the result of a carefully considered legislative choice driven by negotiating States' insistence that nuclear safety remain a national responsibility, that is, that of the installation State.[39] In this vein, the CNS's preamble recognises that 'the Convention entails a commitment to the application of fundamental safety principles … rather than of detailed safety standards'.[40] It thereby gives expression to a view, prevailing then as well as now in industry and the regulatory community, which holds that '[n]uclear safety worldwide will not be enhanced if one insists on

---

In general, key CNS safety obligations are heavily qualified by such words as 'reasonably practical' and 'as soon as practically possible'. Corresponding provisions of the Joint Convention, namely Articles 5 and 12, are couched in similarly conditional language. By the same token, obligations under Articles 4 and 11 of the Joint Convention that relate to 'general safety requirements' applicable to spent fuel and radioactive waste management, respectively, are qualified by such words as 'appropriate' or 'adequate'. Other important safety objectives are circumscribed in terms of Contracting Parties' obligation to 'strive to' or 'aim to' avoid proscribed impacts from management operations. CNS, *supra* note 14; Joint Convention, *supra* note 15.

[37] The only exceptions in the CNS are obligations established by Article 5 (reporting on implementation measures), Article 7 (establishment and maintenance of a legislative and regulatory framework) and Article 8 (establishment of a regulatory body). CNS, *supra* note 14.

[38] An obligation of conduct represents a legal commitment, not to guarantee a certain outcome, but to employ best possible efforts to secure the realisation of the objective concerned, the State being at liberty to choose the specific means of doing so.

[39] For details, see Odette Jankowitsch-Prevor, 'The Convention on Nuclear Safety' in OECD NEA/IAEA, *International Nuclear Law in the Post-Chernobyl Period*, Joint Report by the OECD Nuclear Energy Agency and IAEA (2006), 155 at 159–60; and Tammy de Wright, 'The "Incentive" Concept as Developed in the Nuclear Safety Conventions and its Possible Extension to Other Sectors' (2007) 80 *Nuclear Law Bulletin* 29.

[40] CNS, *supra* note 14 at para viii.

establishing classical and legally binding international instruments because states do not accept them'.[41]

In other words, both the CNS and the Joint Convention are designed as 'incentive conventions'. They rely on contracting parties' rational self-interest in policing, maintaining and improving upon nuclear safety through a process of mandatory national safety reporting and periodic peer reviews.[42] This process is given teeth by the Conventions' incorporation by reference of internationally formulated or endorsed standards and criteria, as either capable of providing 'guidance on contemporary means of achieving a high level of safety'[43] or, of informing the State's obligation to adopt legislation for the effective protection of individuals, society and the environment against radiological hazards.[44] Thus, even though they are not legally binding per se,[45] extra-conventional international safety standards or criteria compensate for the Conventions' own relative normative weakness. In this scheme, IAEA safety standards, which include the so-called safety fundamentals, safety requirements and safety guides, play a crucial, central role.[46] They either inform the interpretation of the installation State's conventional obligations because they must be deemed expressly incorporated as such or, alternatively, because they generally reflect the degree of due diligence the installation

---

41 Norbert Pelzer, 'Safer Nuclear Energy through a Higher Degree of Internationalisation? International Involvement versus National Sovereignty' (2013) 91 *Nuclear Law Bulletin* 43 at 76.

42 See CNS, *supra* note 14, Arts 5 and 20; Joint Convention, *supra* note 15, Arts 30 and 32. For a detailed analysis, see e.g., Günther Handl, 'The IAEA Safety Conventions: An Example of Successful Treaty Management?' (2003) 72 *Nuclear Law Bulletin* 7.

43 CNS, *supra* note 14 at para viii.

44 See Joint Convention, *supra* note 15, Arts 4 and 11.

45 For example, with regard to IAEA safety standards, the Agency itself is careful to point out that, '[u]nder the terms of Article III of its Statute, the IAEA is authorized to establish standards of safety for protection of health, life and property against ionizing radiation'. IAEA, IAEA Safety Standards Series, *Safety of Nuclear Power Plants: Operation,* IAEA Doc No NS-R-2 (2000). While IAEA safety standards are not legally binding on Member States, they may be, and very often are, adopted by them for use in national regulations. Moreover, they are binding on the IAEA in relation to its own operations and on States in relation to operations assisted by the IAEA.

46 For an overview of these standards, see IAEA, *Long Term Structure of the IAEA Safety Standards and Their Current Status*, accessed 9 January 2015 at http://www-ns.iaea.org/committees/files/CSS/205/status.pdf.

State will have to apply in a particular situation.[47] Besides, their normative guidance function is further validated as a result of their adoption – in whole or in part – by States, regional[48] and industry organisations[49] alike, quite apart from their applicability to the organisation's own safety services.[50]

In short, IAEA safety standards can provide a solid authoritative basis for evaluating installation States' safety-related activities. Indeed, as the Agency itself proclaims, IAEA standards represent 'a consistent, reliable means of ensuring the effective fulfilment of obligations of States under the Conventions'.[51] Nevertheless, Fukushima has rekindled doubts even among the conventional regime's supporters about the CNS's continuing effectiveness.[52] Moreover, as recent proposals to formally amend the

---

[47] Thus, unlike other transboundary impact situations in which the task of ascertaining the standard of due diligence applicable in a particular situation can be challenging, this need not be the case in the context of nuclear power activities.

[48] See e.g., European Council Directive, *supra* note 3 at para 2.

[49] For example, the Western European Nuclear Regulators Association's new reference levels for existing nuclear reactors directly take into account revisions to IAEA safety requirements following the Fukushima accident. See e.g., Western European Nuclear Regulators Association Reactor Harmonisation Working Group (WENRA RHWG), *Updating WENRA Reference Levels for Existing Reactors in the Light of TEPCO Fukushima Dai-ichi Accident Lessons Learned* (November 2013) at 7.

[50] For example, the Operational Safety Review Team (OSART) programme, the IAEA's longest existing service, provides advice and assistance to Member States to enhance the safety of nuclear power plants during commissioning and operation. 'The criteria used by the teams as they formulate their conclusions are based on IAEA Safety Standards and the best prevailing international practices …'. See IAEA, *OSART Mission Highlights 2010–2012: Operational Safety Practices in Nuclear Power Plants* (2012) at 9.

[51] Mohamed ElBaradei, Foreword, IAEA Safety Standards, *Fundamental Safety Principles: Safety Fundamentals,* IAEA Series No. SF-1 (2006) (*Fundamental Safety Principles*).

[52] Concerns of this kind were expressed at the 2011 IAEA Ministerial Conference. See 'Chairpersons' Summaries', IAEA Ministerial Conference on Nuclear Safety (20–24 June 2011) at 7, accessed 9 January 2015 at http://www-pub.iaea.org/MTCD/Meetings/PDFplus/2011/cn200/documentation/chairpersons-summaries.pdf: 'In order to ensure that all safety issues are fully considered, it was suggested that there be a review of the effectiveness of the Convention on Nuclear Safety and its associated review mechanisms, both of which form an important part of the global nuclear safety framework'; see also Peri Johnson, 'The post-Fukushima Daiichi Response: The Role of the Convention on Nuclear Safety in Strengthening the Legal Framework for Nuclear Safety' (2013) 91

CNS clearly confirm, there exists a perception, certainly among some States, that IAEA safety standards' *de facto* normativity warrants further strengthening.[53]

III. THE NEED FOR STRENGTHENED INTERNATIONAL SAFETY REVIEW PROCESSES

While an enhanced formal linkage between conventional safety obligations and IAEA standards and criteria is merely desirable, improvements in nuclear safety assessment and peer review processes are simply indispensable to ensuring the global effectiveness of the international nuclear safety regime.[54] There is no mistaking the fact that the nuclear safety Conventions' regularly scheduled or extraordinary review meetings, which are organised as peer review exercises, represent the linchpin of the two conventional regimes.[55] Similar review processes are a hallmark of the wider IAEA safety framework, namely as part of the Agency's safety review services,[56] as well as industry efforts at bolstering nuclear power plant safety, such as through the peer review services of

Nuclear Law Bulletin 7 at 8; Matthew Bunn and Olli Heinonen, 'Preventing the Next Fukushima' (2011) 333 *Science* 1580; Norbert Pelzer, 'Does the Fukushima Incident Require a Revision of the International Legal Regime on Nuclear Safety?' (Paper delivered at the IAEA Ministerial Conference on Nuclear Safety, Vienna, 2011) accessed 9 January 2015 at http://www.xmuenergy.info:1235/w_d4_N.Pelzer.pdf.

53 See e.g., the Proposal for Amendments by the Russian Federation, in CNS 2nd Extraordinary Meeting, *supra* note 9. Russia's proposal would inter alia have pegged expressly an installation State's obligations bearing on assessment and verification of the safety of the nuclear installation to 'IAEA safety standards'. Proposals of the Russian Federation for amendments to the Convention on Nuclear Safety, Proposal 1 to Article 6.

54 See e.g., *Fundamental Safety Principles*, *supra* note 51 at paras 3.15–3.16.

55 See *supra* text at notes 42–43; Handl, *supra* note 42 at 18–26; Patrick Reyners, 'The Convention on Nuclear Safety of 1994' (1995) 5 *Review of European Community and International Environmental Law* 231 at 235–6.

56 Peer reviews (PRs) are an intrinsic feature of safety services provided through, for example, the Operational Safety Review Team (OSART) programme, Integrated Regulatory Review Service (IRRS), Integrated Nuclear Infrastructure Review (INIR), Safety Aspects of Long Term Operation (SALTO) and Emergency Preparedness Review (EPREV) missions.

the World Association of Nuclear Operators.[57] Self-evidently, periodic peer reviews are deemed essential to achieving and maintaining a high degree of nuclear safety worldwide, let alone to ensuring 'the continuous improvement of safety of nuclear installations'.[58]

Post-Fukushima, recognition of this fact has strengthened to the point of becoming an outright safety mantra.[59] Thus, one of the central findings to emerge from the various reviews of the accident concerned the need for strengthened and expanded international safety assessment and periodic peer reviews. For example, the 2011 IAEA Ministerial Conference on Nuclear Energy called special attention to the 'benefits of strengthened and high quality independent international safety expert assessments'[60] and suggested 'a review of the effectiveness of the Convention on Nuclear Safety and its associated review mechanisms, both of which form an important part of the global nuclear safety framework'.[61] The need for improved peer reviews also figured prominently in the IAEA Action Plan on Nuclear Safety.[62] In its report on progress in implementing the Action Plan, the Agency identified specific safety areas in relation to which IAEA peer review services needed to be strengthened: the need for enhanced transparency of peer reviews; and the need to encourage Member States to voluntarily host peer reviews, including follow-up

---

57 Since 1992, World Association of Nuclear Operators (WANO) has conducted over 500 PRs in 31 countries. Its reviews include so-called pre-startup station PRs, corporate PRs and follow-up PRs. US Nuclear Safety, *supra* note 3 at 9.

58 Declaration by the IAEA Ministerial Conference, *supra* note 7 at para 12.

59 See e.g., European Council Directive, *supra* note 3 at paras 22–23. See also European Nuclear Safety Regulators Group (ENSREG), *Post-Fukushima Accident Peer Review Report: Stress Tests Performed on European Nuclear Power Plants* (2012) at 3–4, accessed 9 January 2015 at http://www.ensreg.eu/sites/default/files/EU%20Stress%20Test%20Peer%20Review%20Final%20Report.pdf; WENRA RHWG, *Position Paper on Periodic Safety Reviews (PSRs) taking into account the Lessons Learnt from the TEPCO Fukushima Dai-ichi NPP Accident* (March 2013) at 4, accessed 9 January 2015 at http://www.wenra.org/media/filer_public/2013/04/05/rhwg_position_psr_2013-03_final_2.pdf; Foro iberamericano de Organismos Reguladores Radiológicos y Nucleares, *Evaluación de Resistencia de las Centrales Nucleares de los Países Miembros del Foro, Informe Final* (June 2012) at 15.

60 See Declaration by the IAEA Ministerial Conference, *supra* note 7 at para 12.

61 See e.g., Chairpersons' Summaries, *supra* note 52 at 7.

62 IAEA, *IAEA Action Plan on Nuclear Safety*, IAEA Doc GC(55)/14 (5 September 2011) at 3.

reviews, on a regular basis.[63] In 2012, finally, the Second Extraordinary Meeting of the Contracting Parties to the CNS agreed to a revision of IAEA guidance documents related to the review process and national reports under the CNS.[64] Additionally, the Meeting decided to establish a 'Working Group on Effectiveness and Transparency' with the task of reporting to the next, the sixth (ordinary) CNS review meeting in 2014 on a list of actions to strengthen the peer review process and on proposals to amend, when necessary, the Convention.[65]

Generally speaking, for safety assessments to play their assigned critical role in the nuclear safety system, they must be embedded in a robust procedural framework. A nuclear operator's self-assessment, as the obvious first step in this process, ought to be complemented by periodic external reviews[66] involving independent experts. It ought to be able to draw on timely, comprehensive and accurate information whose conveyance is structured in such a way to make the information readily accessible and permit meaningful analysis of discrete safety issues. All safety assessments, whether undertaken within the ambit of the Conventions themselves or as part of IAEA safety services, organised regionally,[67] or through industry organisations,[68] should be reasonably transparent to facilitate the sharing among States, regulators and industry of nuclear safety-relevant 'lessons to be learned',[69] to boost the process's credibility and thereby public confidence in the safety of nuclear power. Finally, safety reviews should allow for specific follow-up action and, as

---

63 IAEA, *Progress in the Implementation of the IAEA Action Plan on Nuclear Safety: Supplementary Information* (2013) at 8, accessed 9 January 2015 at http://www.iaea.org/About/Policy/GC/GC57/GC57InfDocuments/English/gc57inf-5-att1_en.pdf.

64 See CNS 2nd Extraordinary Meeting, *supra* note 9 at 8.

65 *Ibid* at 8–9.

66 Such reviews ought to occur sufficiently frequently to permit timely assessment of, and responses to, relevant safety-related developments.

67 As to EU-wide assessments, see European Council Directive, *supra* note 3.

68 One justified criticism of WANO's peer review services is that results are not publicised.

69 One important step in this direction is, of course, a deeper cooperation between IAEA and other organisations providing peer review services. Note in this respect the agreement between the IAEA and WANO to enhance information exchange on nuclear power plants' operating experience and other relevant areas. IAEA, Press Release, 2012/22, 'IAEA and WANO Sign New Memorandum of Understanding' (17 September 2012) accessed 9 January 2015 at http://www.iaea.org/newscenter/pressreleases/2012/prn201222.html.

part of this arrangement, ought to include mandatory rather than merely voluntary on-site visits by external safety experts.[70]

Measured against these parameters, it is easy to see why the CNS periodic safety reviews might have been found wanting, thus to be in need of improvement. First, the CNS process is blemished by the fact that some contracting parties fail to file national reports altogether or in time for proper consideration at the review meetings. It is also hampered by a lack of preparation and unwillingness or inability of some States to critically discuss the safety issues under review. Thus, full and effective participation by States Parties in the CNS review meetings remains an unrealised objective.[71] Second, the review meetings' focal point itself has come in for criticism as being less than optimal. Specifically, it has been suggested that the conventional processes established under the CNS and Joint Convention ought to 'be more probing, perhaps by focussing on the most important safety issues, rather than by emphasizing a wide (and necessarily superficial) survey that is today's norm'.[72] Third, the CNS process suffers from a lack of transparency as evidenced by the fact that States are free to decide on whether to make public the national reports they submit pursuant to Article 5 of the Convention, as well as any questions and comments they receive during the review meetings or their responses thereto.[73] Fourth and perhaps most importantly, CNS safety reviews remain in essence desk-top/paper exercises. Although States' nuclear safety performance is subject to mandatory self-reporting and carefully orchestrated scrutiny in the review meetings, the process does

---

[70] In this sense, nuclear safety reviews might undergo a metamorphosis not unlike that experienced by the International Maritime Organization's (IMO) Member State audit scheme. See IMO, Assembly, 28th Session, *Resolution on the IMO Instrument Implementation Code (III Code)*, IMO Doc A 28/Res.1070 (10 December 2013).

[71] As the Summary Report of the 6th Review Meeting critically observes, 'three years after the Fukushima accident, a number of the 76 Contracting Parties either did not submit their national report or did not submit their national report to the Secretariat in time for effective review by other Contracting Parties'. *Summary Report*, 6th Review Meeting of the Contracting Parties to the Convention on Nuclear Safety, Doc CNS/6RM/2014/11_Final (24 March–4 April 2014) (CNS 6th Review Meeting Summary Report) at 4.

[72] See Richard Meserve, 'The Global Nuclear Safety Regime' (2009) 138 *Daedalus* 100 at 109.

[73] See e.g., IAEA, Information Circular, 'Annex to Guidelines for National Reports: Voluntary Practices Regarding Publicly Available Information' in *Guidelines regarding National Reports under the Convention on Nuclear Safety*, IAEA Doc INFCIRC/572/Rev.4 (16 April 2013) at 19.

not include on-site, periodic follow-up reviews[74] of contracting parties' actual compliance with safety obligations or their implementation of suggestions offered at the review meetings, unless the State concerned voluntarily submits to such an outside assessment. To be sure, many States have done so by inviting IAEA and/or industry review missions.[75] But the fact remains that, in the final analysis, contracting States retain a significant margin of appreciation in interpreting and implementing safety obligations notwithstanding the CNS review process. In the absence of mandatory in-country international safety reviews, there is simply no guarantee that national practices might not end up compromising global nuclear safety. In other words, it is uncertain whether the CNS's review meetings as presently administered can truly ensure installation States' timely and effective adjustment of domestic policies and practices in response either to evolving international best safety practices or the identification of national shortcomings or deficiencies.

It is partially in response to these perceived shortcomings that two amendment proposals had been tabled at the CNS's Second Extraordinary Meeting.[76] One, by Switzerland, drawing on a working paper submitted to the 2011 Ministerial Conference,[77] is especially noteworthy in the present context as it aimed at enhancing transparency in matters of nuclear safety. It also called for national nuclear regulatory bodies to undergo periodically a compliance-with-IAEA-requirements audit, and for the design, construction and operational safety of nuclear installations to be subject to mandatory outside expert reviews, pegged similarly to IAEA safety standards.[78] However, over four sessions of the Working Group on Transparency and Effectiveness, the original Swiss proposals

---

[74] For example, the European Nuclear Safety Regulators Group's peer review Board recommends periodic safety reviews 'as often as appropriate but at least every 10 years'. See ENSREG, *supra* note 59 at 4.

[75] One reason for this, as the US GAO notes, is the fact the IAEA does not systematically track the status of the recommendations made by IRRS peer review missions; hence, it does not know the extent to which the recommendations have been implemented by the host countries. See US Nuclear Safety, *supra* note 3 at 39.

[76] The proposal by the Russian Federation called inter alia for mandatory, regular assessments of the safety of nuclear power plants. See *supra* note 53.

[77] 'Working Paper submitted by Switzerland', IAEA Ministerial Conference on Nuclear Safety (20–24 June 2011) accessed 9 January 2015 at http://www-pub.iaea.org/MTCD/meetings/PDFplus/2011/cn200/cn200_Swiss_Working_Paper.pdf.

[78] The Swiss proposal would have amended Arts 8, 18, 19 and 25 of the CNS. See CNS 2nd Extraordinary Meeting, *supra* note 9.

either faded away or were scaled back rather dramatically. In the end, the 6th CNS Review Meeting endorsed only a single amendment proposal unrelated to the review process as such.[79] It also adopted, however, a recommendation for a review of the present Guidelines regarding National Reports[80] with the aim of strengthening the normative effects of IAEA safety standards. Specifically, the recommendation called for the development of a national reporting template through which the installation State's article-by-article review of its CNS obligations would 'address and, if and where necessary, refer to relevant to IAEA safety requirements'.[81] Additionally, the Review Meeting recommended several measures to increase the transparency of the peer review process itself and the use of existing nuclear safety databases 'as a platform for sharing of operating experience on safety culture'.[82]

These steps, if and when they are fully implemented, are measures that undoubtedly can help strengthen nuclear safety globally. In this vein, a major step in the right direction is the European Council Directive No. 2014/87/EURATOM.[83] Although the European Commission's initiative to revise the European nuclear safety directive initially had run into strong headwinds from industry and some Member States expressed concerns and misgivings,[84] the amended version of the European Council Directive now calls for periodic mandatory, more focused peer reviews to address specific nuclear safety topics 'selected by the Member States jointly or in close co-ordination with the Commission'.[85] However, there is no hiding

---

[79] The only formal amendment proposal to be submitted to a diplomatic conference for further consideration aims to make avoidance of off-site contamination a legally binding safety objective. See CNS 6th Review Meeting Summary Report, *supra* note 71 at Annex 1.

[80] IAEA, Information Circular, *Guidelines Regarding National Reports under the Convention on Nuclear Safety,* IAEA Doc INFCIRC/572/Rev.4 (16 April 2013).

[81] See CNS 6th Review Meeting Summary Report, *supra* note 71 at 10 and Annex 3 at para 1.

[82] *Ibid* at paras 2–4, 6.

[83] See European Council Directive, *supra* note 3.

[84] For example, it is being claimed that the proposal would impose unacceptable constraints on Member States' freedom of action. See e.g., UK, House of Commons, European Scrutiny Committee, *Twenty-Third Report of Session 2013–14* (4 December 2013) at 8.

[85] See EC, *Proposal for a Council Directive amending Directive 2009/71/ EURATOM establishing a Community framework for the nuclear safety of nuclear installations* (17 October 2013) at 10; European Council Directive, *supra* note 3, Art 8(e).

the fact that today, in contrast to the immediate aftermath of the Fukushima accident, States' support for bold reforms of the nuclear safety regime at the global level has declined.[86] Regrettably, the net result of this slowdown in global reform efforts is that the problematic combined effect on nuclear safety of the CNS review process's still inadequate transparency and the absence of in-country mandatory safety assessments remains essentially unaddressed.

In conclusion, therefore, two basic observations might be offered. First, Norbert Pelzer's quasi-rhetorical question of whether safer nuclear energy can be achieved through a higher degree of internationalisation[87] can be answered with an emphatic yes. Although the principle of national responsibility may still dominate the global discourse about nuclear safety, in reality the increasing internationalisation of key aspects of nuclear safety as evidenced, in particular, by calls for the strengthening of international safety reviews in the wake of Fukushima, has become an essential underpinning to the very credibility of our convention-based global nuclear safety system. Second, although the international nuclear safety regime will undoubtedly emerge improved in the wake of the accident, one of its critical cornerstones, the international peer review process, continues to suffer from some evident structural deficiencies that are capable of affecting nuclear power plant safety and of eroding public support of nuclear power worldwide.

IV. TERRITORIAL SOVEREIGNTY AND ENHANCED LEGAL CLAIMS FOR THE PREVENTION OF TRANSBOUNDARY NUCLEAR POLLUTION

A. Border Siting and Transboundary Emergency Planning Zones

One factor that bears critically on the transboundary impact potential of a nuclear power installation is a facility's location relative to the

---

[86] Critics have also noted that while 'IAEA nuclear safety services suffer from limited resources … members blocked efforts to expand the agency's nuclear safety responsibilities during the Board of Governors meeting in September 2011'. See Michelle Dover and Miles Pomper, 'Nuclear Security, Post-Fukushima: Expectations for the Seoul Summit' *Nuclear Threat Initiative* (12 March 2012) accessed 9 January 2015 at http://www.nti.org/analysis/articles/nuclear-security-post-fukushima-expectations-seoul-summit/.

[87] See *supra* note 41.

international boundary line. While a nuclear installation's source term[88] is the single most important parameter, geographical proximity to the plant – apart from such factors as the potentially affected area's topography, demography, economic activity and environmental resources – co-determines the risk for the neighbouring State(s) concerned. This is so despite the installation State's traditional defence-in-depth approach to nuclear safety[89] and the post-Fukushima prospect of enhanced on-site emergency response capabilities,[90] all aimed at preventing uncontrolled releases of radioactivity into the environment. For the simple truth is that for the foreseeable future major accidents entailing off-site releases cannot be ruled out.[91] To be sure, both Switzerland's formal proposal to amend the CNS and the European Council Directive introduce, as a new legally binding safety objective, the avoidance of off-site contamination.[92] However, such an outcome cannot, of course, be mandated legislatively.

Given that a transboundary nuclear risk is inter alia a function of an installation's distance from the border, the siting of nuclear facilities in border areas, especially of nuclear power plants, has, not surprisingly, been a contentious international issue in the past.[93] While admittedly, a new generation of small and medium-sized nuclear power reactors (of less than

---

[88] 'Source term' refers to the types, quantities, and chemical forms of the radionuclides that encompass the source of potential for exposure to radioactivity. 'Radiation Terms and Definitions' *Health Physics Society*, accessed 9 January 2015 at http://hps.org/publicinformation/radterms/Sourceterm.html.

[89] Article 18(i) of the CNS requires installation States to ensure that 'the design and construction of a nuclear installation provides for several reliable levels and methods of protection (defense in depth) against the release of radioactive materials'. See CNS, *supra* note 14, Art 18(i). Thus, '[a] key to a defense-in-depth approach is creating multiple independent and redundant layers of defense to compensate for potential failures and external hazards so that no single layer is exclusively relied on to protect the public and the environment'. See US, Nuclear Regulatory Commission, *Recommendations for Enhancing Reactor Safety in the 21st Century: The Near-Term Task Force Review of Insights from the Fukushima Dai-Ichi Accident* (12 July 2011) at 15.

[90] See e.g., CNS 2nd Extraordinary Meeting, *supra* note 9 at para 17: 'The displacement of people and the land contamination after the Fukushima Daiichi accident calls for all national regulators to identify provisions to prevent and mitigate the potential for severe accidents with off-site consequences.'

[91] Of course, so-called inherently safe nuclear reactors, if and when they come on-line, might change the situation.

[92] See CNS 2nd Extraordinary Meeting, *supra* note 9 at 5; European Council Directive, *supra* note 3, Art 8a, respectively.

[93] See e.g., Handl, *supra* note 20 at 27–30.

700 MW output) may eventually alleviate the siting issue,[94] recent evidence confirms that the 'border siting' of new power plants continues to engender controversy between installation and the risk-exposed States.[95]

---

[94] The size of the required emergency planning zone for some of these plants is claimed to have no more than a 300 metre radius. See World Nuclear Organization, 'Small Nuclear Power Reactors' (July 2014) accessed 9 January 2015 at http://www.world-nuclear.org/info/Nuclear-Fuel-Cycle/Power-Reactors/Small-Nuclear-Power-Reactors/. However, for the time being, traditional, large-scale nuclear power plants with their attendant risk potential, dominate the field. See generally IAEA, *Status of Small and Medium Sized Reactor Designs: A Supplement to the IAEA Advanced Reactors Information System (ARIS)* (September 2012).

[95] Examples in point include controversies between: Armenia-Turkey over the safety of Metsamor, an old Armenian Nuclear Power Plant, just 10 miles from the Turkish border (see Lily Torosyan, 'Revisiting the Metsamor Nuclear Power Plant' *The Armenian Weekly* (7 December 2012) accessed 9 January 2015 at http://www.armenianweekly.com/2012/12/07/revisiting-the-metsamor-nuclear-power-plant/); Austria-Czech Republic over a proposal for a nuclear waste final repository in Boletice, 20 kilometres from the Austrian border (see Kerstin Scheller, 'Tschechien will Atommüll-Endlager nahe Oberösterreich' *Der Standard* (26 April 2013)); Bulgaria-Turkey over Turkish plans for a power plant at Igneada, on the Black Sea coast, just 5 kilometres south of the Rezovo River marking the border with Bulgaria (see 'Turkey to Build Its 3rd Nuclear Power Plant on Bulgarian Border' *Sofia News Agency* (6 April 2011) accessed 9 January 2015 at http://www.novinite.com/view_news.php?id=127024); Albania-Montenegro over Albanian plans for a Nuclear power plant in the Shkoder region (see John Daly, 'Energy-Poor Albania Wants Nuclear Power Plant, Neighboring Montenegro Opposes' *Oilprice.com* (1 February 2012) accessed 9 January 2015 at http://oilprice.com/Alternative-Energy/Nuclear-Power/Energy-Poor-Albania-Wants-Nuclear-Power-Plant-Neighboring-Montenegro-Opposes.html); Russia-Lithuania over a nuclear power plant in Kaliningrad, within 6 miles of the Lithuanian border whose construction was suspended in 2013 (see Bryan Bradley, 'Lithuania Demands Russia Clarify Safety of Baltic Nuclear Plant' (27 August 2013) accessed 9 January 2015 at http://www.businessweek.com/news/2013-08-27/lithuania-demands-russia-clarify-safety-of-baltic-nuclear-plant); Belarus-Lithuania, over a proposed NPP to be located in Ostrovets, Belarus, 12 miles east of the Lithuanian border (see Lubomir Mitev, 'Belarus Nuclear Plant In Non-Compliance With Espoo Convention' *NucNet* (10 July 2014) accessed 9 January 2015 at http://www.nucnet.org/all-the-news/2014/07/10/belarus-nuclear-plant-in-non-compliance-with-espoo-convention); Paraguay-Argentina (see 'Paraguay concerned with Argentine plans to move nuclear facilities to border area' *Mercopress* (11 February 2014) accessed 9 January 2015 at http://en.mercopress.com/2014/02/11/paraguay-concerned-with-argentine-plans-to-move-nuclear-facilities-to-border-area); Saudi Arabia-Jordan (see 'Saudi Arabia-Jordan Agree to Cooperate' *World Nuclear News* (23 January 2014) accessed 9 January 2015 at

Indeed, Fukushima might have brought the issue into even sharper relief. As is well-known, under deteriorating conditions at the stricken Fukushima-Daiichi plant, Japanese authorities were repeatedly forced to expand the initial evacuation area, from a 3 kilometre-radius to 10, and then again to 20 kilometres.[96] Eventually, protective measures were ordered also for residents in an area between 20 and 30 kilometres from the plant: although at first only ordered to shelter-in-place, these residents were eventually urged to voluntarily evacuate.[97] However, the accident also produced radioactive hotspots further afield,[98] thus accentuating doubts about the adequacy of Japan's emergency planning and preparedness.[99] In light of these experiences, a number of countries began to

---

http://www.world-nuclear-news.org/NP-Saudi-Arabia-Jordan-agree-to-cooperate-2301147.html); and notwithstanding the proposed sites' considerable distance from the border, India-Sri Lanka over the Kudankulam Nuclear Power Plant site, 250 kilometres from the neighbouring country (see M P Ram Mohan, 'A Nuclear Agenda for SAARC: Establishing a South Asian Nuclear Risk Community' *Indian Nuclear Law Association* (26 May 2012) accessed 9 January 2015 at http://www.nlain.org/blog/anuclearagendaforsaarcestablishingasouthasiannuclear riskcommunity); Finland-Sweden over the planned Pyhäjoki NPP, at a distance of 150–160 kilometres from Sweden (see Thomas Nilsen, 'New nuclear power plant triggers concern' *Barents Observer* (January 2014) accessed 9 January 2015 at http://barentsobserver.com/en/energy/2014/01/new-nuclear-power-plant-triggers-concern-10-01).

[96] See The National Diet of Japan, *supra* note 4 at 38.

[97] See *ibid.*

[98] For example, at Iitate, 40 kilometres from the plant, levels of caesium-137 were as high as 18 MBq/m2, well above the level at which evacuation would be deemed advisable. See Debora MacKenzie, 'IAEA says Fukushima fallout warrants more evacuation' *New Scientist* (31 March 2011) accessed 9 January 2015 at http://www.newscientist.com/article/dn20324-iaea-says-fukushima-fall out-warrants-more-evacuation.html#.U8BCsbEo4dV.

[99] See e.g., Norimitsu Onishi and Martin Fackler, 'Japan Held Nuclear Data, Leaving Evacuees in Peril', *The New York Times* (8 August 2011) accessed 9 January 2015 at http://www.nytimes.com/2011/08/09/world/asia/09japan.html?pagewanted=all&_r=0?. These doubts were reinforced by a US government recommendation that American citizens in Japan evacuate from an area of up to 50 miles from the plant. See Brian Vastag, Rick Maese and David A Fahrenthold, 'U.S. urges Americans within 50 miles of Japanese nuclear plant to evacuate; NRC chief outlines dangerous situation', *The Washington Post* (6 March 2011) accessed 9 January 2015 at http://www.washingtonpost.com/national/us-urges-americans-within-50-miles-of-japanese-nuclear-plant-to-evacuate/2011/03/16/AB wTmha_story.html.

re-evaluate the size and nature of their emergency planning zones.[100] Moreover, an IAEA group of experts concluded that '[a]n internationally agreed calculation methodology … [was] needed for determining the optimal size of emergency planning zones. Emergency planning zones … [were] to be redefined to take into account the experience from the accident at the Fukushima Daiichi nuclear power plant'.[101]

As part of their off-site emergency preparations, nuclear installation States are expected to designate two planning zones: a 'precautionary action zone' (PAZ), an area in the immediate vicinity of the facility in which precautionary protective measures, such as the evacuation, sheltering in place and the administration of stable iodine to the population,[102] would be taken before or immediately after an off-site release of radioactivity to minimise exposure to the radioactive plume or the effects thereof; and an adjoining 'urgent protective action planning zone' (UPZ), in which, following an environmental release, measures would be taken to minimise radiation exposure along the expected ingestion pathway.[103] While the IAEA suggests a mere 3–5 kilometre radius for the former and a 5–30 kilometre radius for the latter,[104] many nuclear installation States, including the United States,[105] have opted for somewhat larger PAZs covering areas of a 10–16 kilometre radius.[106]

---

[100] See generally OECD/NEA Nuclear Safety Response, *supra* note 4 at 28. See further UK, Health and Safety Executive (HSE) Office for Nuclear Regulation, *Japanese Earthquake and Tsunami: Implications for the UK Nuclear Industry,* Final Report of HM Chief Inspector of Nuclear Installations (September 2011) at 144–5.

[101] IAEA, *IAEA Report on Preparedness for a Nuclear or Radiological Emergency in the Light of the Accident at the Fukushima Daiichi Nuclear Power Plant*, Annex A: 'Main Conclusions from the Workshop on Sharing Lessons Identified from Past Responses and Exercises, 23–27 April 2012, Vienna' (2013) at 40.

[102] See IAEA, IAEA Safety Standards, *Arrangements for Preparedness for a Nuclear or Radiological Emergency*, IAEA Safety Guide No GS-G-2.1 (2007), Appendix V at 95–103.

[103] IAEA, IAEA Safety Standards Series, *Preparedness and Response for a Nuclear or Radiological Emergency,* IAEA Requirements No GS-R-2 (November 2002) at 21–23. Measures to be taken in the UPZ aim at averting off-site radiation doses 'in accordance with international standards'. *Ibid*, Annex III at 50.

[104] *Ibid.*

[105] See US Nuclear Regulatory Commission (NRC), *NRC Regulations Title 10*, 10 CFR § 50.47(c)2: 'Generally, the plume exposure pathway EPZ for nuclear power plants shall consist of an area about 10 miles (16 km) in radius and the ingestion pathway EPZ shall consist of an area about 50 miles (80 km) in radius.'

[106] See e.g., Handl, *supra* note 21 at 30–5; Jozef Kubanyi, Ricardo Bolado Lavin, Dan Serbanescu, Bela Toth and Heinz Wilkening, *Risk Informed Support*

Whatever the most appropriate size of an emergency planning zone might be in a given situation, it seems reasonable to postulate that generically a 16 kilometre (10 mile) PAZ might be advisable for a large commercial nuclear power plant.[107] Thus, if the plant's location is at a distance to the border at which the State's PAZ or exclusion zone would extend into the territory of the neighbouring State, were it not for the existence of the international boundary, this fact raises the obvious question of whether such siting is consistent with the present-day allocation of fundamental rights among States in the international legal system. For unlike other typical transboundary environmental interferences which may require a neighbouring State to adopt a passive conduct, that is, it requires the State to merely abstain from utilising or fully utilising the internationally shared natural resource concerned, a border power plant in the above sense effectively forces the exposed State to become intensively active: it will have to take precautionary measures to ensure its population of a degree of protection against the risks of a nuclear accident that not only the installation State deems essential for its own residents, but that internationally reflects also a standard of good practice. Arguably, therefore, the transboundary nuclear risk imposes a burden that prima facie is at odds with the exposed State's territorial sovereign rights.

---

of Decision Making in Nuclear Power Plant Emergency Zoning: Generic Framework towards Harmonizing NPP Emergency Planning Practices, JRC Scientific and Technical Reports (European Commission Joint Research Centre Institute for Energy, January 2008) at 21–4.

The US Nuclear Regulatory Commission recently rejected a petition for rulemaking which sought an increase in the size of the present plume exposure pathway zone from 10 to 25 miles, establishing a new zone but with less stringent requirements from 25–50 miles around reactors, and expanding the existing ingestion pathway zone from 50 to 100 miles. See US Nuclear Regulatory Commission Before the NRC Staff, *Petition for Rulemaking to Improve Emergency Planning Regulations,* 10 CFR §50.47 (2012). The Commission maintained that 'the current size of the emergency planning zones … [was] appropriate for existing reactors and that emergency plans … [would] provide an adequate level of protection of the public health and safety in the event of an accident at a nuclear power plant'. See *Emergency Planning Zones: A Proposed Rule by the Nuclear Regulatory Commission on 04/09/2014*, 79 Fed Reg 19501 (2014).

[107] E.g., for so-called threat category I facilities, including in particular reactors with a larger than 1,000 MW(th) output, for which on-site events (including very low probability events) are postulated that could give rise to severe deterministic health effects off the site, or for which such events have occurred in similar facilities.

As a general principle, any such derogation from the sovereignty of another State cannot be lightly presumed to be internationally permissible.[108] However, as already intimated before, the risk-exposed State does not have a right to veto a border nuclear facility.[109] Not only does the present existence of several 'border power plants' represent an important datum in the practice of States to this effect, but on several occasions international tribunals have also expressly denied or declined to infer a risk-exposed State's customary international right to a veto, albeit in contexts in which the latter was not *de facto* called upon to take affirmative action in order to minimise a transboundary risk.[110] Additionally, the Draft Articles of Prevention lend strong support to the notion that under international law a risk-exposed neighbouring State does not have a right to veto a proposed border site, whatever the nature and scope of the transboundary risk involved.[111]

Although the burdened neighbouring State may not be able to legally stop a border plant, the question still arises whether it is or should be entitled to an enhanced level of protection against the transboundary nuclear risks emanating from the border facility. After all, the necessity for emergency measures in the risk-exposed State represents not only an unusually massive interference with the exclusivity of that State's powers

---

108 See e.g., *Dispute regarding Navigational and Related Rights (Costa Rica v Nicaragua)*, [2009] ICJ Rep 213 at para 48: '[I]t is certainly true that limitations of the sovereignty of a State over its territory are not to be presumed.'

109 See *supra* text at notes 21–24.

110 Thus, in the *Lake Lanoux* case, the arbitration tribunal famously rejected the idea of a risk-exposed riparian's right of veto under customary international law. *Lake Lanoux Arbitration (France v Spain)* (1957) 12 RIAA 281 (French) at paras 12–23, with English translations at (1961) 24 ILR 101; (1959) 53 *American Journal of International Law* 156 (*Lake Lanoux*). In the *Pulp Mills* case, the ICJ in construing Argentina's right to stop construction of controversial pulp mills in Uruguay declined to infer a 'no construction obligation' incumbent upon Uruguay under customary international law. The Court argued instead that the bilateral treaty conferring jurisdiction did not authorise the Court to decide the matter in the event the parties failed to reach an agreement on the planned activities at the end of a mandatory negotiation period. See *Pulp Mills*, *supra* note 27 at paras 152–154; see also *Case concerning Land Reclamation by Singapore in and around the Straits of Johor (Malaysia v Singapore),* Provisional Measures, Order of 8 October 2003, ITLOS Case No 12, Joint Declaration of Judges ad hoc Hossain and Oxman (noting that 'Malaysia accepts the importance of land reclamation and does not claim a veto over Singapore's activities …').

111 See Draft Articles on Prevention, *supra* note 17 at 411–12.

in its own territory – the very hallmark of sovereignty[112] – but, absent special circumstances,[113] raises also a question of fundamental fairness and equity between installation and neighbouring States as regards the distribution of costs and benefits associated with the border facility.

Both Conventions expressly recognise the international sensitivity of the siting issue. In virtually identical language, they call upon the installation State to consult with, and to provide relevant information upon request to, any contracting party 'in the vicinity' of the nuclear installation concerned.[114] Moreover, under customary international law today, any installation State arguably has a legal obligation to assess the transboundary impact potential of a nuclear facility,[115] as well as to inform and consult with the neighbouring State(s) about the siting (as well as the construction, operation and decommissioning) of a border nuclear facility. There is no need here to elaborate further on procedural safeguards,[116] except perhaps to point out that since Chernobyl, international practice appears to indicate an evolution of the risk-exposed State's right to be consulted in nuclear facility siting matters towards a

---

[112] See e.g., Max Huber in the *Island of Palmas case (Netherlands v USA)*, 4 April 1928, 4 RIAA 829 at 838.

[113] Such circumstances include the case of reciprocal risk-creation, i.e., when the exposed neighbouring State in turn operates a nuclear power plant in the border area, so that the States' mutually operative sovereign rights cancel each other out. Second, a risk-exposed neighbouring State might benefit directly from the border plant, be that it or one of its public entities has a financial stake in the plant, or that it has entered into a power supply arrangement with the installation State that involves the border plant.

[114] See CNS, *supra* note 14, Art 17, para (iv); Joint Convention, *supra* note 15, Art 13, para 1(iv).

[115] See generally *Pulp Mills*, *supra* note 27 at para 204. As to multilateral treaty-based requirements of environmental assessments of nuclear installations specifically, see, for example, Convention on Environmental Impact Assessment in a Transboundary Context, UNECE, *Background Note on the Application of the Convention to Nuclear Energy-Related Activities,* UN Doc ECE/MP.EIA/2011/5 (2 April 2011).

[116] See generally, Arts 7–15 of the Draft Articles on Prevention, *supra* note 17 at 373–5; Günther Handl, 'Transboundary Impacts' in Daniel Bodansky, Jutta Brunnée and Ellen Hey (eds), *Oxford Handbook of International Environmental Law* (Oxford: Oxford University Press, 2007) 531 at 540–44. But see Phoebe Okowa, 'Procedural Obligations in International Environmental Law' (1966) 67 *British Yearbook of International Law* 275 at 317, 336.

right to inspect the facility concerned, either jointly with the installation State or as a member of an international review team.[117]

The purpose of these procedural obligations is, of course, to protect states' substantive rights[118] and, more specifically, to guide the installation and risk-exposed State(s) towards a mutually agreeable accommodation of their national interests. Thus, analogous to other actual or potential international conflicts over natural resources, each State will be under a good faith obligation to 'take into consideration the various interests involved, to seek to give them every satisfaction compatible with the pursuit of its own interests, and to show that in this regard it is genuinely concerned to reconcile the interests of the other ... [state(s)] with its own'.[119] Nevertheless, if this process does not lead to a consensual solution the installation state will be 'permitted to go ahead with the activity, for the absence of such an alternative [solution] would, in effect, create a right of veto for the [risk-exposed neighbouring State(s)]'.[120]

---

[117] An incipient trend to this effect had been noticeable in Western Europe and North America and seems to have become more pronounced in the wake of the dissolution of the Soviet Union and the subsequent enlargement of the EC/EU. For details, see Handl, *supra* note 21 at 84–7. Examples in point include the so-called Mochovce 'walk-down', an on-site review of the safety level of the Slovak nuclear power plant by an international team of experts which included representatives of the risk-exposed neighbouring State of Austria; and expert safety missions to review the Temelin nuclear power plant in the Czech Republic under the so-called 'Melk Protocol Process', an arrangement between the installation State and Austria as well as the EC. See Austrian Aid Memoire, 'Mochovce Nuclear Power Plant' (18 May 1998) (on file with the author); Nuclear Power Plant Temelín, 'Austrian Technical Position Paper' (July 2001) accessed 9 January 2015 at http://www.umweltbundesamt.at/fileadmin/site/umweltthemen/kernenergie/temelin/Melk/Trialog/positionpaper/etetrialogreport_1.pdf. Of course, on-site safety inspections of nuclear facilities are routine events within, for example, the framework of IAEA's – voluntary – OSART mission programme, but can be mandatory under the EURATOM Treaty and will be so under the revised European Council Directive, *supra* note 3.

[118] As to the interrelationship of procedural and substantive obligations in the context of international environmental disputes, see *Pulp Mills*, *supra* note 27, Joint Dissenting Opinion of Judges Al-Khasawneh and Simma at para 26.

[119] *Lake Lanoux*, *supra* note 110 at para 22; see also *Case concerning the Gabčíkovo-Nagymaros Project (Hungary v Slovakia),* [1997] ICJ Rep 7 at para 112; Draft Articles on Prevention, *supra* note 17 at 409–12.

[120] Draft Articles on Prevention, *supra* note 17 at 412.

As the ILC reminds us, the installation State, 'while permitted to go ahead with the activity, is still obligated, as [a] measure of self-regulation, to take into account the interests of the states likely to be affected'.[121] Such an outcome, however, appears inherently problematic: even from a 21st century perspective of diminished legal expectations, the end result appears out of line with the respect due to neighbouring States' territorial sovereign rights. Nor does the result seem equitable. Indeed, as a matter of international public policy, the risk-exposed State's right to have its justified interests taken into account by the installation State deserves stronger procedural protection. This would mean in essence that, if the two sides fail to reach a mutual accommodation, recourse to an international third-party decision-making procedure should be available. The disputed issue might then be submitted to a conciliation proceeding or be decided in binding fashion by a committee of technical experts or a judicial panel. In either event, while such proceedings are under way, the installation State should not be permitted to proceed with its proposed activity.

The need for formal adoption of such an additional safeguard seems obvious.[122] Certainly, improvements in bilateral cooperation between installation and risk-exposed States have been authoritatively identified as part of the post-Fukushima agenda. For example, the 5th CNS Review Meeting specifically recognised the need for the installation State to closely cooperate with neighbouring countries with regard to the siting of new nuclear power plants, and to evaluate potential sites for new facilities in accordance with IAEA safety standards.[123] Similarly, at the 6th CNS Review Meeting several contracting parties re-emphasised the importance of transboundary consultations and the provision of information to the other parties in the vicinity of proposed new nuclear installations.[124] Still, it is unlikely that the continuing global nuclear safety discourse will lead

---

[121] *Ibid.*

[122] For example, Lithuania pointedly complained recently that after several futile attempts at bilateral consultations, the '[n]ecessity to accomplish comprehensive site evaluation … before the start of a new NPP construction … [was] one of the issues, which currently remain[ed] unsolved, in regard to sites in close vicinity to the Lithuanian border selected for construction of nuclear power plants in Belarus and the Kaliningrad region of the Russian Federation'. See Republic of Lithuania, *Convention on Nuclear Safety Sixth National Report* (2013) at 103.

[123] *Summary Report*, 5th Review Meeting of the Contracting Parties to the Convention on Nuclear Safety, Doc CNS/RM/2011/6/FINAL (4–14 April 2011) at paras 38–39.

[124] CNS 6th Review Meeting Summary Report, *supra* note 71 at para 29.

to a formal endorsement of a mandatory third-party role in the border site selection and evaluation process anytime soon. Nevertheless, it should be noted that a normative change of this kind would only signal a small step beyond the evolving expanded transboundary consultation obligation.[125] It would also fit easily with growing normative expectations that in-country international nuclear safety reviews be made mandatory.[126]

B. Transportation of Radioactive Materials through the Territorial Sea

Finally, the carriage at sea of certain nuclear materials can give rise to another scenario that fits in with the transboundary nuclear pollution template.[127] To the extent such transports involve the territorial sea of a foreign State, any transport accident resulting in an environmental release of radioactivity by definition affects the sovereign territory of the coastal State.[128] This fact in and of itself raises the question of whether the risk-exposed coastal State might not be the beneficiary of special international legal rules aimed at the prevention of nuclear harm.

While it is notoriously difficult to assess the characteristics and frequencies of shipments of nuclear materials by sea globally,[129] the carriage of radioactive materials through waters in the proximity of coastal States is probably an infrequent occurrence.[130] Nevertheless, it is one that has riled coastal States in the past.[131] It is, moreover, an issue

---

[125] See *supra* text at note 117.

[126] See *supra* text at notes 76–85.

[127] See *supra* text at note 11.

[128] 'The sovereignty of the coastal state extends … to the territorial sea'. United Nations Convention on the Law of the Sea, adopted 10 December 1982, 1833 UNTS 3 (entered into force 16 November 1994) (UNCLOS), Art 2(1).

[129] See IAEA, Information Circular, Report of the Working Group, Annex 1, in *Communication dated 15 April 2014 received from the Resident Representative of Norway to the Agency regarding the Working Group on Best Practice Guidelines for Voluntary and Confidential Government-to-Government Communications on the Transport of MOX Fuel, High Level Radioactive Waste and, as appropriate, Irradiated Nuclear Fuel by Sea*, IAEA Doc INFCIRC/863 (5 May 2014) (Report of the Working Group on Best Practice Guidelines).

[130] See *ibid* (noting that in 2012 only 64 of 14,545 transits through the Panama Canal involved transports of radioactive materials, and only two of these shipments related to the back-end of the fuel cycle). See also Paul Gaffney, *The Transport of Nuclear Materials by Sea in Northern Europe* (Kommunenes Internasjonale Miljøorganisasjon, 2011).

[131] See e.g., Caricom Heads of Government, 'Statement by Caricom Heads of Government on the Shipment of Nuclear Waste Through the Caribbean Sea'

that may have gained renewed prominence in the wake of the Fukushima accident. Although the latter may not have obvious implications for maritime shipping,[132] it certainly holds some lessons for nuclear transport safety generally,[133] therefore also indirectly for the maritime transportation of nuclear materials. At any rate, it seems fair to say that Fukushima has re-animated the international debate about maritime transport safety[134] and rekindled some coastal States' long-standing or latent concerns about the routing of nuclear materials shipments through

(16 July 1999) accessed 9 January 2015 at http://www.nci.org/c/caricom716.htm; Nadi Declaration on ACP Solidarity in a Globalised World, Doc ACP/28/029/02 [Final], adopted by the 3rd Summit of African, Caribbean and Pacific Groups of States, Heads of State and Government (Denaru, Nadi-Fiji Islands, 19 July 2002) at para 50: 'We express our strong objection to the transport of nuclear and other hazardous materials through the waters around ACP States. We call for the immediate cessation of such practice in order to, inter alia, prevent any occurrence of accidents that could seriously threaten their sustainable development and the health of their peoples.'

[132] Of course, the temporary shutdown of all nuclear reactors in Japan following the accident is also likely to affect maritime transportation of radioactive materials between Japan and Europe.

[133] See e.g., Gary Owen and Simon Chan, 'The Impact of the Fukushima Disaster on International Nuclear Transport Safety – a UK Perspective' (presented in proceedings of the 17th International Symposium on the Packaging and Transportation of Radioactive Materials (PATRAM) 2013, 18–23 August 2013) at 10–11, who specifically identify transport safety, human factors, communications, and emergency response preparations as topics for which Fukushima offers directly relevant lessons. See also André-Claude Lacoste, 'Safety, Security and Response: A View of the IAEA Today' (Presentation delivered at the International Conference on the Safe and Secure Transport of Radioactive Material: The Next Fifty Years of Transport – Creating a Safe, Secure and Sustainable Framework, Vienna, 17–21 October 2011) (discussing, for example, the need for re-assessing transport package integrity in beyond-design-accident situations) accessed 9 January 2015 at http://www-pub.iaea.org/MTCD/Meetings/cn187 presentations.asp.

[134] See e.g., Joseph Ballard, New Zealand Deputy Permanent Representative (Statement to the Conference on Disarmament in the Preparatory Committee for the 2015 Review Conference of the Parties to the Treaty on the Non-Proliferation of Nuclear Weapons, 5 May 2014) accessed 9 January 2015 at http://www.converge.org.nz/pma/NPT-NZ-5MAY2014.pdf: 'As Fukushima clearly demonstrated, nuclear accidents do not respect national boundaries. As a country that has not chosen nuclear power for itself, New Zealand is nevertheless still susceptible to suffering damage as a result of a nuclear accident elsewhere. This is particularly true for an accident occurring during the maritime transport of nuclear material.'

offshore waters.[135] Two basic types of claims have generally been advanced by coastal States when objecting to the carriage of nuclear materials of special concern[136] through their territorial seas.[137] The first one amounts to an outright denial of the right of innocent passage on the part of foreign nuclear transport vessels; the second conditions the exercise of innocent passage on the coastal State's prior consent or in advance notification. To this day, both these claims have remained unfounded as a matter of law.

The Conventions themselves essentially remain silent on this matter. While Article 27 of the Joint Convention does address the transboundary movement of spent fuel and radioactive waste, it does not establish an obligation to notify, let alone to obtain the consent from a State of transit, that is, any State other than the State or origin or final destination.[138] Rather, paragraph 3 of the Article contains an express savings clause for navigational rights and freedoms under international law.[139] Thus, absent any special *inter se* regulation,[140] the right of passage of nuclear cargo

---

[135] See e.g., Caribbean Community Secretariat, Press Release, 287/2011, 'CARICOM Demands Halt to Passage of Hazardous Waste through Caribbean Sea' (20 July 2011) accessed 9 January 2015 at http://www.caricom.org/jsp/pressreleases/press_releases_2011/pres287_11.jsp; Caribbean Community Secretariat, Press release 338/2012, 'Stop Hazardous Waste Shipment through Caribbean Sea – CARICOM Chairman' (28 December 2012) accessed 9 January 2015 at http://www.caricom.org/jsp/pressreleases/press_releases_2012/pres338_12.jsp.

[136] From a risk perspective, the most significant radioactive materials are associated with the back-end of the fuel cycle, i.e., MOX fuels, high-level radioactive waste, plutonium and spent nuclear fuel.

[137] Most often coastal States object not only to transportation through their territorial seas but also through their exclusive economic zones. However, the focus here is exclusively on the legal regime applicable to transport-related nuclear risks associated with innocent passage through the territorial sea as the coastal State's position resembles more closely that of the risk-exposed State in an on-land transboundary nuclear risk setting. In any event, this chapter's largely negative conclusions regarding the coastal State's claims in respect of enhanced protections in the territorial sea would apply a fortiori to the EEZ in which the coastal State enjoys limited 'sovereign rights', while foreign nuclear transporters enjoy freedom of navigation.

[138] See Joint Convention, *supra* note 15, Art 27(1)(ii) in conjunction with Art 2(s).

[139] 'Nothing in this Convention prejudices or affects: (i) the exercise by ships and aircraft of all States, of maritime … rights and freedoms, as provided for in international law.' *Ibid*, Art 27(3).

[140] See *infra* note 165.

through the territorial sea is governed by UNCLOS or customary international law.[141] Within this normative framework, a categorical denial of a right of innocent passage on the part of a foreign flagged vessel carrying nuclear materials of interest here is already implausible on account of UNCLOS Articles 22 and 23 whose reference to ships carrying nuclear substances or materials[142] proves the very existence of the right whose denial is being advocated. By the same token, any attempt at characterising nuclear material trans-shipments as inherently non-innocent, therefore illegal, is unpersuasive. To begin with, the very notion of non-innocent passage is tied to a vessel's activities or conduct during transit in the territorial sea, rather than the vessel's flag, origin, destination or indeed cargo.[143] Moreover, the argument that any carriage, in the territorial sea, of radioactive materials of special concern amounts to the creation of an exorbitant risk of harm to the coastal State's interests and thus can be equated to an 'act of wilful and serious pollution' rendering passage non-innocent[144] is difficult to support.[145]

---

[141] Thus, UNCLOS, *supra* note 128, Art 311(3) permits inter se agreements modifying provisions of the Convention 'not incompatible with "the effective execution of the object and purpose" of the Convention and provided they do not affect the application of its basic principles'.

[142] *Ibid* Article 22(2) specifically refers to 'ships carrying nuclear or other inherently dangerous or noxious substances or materials' as being required to follow coastal States' sea lanes and traffic separation schemes in the territorial sea. Article 23 provides: 'Foreign nuclear-powered ships and ships carrying nuclear or other inherently dangerous or noxious substances shall, when exercising the right of innocent passage through the territorial sea, carry documents and observe special precautionary measures established for such ships by international agreements.'

[143] See e.g., *Message from the President of the United States Transmitting United Nations Convention on the Law of the Sea*, 103rd Cong, 2nd Sess (1994) at 15: '[UNCLOS] Article 19(2) refers to activities that occur in the territorial sea. This means that any determination of non-innocence of passage by a transiting ship must be made on the basis of acts it commits while in the territorial sea. Thus, cargo, means of propulsion, flag, origin, destination or purpose of voyage cannot be used as criteria in determining that the passage is not innocent.'

[144] Thus, Article 19(2)(h) lists among non-innocent vessel activities – because prejudicial to the peace, good order or security of the coastal State – 'any act of wilful or serious pollution contrary to this Convention'. UNCLOS, *supra* note 128, Art 19.

[145] See also Masahiro Miyoshi, 'Ocean Transport of Radioactive Fuel and Waste: A Japanese Perspective' in David Caron and Harry Schreiber (eds), *The Oceans in the Nuclear Age: Legacies and Risks* (Leiden: Martinus Nijhoff

First, coastal States' espousal of a modified-burden-of-proof approach[146] in challenging the legality of such risk-bearing passage[147] would be difficult to square with the fundamental thrust of the Draft Articles on Prevention, as discussed.[148] Second, and more importantly, international practice points in the opposite direction as a majority of States simply continues to support an unrestricted right of innocent passage.[149] Admittedly, on a number of occasions States engaged in the maritime transport of nuclear materials may have heeded coastal States' warnings or protests and re-routed such transports to avoid the territorial waters of third States.[150] However, it has also been noted that this practice, limited as it is, is ambiguous in terms of its evidentiary value for inferring a new customary norm requiring foreign flagged nuclear transport vessels to avoid third States' territorial seas altogether or to obtain a coastal State's prior authorisation.[151] The simple fact is, as Tullio Treves puts it, that

---

Publishers, 2010) 197 at 204; Patricia Birnie, Alan Boyle and Catherine Redgwell, *International Law and the Environment*, 3rd edn (Oxford: Oxford University Press, 2009) at 415.

[146] See *supra* note 19.

[147] Especially Caribbean countries have characterised transshipments of nuclear materials as legally unacceptable on account of the extraordinary risk of harm, albeit of an extremely low probability, associated with a transport accident in light of 'the economic importance and ecological fragility of the Caribbean Sea and the well-being of the millions of people who depend on this unique resource for their very existence'. See Canute James, 'Nuclear Waste Raises the Heat on its Way through Caribbean', *The Financial Times* (30 April 1999) at 3.

[148] See *supra* text at notes 25–26.

[149] Of course, coastal States have the right to regulate innocent passage in accordance with Articles 21, 22 and 24 of UNCLOS, *supra* note 128. However, this authority does not cover also the jurisdictional claims here under review.

[150] Examples on point include the voyages of the Akatsuki Maru (see Jon Van Dyke, 'The Legitimacy of Unilateral Actions to Protect the Ocean Shipment of Ultrahazardous Radioactive Materials' (13 December 1996) accessed 9 January 2015 at http://www.nci.org/i/ib121396.htm), as well as the Pacific Swan, Pacific Teal and the Pacific Pintail. For details, see David Dixon, 'Transnational Shipments of Nuclear Materials by Sea: Do Current Safeguards Provide Coastal States a Right to Deny Innocent Passage?' (2006) 16 *Journal of Transnational Law and Policy* 73 at 76–81.

[151] Thus, Miguel Sousa Ferro points out that, on occasion, flag States promising to keep national vessels out of the waters of objecting coastal States simply failed to keep their promises. Miguel Sousa Ferro, 'Right of Innocent Passage of Ships Carrying Ultra-hazardous Cargoes' (2006) 78 *Nuclear Law Bulletin* 5 at 12. Besides, these instances of practice – whatever their relevance – are undermined by incidents in which flag States rejected costal States' insistence on prior authorisation and/or notification as legally impermissible, an example of

'flag states of such [nuclear cargo carrying] ships refuse to accept formally such requirements'.[152] Other evidence supposedly supportive of a coastal State's right to ban altogether or to subject to prior authorisation the passage of vessels carrying nuclear materials, is at best inconclusive.[153] In the final analysis, therefore, given that the burden of proof

---

which is the official US position in relation to a nuclear reactor vessel shipment from California to South Carolina around Cape Horn. See e.g., the 'U.S. Diplomatic Note of 23 March 2003', protesting restrictive provisions of Chile's new Nuclear Safety Law, in Sally Cummins and David Stewart, (2003) *Digest of Practice in International Law* 759 (the 'requirement of authorization prior to transiting a country's territorial sea … clearly exceeds the provisions of international law as reflected in the LOS Convention'); and Sally Cummins (ed), *Digest of United States Practice of International Law* (New York: Oxford University Press, 2004) at 713. Finally, in documented cases of abidance by some operators/flag States by coastal States' restrictions, the parties' conduct may not, or at least not always, reflect a sense of legal obligation. See *ibid*; Ferro, *supra* note 151 at 12.

152 Tullio Treves, 'Navigation of Ships with Nuclear Cargoes: Dialogue between Flag and Coastal States as a Method for Managing the Dispute' in Caron and Schreiber, *supra* note 145 at 217, 225.

153 Admittedly, a number of States may have adopted national regulations to this effect. Specifically, some of them have argued that Article 23 permits coastal States to temporarily adopt special rules. Thus, upon ratification of UNCLOS, Malaysia offered the following observation: 'In view of the inherent danger entailed in the passage of nuclear-powered vessels or vessels carrying nuclear material or other material of a similar nature … the Malaysian Government … requires the aforesaid vessels to obtain prior authorization of passage before entering the territorial sea of Malaysia until such time as the international agreements referred to in article 23 are concluded and Malaysia becomes a party thereto.' Government of Malaysia, Declaration of 14 October 1996 to the UNCLOS, accessed 9 January 2015 at http://www.un.org/Depts/los/convention_agreements/convention_declarations.htm. By contrast, the declaration by the Netherlands (28 June 1996, *ibid*) takes the opposite view: 'The Convention permits innocent passage in the territorial sea for all ships, including … nuclear-powered ships and ships carrying nuclear or hazardous waste, without any prior consent or notification, and with due observance of special precautionary measures established for such ships by international agreements.' Many other countries, upon signing or ratifying UNCLOS, have similarly rejected a coastal State's right to impose conditions. For example, the US (see *supra* note 143) maintains that 'the Convention provides no authority for a coastal State to condition the exercise of the right of innocent passage by any ships, including warships, on the giving of prior notification to or the receipt of prior permission from the coastal State'. Finally, Erik Molenaar makes the point that, in any event, foreign flag vessels would not be required to observe a coastal State's special national standards applicable to nuclear cargo carriers, as Art 23 expressly refers

regarding the existence of a general practice accepted as law lies with the proponents of such a (new) limited or conditional customary right of innocent passage, the conclusion must be that 'international law, as it stands today, does not allow for such a restriction on the right of innocent passage'.[154]

At first glance, the claim that a coastal State is entitled to prior notification of a nuclear material transshipment through its territorial sea might look more persuasive. After all, the effectiveness of UNCLOS Article 22, which authorises the coastal State to establish sea lanes and traffic separation schemes in the territorial sea for, inter alia, ships carrying nuclear substances or materials, seems to be premised on the State having advance knowledge of the arrival of such vessels in its waters.[155] However, while such an inference may seem logical, it is not supported in international practice.

Tellingly, for many years now, the annual attempt by small island developing countries to persuade a majority of States in the UN General Assembly to support their call for international recognition of coastal States' right to condition passage through the territorial sea on prior notification, if not consent, have foundered because of strong opposition by a coalition of maritime and other States.[156] Similarly, in the debates in the 1990s regarding revisions of the International Maritime Organization (IMO) Code for the Safe Carriage of Packaged Irradiated Nuclear Fuel, Plutonium and High-Level Radioactive Wastes on Board Ships (INF Code),[157] a proposal to require notification to the coastal State failed to garner acceptance.[158] Admittedly, the IAEA (voluntary) Code of Practice

---

to 'international agreements'. See Erik Molenaar, *Coastal State Jurisdiction over Vessel-Source Pollution* (The Hague: Kluwer Law International, 1998) at 205.

154 Ferro, *supra* note 151 at 14.

155 To this effect, see Tullio Scovazzi, 'The Transboundary Movement of Hazardous Waste in the Mediterranean Regional Context' (2000) 19(1) *UCLA Journal of Environmental Law* 231 at 243.

156 For the most recent such attempt during the 2013 session, see *Oceans and the Law of the Sea*, Res A/68/70, UNGAOR, 68th Sess, UN Doc A/68/L.18 and Add.1 (2014) at paras 130–131.

157 The originally non-binding Code was amended in 1997 by International Maritime Organisation Resolution A.853(20), Amendments to the Code for the Safe Carriage of Irradiated Nuclear Fuel, Plutonium and High-Level Radioactive Wastes in Flasks on Board Ships (27 November 1997) and became mandatory on 1 January 2001 as part of Chapter VII – Carriage of Dangerous Goods of the International Convention for the Safety of Life at Sea.

158 The Irish government had led the effort to have the IMO incorporate into the INF Code 'a notification structure … whereby coastal states would be

on the International Transboundary Movement of Radioactive Waste does recommend prior notification of transit States.[159] However, the INF Code features a disclaimer according to which none of its provisions 'prejudices or affects in any way' navigational rights and freedoms under customary international law, UNCLOS and under other relevant international legal instruments.[160] None of the multilateral legal instruments that specifically cover the transport of nuclear materials can be deemed to endorse prior notification as a binding legal obligation associated with innocent passage.[161]

Some commentators point to treaty practice related to the transboundary movement of hazardous wastes as evidence of the emergence of a customary obligation of prior notification applicable to all types of waste,

---

informed of the passage of INF cargoes past our coast'. See Seanad Éireann, *Nuclear Free Zone: Motion*, vol 150, No 15 (9 April 1997) at 12. In the end, however, IMO's Maritime Safety and Marine Environment Protection Committees agreed not to pursue the matter further noting that a 'prior notification' requirement might result in contravention of UNCLOS and the Convention on Protection of Nuclear Material [,] … an undesired precedent for the transportation of other dangerous goods and … interference by terrorists.' See IMO, *Report of the Maritime Safety Committee on its Sixty-Ninth Session*, IMO Doc MSC 69/22 (29 May 1998) at para 15.5.

[159] See IAEA, Information Circular, *Code of Practice on the International Transboundary Movement of Radioactive Waste*, IAEA Doc INFCIRC/386 (13 November 1990), Art 3 at para 5.

[160] *Ibid*, Art 1.

[161] The IAEA, IAEA Safety Standards Series, *Regulations for the Safe Transport of Radioactive Material, Specific Safety Requirements*, IAEA Doc No SSR-6 (2012) at para 558, call for notification to the competent authority of a country 'through (…) which' a consignment might be shipped, thus conceivably cover also coastal transit States. However, the provision, which is not per se legally binding, is not maritime transport-specific. Although both the INF Code and the International Maritime Dangerous Goods (IMDG) Code establish binding regulations for the ocean transport of nuclear materials, neither addresses the issue of coastal States' prior notification. Finally, the system established by EC, *Council Directive 2006/117/EURATOM of 20 November 2006 on the Supervision and Control of Shipments of Radioactive Waste and Spent Fuel*, [2006] OJ L 337/21 (2006/117 EURATOM Council Directive) is according to Art 1(6) to be 'without prejudice to rights and obligations under international law'. However, in Art 11(2), the 2006/117 EURATOM Council Directive calls upon any EU Member State that is the State of destination for such a shipment to notify any third country of transit upon receipt of each shipment, but does not call for prior notification of such a country of transit.

including nuclear wastes.[162] Indeed, the Basel Convention on the Transboundary Movement of Hazardous Wastes and Their Disposal (Basel Convention),[163] as a global instrument, and the Izmir Protocol,[164] as an example of a regional agreement,[165] both establish an obligation to notify the coastal State that applies also to hazardous waste shipments through the territorial sea.[166] However, apart from the fact that the Basel Convention does not cover radioactive waste,[167] the Convention includes an express savings clause as regards foreign flag vessels' navigational

---

[162] See e.g., Laura Pineschi, 'The Transit of Ships Carrying Hazardous Wastes through Foreign Coastal Zones' in Francesco Francioni and Tullio Scovazzi (eds), *International Responsibility for Environmental Harm* (London: Graham & Trotman, 1991) 299 at 315, who invokes 'the spirit of the 1989 Basel Convention' as evidence of the emergence of a general legal obligation of prior notification. *C.f.* Jon van Dyke, 'Sea Shipment of Japan Plutonium under International Law' (1993) 24 *Ocean Development and International Law* 399 at 412.

[163] Basel Convention on the Transboundary Movement of Hazardous Wastes and Their Disposal, adopted 22 March 1989, 1673 UNTS 57 (entered into force 5 May 1992) (Basel Convention).

[164] Protocol on the Prevention of Pollution of the Mediterranean Sea by Transboundary Movements of Hazardous Wastes and their Disposal, adopted 1 October 1996 (entered into force 18 January 2008) accessed 9 January 2015 at http://www.unep.ch/regionalseas/main/med/medhaz.html (Izmir Protocol).

[165] Other regional agreements cover nuclear materials as well but establish far from uniform requirements. For example, while the Bamako Convention on the Ban of the Import into Africa and the Control of Transboundary Movement and Management of Hazardous Wastes within Africa, adopted 30 January 1991 (entered into force 22 April 1998) accessed 9 January 2015 at http://www.unep.org/delc/Portals/119/Bamako%20Convention%20-%20Text%20English.pdf, establishes no exception for nuclear transit shipments through the territorial sea from its notification requirements, the Waigani Convention and the Tehran Protocol feature a savings clause as regards international navigational rights in the territorial sea. See Art 2(4)–(5) of the Convention to Ban the Importation into Forum Island Countries of Hazardous and Radioactive Wastes and to Control the Transboundary Movement and Management of Hazardous Wastes within the South Pacific Region, adopted 16 September 1995 (entered into force 21 October 2001) accessed 9 January 2015 at http://www.sprep.org/attachments/legal/WaiganiConvention.pdf (Waigani Convention); and Art 4(14) of the Tehran Protocol of Marine Transboundary Movements and Disposal of Hazardous Waste and Other Wastes, adopted 17 March 1998 (entered into force 21 August 2005) accessed 9 January 2015 at http://ropme.org/uploads/protocols/hazardous_wastes_protocol.pdf (Tehran Protocol).

[166] See Basel Convention, *supra* note 163, Art 2(3); Izmir Protocol, *supra* note 164, Art 6(4).

[167] See Basel Convention, *supra* note 163, Art 1(3).

rights under international law in the territorial sea.[168] In light of this and various 'clarifying statements' offered by States at the time of the adoption of the Basel Convention,[169] it should be evident that the Convention does not purport to derogate from the customary legal concept of 'innocent passage' through the territorial sea.[170] By contrast, the Izmir Protocol applies to nuclear materials and indisputably establishes an obligation on the part of the exporting State to notify any transit State.[171] However, any claim to a broader significance of this instrument is undermined by the fact that States' participation in the Izmir Protocol is limited.[172] Equally importantly, at the conclusion of the Izmir Conference, the European Community and France strongly objected to the prior notification provision, while both of them as well as Israel opposed the inclusion of radioactive wastes among the waste substances covered by the instrument.[173]

---

[168] Thus, *ibid*, Article 4(12) stipulates: 'Nothing in this Convention shall affect in any way the sovereignty of States over their territorial sea established in accordance with international law, and the sovereign rights and the jurisdiction which States have in their exclusive economic zones and their continental shelves in accordance with international law, and the exercise by ships and aircraft of all States of navigational rights and freedoms as provided for in international law and as reflected in relevant international instruments.' See further Iwona Rummel-Bulska, 'The Basel Convention and the UN Convention on the Law of the Sea' in Henrik Ringbom, *Competing Norms in the Law of Marine Environmental Protection* (The Hague: Kluwer Law International, 1997) 83 at 89–92.

[169] For example, Japan declared its understanding 'that nothing in this Convention sh[ould] be deemed to require notice to or consent of any state for the mere passage of hazardous wastes on a vessel.' See Japan's Declaration to the Basel Convention, accessed 9 January 2015 at https://treaties.un.org/pages/ViewDetails.aspx?src=TREATY&mtdsg_no=XXVII-3&chapter=27&lang=en. Similarly, Germany upon ratifying the Convention expressed the view that 'nothing in this Convention sh[ould] be deemed to require notice to or consent of any State for the passage of hazardous wastes on a vessel under the flag of a Party exercising its right of innocent passage through the territorial sea or the freedom of navigation in an exclusive economic zone under international law.' Basel Convention, *supra* note 163 at 325.

[170] See also Tullio Treves, 'Codification du droit international et pratique des Etats dans le droit de la mer' (1990) 223 *Recueil des Cours* 9 at 123.

[171] Izmir Protocol, *supra* note 164, Art 6(4).

[172] As of 9 May 2014, only 7 out of 22 contracting parties to the framework agreement – the Barcelona Convention – had ratified the Protocol.

[173] The respective declarations are reproduced in Tullio Scovazzi, *Marine Specially Protected Areas: The General Aspects and the Mediterranean Regional System* (The Hague: Kluwer Law International, 1999) at 216–17.

In sum, the prospect of radioactive materials of interest transiting the territorial sea does not in and of itself authorise a coastal State to take special protective measures beyond those envisaged in UNCLOS to minimise or eliminate any risk of harm. Of course, such shipments are subject to various binding safety regulations, such as those of the INF or the IMDG Codes, quite apart from security-related international requirements.[174] However, neither they, nor UNCLOS or general international law support a coastal State's claim to be entitled to prior notification of such transits, however desirable such notification might be. There is no need to review in detail the obvious merits – in terms of better emergency preparedness and enhanced public confidence in the coastal State in the safety of such transports – associated with improved communications between shipping/flag States and coastal States.[175] Pitted against these real benefits are, of course, the real risks of an increase in denial of nuclear shipments,[176] of potential security threats from criminal or terrorist organisations,[177] as well as a concern among many States of creating a precedent for additional jurisdictional claims by coastal States. Looking at this balance of interests, some commentators have urged a compromise. For example, Tullio Treves has called for envisaging nuclear transits through the territorial sea 'as an exercise of a right that requires cooperation between flag and the coastal State in order to protect the rights and interest of both sides'.[178] And indeed, today the international political and legal discourse is less about the basic advisability of a prior notification scheme than about its operational modalities.

In October 2011, the International Conference on the Safe and Secure Transport of Radioactive Material[179] identified improved communications

---

174 See e.g., IAEA, *Convention on the Physical Protection of Nuclear Material*, IAEA Doc INFCIRC/274/Rev.1 (3 March 1980).

175 See also IAEA, General Conference, *Measures to Strengthen International Cooperation in Nuclear, Radiation, Transport and Waste Safety*, IAEA Doc GC(57)/RES/9 (19 September 2013) at para 66.

176 For an acknowledgement of the nature and scope of the problem of denial of (and delays in) the shipment of radioactive materials, see e.g., *ibid* at para 72.

177 See *supra* note 158.

178 Treves, *supra* note 152 at 232. Similarly, Tullio Scovazzi advocates 'notification without authorization'. Tullio Scovazzi, 'The Evolution of International Law of the Sea: New Issues, New Challenges' (2000) 286 *Recueil des Cours* 39 at 160.

179 IAEA, International Conference on the Safe and Secure Transport of Radioactive Material: The Next Fifty Years of Transport – Creating a Safe, Secure and Sustainable Framework (17–21 October 2011) accessed 9 January

between coastal States and shipping/flag States in advance of nuclear shipments through offshore waters as a topic of interest, while noting also some delegates' expressions of concern about the 'appropriateness, practicality and legal aspects of prior notification'.[180] The Conference concluded by recommending the development of 'best practice guidelines' for systematic and timely government-to-government communications to build on the current practice of voluntary exchange of information between the States concerned.[181] It thereby initiated a process which by April 2014, after a technical review meeting of IAEA Member States in 2012 and several sessions of a dedicated working group, resulted in a Report on Best Practice Guidelines.[182] The Report's suggestions are premised on communications being voluntary, confidential and of a government-to-government nature; they cover the transport by sea of only nuclear materials of significance, namely mixed-oxide fuel, high-level radioactive waste and, as appropriate, irradiated fuel; and they define 'coastal State' as any coastal State with a concern about shipments, irrespective of its geographical proximity to the actual route being used, while referring to the State on the other side of the communication exchange as 'shipping State', that is, the State that regulates consigners, consignees and carriers for transports by sea.[183] With regard to the information to be provided, upon 'assurances of confidentiality' ten days before departure, the Working Group proposes the inclusion of, inter alia, details regarding the vessel itself (type, flag, age and name), the nuclear material to be transported, the ship's likely route and emergency preparedness and response plans relevant to the voyage.[184] Annexes 3–5 of the Report contain a 'number of other suggestions' made in a joint submission by the principal nuclear shipping countries, the United Kingdom, France and Japan; by New Zealand on behalf of a group of coastal States; and the response thereto by the shipping countries.

Undoubtedly, the Report of the Working Group is not the final word on the matter of State-to-State communications bearing on nuclear transports

---

2015 at http://www-pub.iaea.org/iaeameetings/38298/International-Conference-on-the-Safe-and-Secure-Transport-of-Radioactive-Material-The-Next-Fifty-Years-of-Transport-Creating-a-Safe-Secure-and-Sustainable-Framework.

[180] *Ibid*, President's Findings.

[181] *Ibid*.

[182] Report of the Working Group on Best Practice Guidelines, *supra* note 129.

[183] *Ibid* at paras 1–2.

[184] *Ibid* at para 4(a)–(i).

by sea. However, it is a milestone in a process of deepening dialogue and consultations between coastal States and nuclear shipping/flag States. This process, strongly supported as it is both in the IAEA[185] and the UN General Assembly,[186] is likely to bring about an accommodation that will broadly recognise, subject to reasonable safeguards and a clear understanding of its exceptional nature, the legitimate interests of coastal States in being informed of risks emanating from nuclear shipments through their waters offshore. It thereby will also give at least *de facto* recognition to the long-standing claim that in its territorial sea the coastal State is entitled to special consideration with regard to the prevention of nuclear harm.

V. SUMMARY AND CONCLUSIONS

By now, seemingly every aspect of the accident at the Fukushima-Daiichi nuclear power plant and its implications for nuclear power safety, at both national and global levels, has been the object of an amazing number of studies, analyses, comments and recommendations of a technical, political and regulatory nature. As noted, some of this activism has already translated into adjustments of specific legal standards bearing on nuclear safety. Other normative changes are bound to follow. There can be little doubt, therefore, that in a formal sense the post-accident global nuclear safety regime will be a different, much improved one. However, the test of whether the lessons of Fukushima will have been truly mastered lies less in the adoption of new safety laws and regulations than in industry's and regulators' ability and willingness to ensure that these prescriptions become and continue to be operational reality.

A critical precondition for achieving this objective is an effective global system of international safety reviews. As discussed, post-Fukushima, there exists a heightened awareness among all interested parties of the need to strengthen existing peer review processes, including those under the Conventions themselves. The aim ought to be to establish periodic, more focused and mandatory reviews that allow for in-country follow-up assessment to ensure that, beyond representing formal undertakings, national safety standards and safety practices actually do conform to installations States' obligations under the Conventions, customary

---

185 See e.g., IAEA, *Measures to strengthen international cooperation in nuclear, radiation, transport and waste safety,* IAEA Doc GC(55)/RES/9 (22 September 2011) at para 48.

186 See e.g., *Oceans and the Law of the Sea*, *supra* note 156 at para 130.

international law or reflect industry best practices. Clearly, there is substantial support for such a reform given independent peer reviews' indispensable function in the global nuclear safety regime. However, it is well worth recalling also that in the wake of Chernobyl initial support for similar, mandatory safety reviews in the end failed to produce legislative change. In other words, while on a regional level, progress is being made in this respect, it is far from certain that the improvements necessary to ensure a high level of nuclear safety worldwide, thereby to reduce also the risk of transboundary nuclear pollution generally, will be achievable also globally.

Other factors that bear more specifically on the risk of transboundary nuclear harm are principally a function of geography: the proximity of a risk-exposed country to on-land nuclear installations or the maritime transport of certain nuclear materials through offshore waters. As noted, despite the fact that nuclear 'border' installations may entail a significant imposition on the risk-exposed State's territorial sovereignty given that the State might have to take proactive emergency planning measures, present-day international law does not provide an enhanced level of protection to the State. It certainly does not support the idea of the State's right to veto such an installation. Nevertheless, considering the severity of the interference with the risk-exposed State's sovereign rights and the balance of equities between risk-creating and exposed States – barring special circumstances – the latter's interest in minimising the nuclear pollution risk ought to be accorded stronger procedural protection. This could be done by allowing recourse to third-party decision-making in the event the parties fail to reach an accommodation.

Finally, despite an outstanding safety record and an extremely low probability of harm, nuclear shipments by sea, especially through territorial seas, have been a matter of contention for many coastal States. Their long-standing claim that the transit of nuclear materials of concern is either incompatible with the very notion of innocent passage, or is impermissible except with the coastal State's prior notification or authorisation is not supported in present-day international law. Nevertheless, a possible compromise might now be in the offing: provided appropriate safeguards can be put in place, a majority of States might be willing to accept informally what has been advocated as a 'notification without authorization' approach. If and when this happens, it will signal finally the *de facto* recognition of coastal States' sovereignty-based claim of enhanced protection against the risk of accidental nuclear pollution emanating from foreign flag vessels in the territorial sea.

9. Transboundary air pollution: a tale of two paradigms

Alan Boyle

I. INTRODUCTION

International law offers two paradigms for addressing transboundary air pollution: the *Trail Smelter* arbitration case of 1941 (*Trail Smelter*)[1] and the Convention on Long-Range Transboundary Air Pollution of 1979 (LRTAP).[2] The first of these paradigms is concerned with an identifiable source of transboundary pollution that could be addressed within the framework of bilateral relations. The Arbitral Tribunal concluded that 'no state has the right to use or permit the use of its territory in such a manner as to cause injury by fumes in or to the territory of another or the properties or persons therein'.[3] It awarded monetary damages to compensate for transboundary harm caused by the smelter's emissions, and it prescribed control measures to avert future transboundary harm.[4]

International environmental law has developed greatly since then, but the basic principle upheld by the arbitrators in the *Trail Smelter* case remains unchallenged. In today's international law, that principle is expressed in the obligation of States to use due diligence to regulate and control potentially harmful activities taking place within their own territory, jurisdiction or control. The two leading cases are *Pulp Mills on the River Uruguay*[5] (*Pulp Mills*), and the Advisory Opinion of the

---

1 *Trail Smelter Case (United States of America, Canada)* (1938 and 1941) 3 RIAA 1905 *(Trail Smelter).*

2 Convention on Long-Range Transboundary Air Pollution, adopted 13 November 1979, 1302 UNTS 217 (entered into force 16 March 1983) (LRTAP).

3 *Trail Smelter*, *supra* note 1 at 716.

4 *Ibid.*

5 *Case concerning Pulp Mills on the River Uruguay (Argentina v Uruguay),* [2010] ICJ Rep 14 (*Pulp Mills*).

International Tribunal for the Law of the Sea (ITLOS) on the *Responsibilities and Obligations of States Sponsoring Persons and Entities with respect to Activities in the Area*[6] (*Advisory Opinion on Seabed Activities*), although neither is specifically about air pollution. The first section of the chapter explores this, using the now settled *Aerial Spraying* case to illustrate some of the key issues of general international law that arise when applied to transboundary air pollution.[7] Its most obvious relevance to Singapore is the problem of smog caused by large-scale forest fires in Indonesia.

The second paradigm locates transboundary air pollution within a broader regional context and addresses it through the mechanism of evolutionary regimes of multilateral regulation exemplified by the 1979 Geneva Convention on Long-Range Transboundary Air Pollution and the 1991 US-Canada Air Quality Agreement.[8] This approach treats regional airmasses as shared natural resources.[9] 'Long-range transboundary air pollution' is defined as pollution having effects at such a distance that 'it is not generally possible to distinguish the contributions of individual emission sources or groups of sources'.[10] Thus, it is not aimed at *Trail Smelter* type cases, but at regional problems of acid rain and other widely dispersed pollutants.

The benefits can be observed in the increasingly complex and largely successful structure of protocols built onto framework conventions and dealing with a range of widely dispersed air pollutants. This paradigm is considered briefly in the second part of the chapter. The point that is most relevant in Asia is the absence of any comparable regional regime in this continent, despite the growing and obvious problem of transboundary

---

6 *Responsibilities and Obligations of States Sponsoring Persons and Entities with respect to Activities in the Area,* Advisory Opinion, 1 February 2011, ITLOS Rep 2011, 10 (*Advisory Opinion on Seabed Activities*).

7 See *Aerial Herbicide Spraying (Ecuador v Colombia),* Order of 13 September 2013, accessed 9 January 2015 at http://www.icj-cij.org/docket/index.php?p1=3&code=ecol&case=138 (*Aerial Spraying*). The author was counsel for Ecuador, but the views expressed here are his own and not those of any government he has represented.

8 Agreement between the Government of the United States of America and the Government of Canada on Air Quality (adopted and entered into force on 13 March 1991) 30 ILM 676 (US-Canada Air Quality Agreement).

9 UNEP has referred to 'airsheds' in this context. See UNEP, Co-operation in the Field of the Environment Concerning Natural Resources Shared by Two or More States: Report of the Executive Director, UN Doc UNEP/GC/44 (1975) at para 86.

10 LRTAP, *supra* note 2, Art 1(b).

air pollution on a regional scale, caused in particular by Chinese and Indian emissions and the continued growth in fossil fuel consumption.

II. THE *AERIAL SPRAYING* CASE AND TRANSBOUNDARY AIR POLLUTION

The *Aerial Spraying* case would have been the first case on transboundary air pollution to come before the International Court of Justice (ICJ) and the first before any international tribunal since *Trail Smelter*. Wisely for the parties (Ecuador and Colombia), it was settled only a few weeks before oral argument was due to begin,[11] but the written pleadings are all on the ICJ website,[12] and they make interesting reading. In particular the case is a textbook example of the application of the due diligence obligation to an activity (aerial spraying of herbicides) that was alleged to cause or to risk causing transboundary pollution in Ecuador, downwind of the spraying operation.

The case arose out of Colombia's efforts, funded by the United States, to eradicate illicit cocaine production by spraying coca plantations with herbicides, using specialist crop spraying aircraft and converted military observation aircraft.[13] The problem was that the herbicide drifted into Ecuador when spraying was carried out in close proximity to the border.[14] Ecuador alleged that the herbicide spray caused illness in people and livestock, polluted water supplies and damaged legitimate crops.[15] At its most extreme, it was claimed that it resulted in abandonment of some villages in border areas.[16]

The case raised a number of legal issues of general importance to transboundary pollution. First, what preventive measures will satisfy the standard of due diligence required by international law in this context? Second, is the precautionary approach relevant or of any value in this kind of case where the risks are provable and well known? Third, in what circumstances must an Environmental Impact Assessment (EIA) be

---

[11] See *Aerial Spraying, supra* note 7.

[12] *Aerial Spraying,* written pleadings are available at http://www.icj-cij.org/docket/index.php?p1=3&case=138&code=ecol&p3=1 (accessed 9 January 2015) (*Aerial Spraying*, Pleadings).

[13] See *ibid*, 'Memorial of Ecuador', accessed 9 January 2015 at http://www.icj-cij.org/docket/files/138/17540.pdf (*Aerial Spraying*, Memorial of Ecuador).

[14] See *ibid* at para 2.6.

[15] See *ibid* at para 2.8.

[16] See *ibid* at para 8.16.

carried out before such activities are permitted? Fourth, does human rights law have any bearing in cases of transboundary pollution affecting the population living on the other side of the border?

Plainly, as the case has now settled without a hearing, it can provide no definitive answers to any of these questions. Nevertheless, the proceedings do illustrate how transboundary air pollution might be addressed by an international court or arbitral tribunal, and thus there are lessons here for other States with comparable problems.

The first and most obvious lesson is that filing a case and intimating that an application for provisional measures will follow can be a very effective way of stopping a risky activity. Colombian spraying within ten kilometres of the border ceased in 2007, soon after the case was initiated. It has never resumed, and the settlement agreed in 2013 makes it unlikely that it ever will resume.[17] To win on provisional measures in an environmental case, the applicant State needs to establish an urgent need to prevent irreparable harm that may otherwise occur in the two or three years it will take for a hearing on the merits to come before the court. The record of applications for provisional measures in environmental cases is mixed: Uruguay successfully resisted an Argentine application in the *Pulp Mills* case in 2006, but it had a good case and Argentina did not.[18] For Colombia to concede in 2007 without an application even being filed is remarkable. It shows that in some cases litigation can provide immediate and effective protection from the risk of transboundary harm.

A. Due Diligence

The second lesson is that due diligence matters and must be taken seriously. Proving a failure by the respondent State to act with due diligence is the core of any case involving transboundary pollution. In *Aerial Spraying*, the evidence showed that Colombia recognised the risk of transboundary pollution. It issued a management plan setting out operating parameters for the spraying operation, including aircraft

---

[17] See *Aerial Spraying*, *supra* note 7, 2013 Agreement on Aerial Eradication of Illicit Crops by Colombia near the Ecuadorian Border (Colombia–Ecuador) (9 September 2013).

[18] *Pulp Mills, supra* note 5 at 113, para 62. See also *Certain Activities carried out by Nicaragua in the Border Area (Costa Rica v Nicaragua)*, Order of 8 March 2011, [2011] ICJ Rep 6 at 21, para 63, and *Construction of a Road in Costa Rica along the San Juan River (Nicaragua v Costa Rica)*, Order of 13 December 2013, [2013] ICJ Rep at 7, para 25.

speeds, height and prohibited areas.[19] These were designed to minimise the risk of transboundary drift and they might have worked had they been enforced. But it seems they were not enforced. The evidence appeared to suggest that the aircraft flew too high and too fast and operated in locations they should have avoided.[20]

If true, would this amount to a failure to act with due diligence? Several preliminary points need emphasis. The obligation of due diligence is an obligation of conduct, not an obligation of result.[21] A violation is established by proving failure to take the measures necessary to prevent, regulate or control the potentially risky activity. A violation is not established by showing actual pollution or a risk of pollution. Proof of pollution or the risk of pollution establishes only that the State has a duty to act. It does not tell us that the State has failed in its duty to act. The point is very clearly elaborated in the International Law Commission's (ILC) Draft Articles on Prevention of Transboundary Harm from Hazardous Activities.[22] These articles require States to control activities which may otherwise cause harm in neighbouring States.[23] They are about the management of risk. They do not presuppose the existence of harm. They do presuppose that harm is foreseeable.[24]

The obligation to act is the State's, not the company's. Showing that the company operating the aircraft had not followed Colombia's own rules does not establish a failure by Colombia to act with due diligence.[25] The question is thus what the State did or failed to do, not what the company did or failed to do. Colombia is not vicariously responsible for the companies it employs. It is responsible for its own actions or inactions. The same is true in *Trail Smelter, Pulp Mills*, and the ITLOS

---

[19] See *Aerial Spraying*, *supra* note 7, 'Counter-Memorial of Colombia' (29 March 2010), ICJ Pleadings at paras 4.67–4.69 (*Aerial Spraying*, Counter-Memorial of Colombia).

[20] See *Aerial Spraying, supra* note 7, 'Reply of Ecuador' (31 January 2011), ICJ Pleadings at paras 2.87–2.107 (*Aerial Spraying*, Reply of Ecuador).

[21] *Pulp Mills*, *supra* note 5 at para 187.

[22] 'International Liability for Injurious Consequences Arising out of Acts Not Prohibited by International Law (Prevention of Transboundary Harm from Hazardous Activities)' in 'Report of the International Law Commission to the General Assembly covering the work of its fifty-third session, with commentaries, 2001' (UN Doc A/56/10), Ch V, in *Yearbook of the International Law Commission 2001*, vol II, Part Two (UN, 2001) (Draft Articles on Prevention) at 336.

[23] See *ibid.*

[24] *Ibid* at paras 14–15, 387 at paras 1–3.

[25] *Advisory Opinion on Seabed Activities*, *supra* note 6 at paras 110–112.

Advisory Opinion on Seabed Activities. All of these cases involved businesses rather than State activity, but the State's engagement arose from its obligation to use due diligence to control the activity, not from any attribution of those activities to the State.

What does the exercise of due diligence entail? In its *Pulp Mills* judgment the ICJ reaffirmed that '[a] State is thus obliged to use *all of the means at its disposal* in order to avoid activities which take place in its territory, or in any area under its jurisdiction, causing significant damage to the environment of another state'.[26] The Court identified what due diligence requires in the following terms:

- 'adoption of appropriate rules and measures';[27]
- 'a certain level of vigilance in their enforcement';[28]
- 'the exercise of administrative control applicable to public and private operators';[29]
- 'careful consideration of the technology to be used';[30] and
- EIA.[31]

Uruguay satisfied the Court on all these points. It could show that the technology used in the mill was of the highest international standard.[32] It had set waste emission limits for the main pollutants which ensured that the river water continued to meet the water quality standards agreed in the Administrative Commission of the River Uruguay (CARU).[33] It had a comprehensive monitoring scheme in place, and where necessary it had shown itself prepared to take enforcement action against the operating company.[34] Finally, it had carried out the required EIA.[35] Argentina failed to prove its case, not because pulp mills do not pose a risk of pollution, but because it could not show that Uruguay had not taken appropriate measures to prevent, regulate and control that pollution.

---

26 *Pulp Mills*, *supra* note 5 at para 101 (emphasis added).
27 *Ibid* at para 197.
28 *Ibid*.
29 *Ibid*.
30 *Ibid* at para 223.
31 *Ibid*.
32 *Ibid* at 14.
33 *Ibid*.
34 *Ibid*.
35 *Ibid*.

The ITLOS *Advisory Opinion on Seabed Activities* cites with approval the ICJ's treatment of due diligence, but it also makes the following additional points:

- due diligence is a 'variable concept', which may change over time and differ in respect of different risks;[36] and
- 'the precautionary approach is also an integral part of the general obligation of due diligence of sponsoring States, which is applicable even outside the scope of the Regulations'.[37]

To illustrate the practical requirements of due diligence with respect to pollution, we can look briefly at the report of the National Commission of Enquiry which reported to the US President on the Deepwater Horizon oil rig disaster in 2011.[38] Although not the subject of international litigation, Deepwater Horizon exemplifies the points made in *Pulp Mills* by the ICJ and by the ITLOS in its *Advisory Opinion on Seabed Activities*. In addition to criticisms of BP (the well operator and lessee), Transocean (the rig operator), and Haliburton (the principal contractor), the report gives a damning account of regulatory failure by Congress and by the Minerals Management Service (MMS). Among a long catalogue of failures, the following stand out:

- *Inappropriate regulations*: drilling regulations were well below industry best practice in the North Sea. They were based on a 'prescriptive regulation with inspection model' characterised as 'fundamentally reactive and therefore incapable of driving continuous improvement in policies and practices'.[39] The report notes 'efforts to adopt a more rigorous and effective risk-based safety regime were repeatedly revisited, refined, delayed, and blocked by industry or sceptical agency political appointees'.[40]
- *Inadequate administrative control*: 'MMS became an agency systematically lacking the resources, technical training or experience in petroleum engineering that is absolutely critical to ensuring

---

36 See *Advisory Opinion on Seabed Activities*, *supra* note 6 at para 117.

37 *Ibid* at para 131.

38 US, National Commission on the BP Deepwater Horizon Oil Spill and Offshore Drilling, *Deep Water – The Gulf Oil Disaster and the Future of Offshore Drilling, Report to the President* (US Government Printing Office, 2011) (Gulf Oil Disaster Report).

39 *Ibid* at 69.

40 *Ibid* at 71.

that offshore drilling is being conducted in a safe and responsible manner.'[41]

- *Lack of vigilance in enforcement*: diminished resources meant fewer and less effective inspections.[42] Inspectors were poorly paid, poorly trained, too few in number, too inexperienced.[43] A culture of inspectors accepting gifts from oil companies was prevalent.[44]
- *Unfamiliarity with the technology*: 'As drilling technology evolved, many aspects of drilling lacked corresponding safety regulations.'[45]
- *No EIA*: MMS lacked resources to carry out 'meaningful NEPA [National Environmental Policy Act][46] review' and in practice it exempted Gulf oil drilling from NEPA.[47] The report refers to a 'culture of complacency'.[48] No other agency had the resources or authority to do the job.[49] Notwithstanding 'layers of required environmental scrutiny' – by NEPA, OPA (Oil Pollution Act), OCSLA (Outer Continental Shelf Lands Act), EDSA (Magnuson-Stevens Act), CWA (Clean Water Act) – '*none* of these laws resulted in site-specific review of the drilling operations of the Macondo well'.[50]

Overall, it is hard to imagine a better example of the failure by a State to act with due diligence. If this accident had resulted in damage to other States, the US could not easily have defended itself if sued in an international forum.[51]

As all these examples make clear, due diligence entails an evolving standard of technology and regulation.[52] In some cases, the appropriate standard is agreed internationally, as for example in the case of nuclear

---

41 *Ibid* at 57.
42 *Ibid* at 74.
43 *Ibid* at 79.
44 *Ibid* at 78.
45 *Ibid* at 73.
46 *Ibid* at 82.
47 *Ibid.*
48 *Ibid.*
49 *Ibid* at 84.
50 *Ibid* (emphasis added).
51 Finding a forum would be the difficult part, given that the US is not subject to compulsory jurisdiction under the ICJ Statute or the 1982 UN Convention on the Law of the Sea.
52 Draft Articles on Prevention, *supra* note 22 at 394, para 11.

installations or oil tankers,[53] and the relevant international agency will retain responsibility for periodically reviewing and updating the regulations and guidelines. There are no internationally agreed standards for any form of air pollution save for emissions from ships, which are regulated by the International Maritime Organization (IMO).[54]

In the absence of internationally agreed standards comparison with standards adopted by other States may be a good guide. This approach allows for the standard of diligence to change as technology and operating techniques develop and for new industrial plants to operate to higher standards than existing plants. Thus, in the *Pulp Mills* case, Uruguay accepted that its newly built mill would have to operate to the highest international standards (in that case the EU's), whereas Argentina's older mills most probably could not be expected to do so.[55]

The same approach could easily be applied to air pollution from power stations or industrial processes such as smelters or blast furnaces. In effect that is what the 1979 Geneva Convention on Long-Range Transboundary Air Pollution and the 1991 US-Canada Air Quality Agreement achieve at a regional level. They do not mandate technology standards as such, but by setting air quality standards, they compel the use of more efficient technology capable of meeting the required air quality standard.[56] Older less efficient plant is then more likely to be phased out.

---

[53] See Convention on Nuclear Safety, adopted 17 June 1994, 1963 UNTS 293, 33 ILM 1514 (entered into force 24 October 1996) (1994 Nuclear Safety Convention); International Convention for the Prevention of Pollution from Ships (MARPOL), as amended by the Protocol relating to the 1973 International Convention on the Prevention of Pollution from Ships, adopted 17 February 1978, 1340 UNTS 61 (entered into force 2 October 1983) and annexes thereto (MARPOL 73/78).

[54] See 2012 amendments to MARPOL Annex VI making the Energy Efficiency Design Index (EEDI) and the Ship Energy Efficiency Management Plan (SEEMP) mandatory subject to certain conditions. International Maritime Organisation, Marine Environment Protection Committee, *Amendments to the Annex of the Protocol of 1997 to Amend the International Convention for the Prevention of Pollution from Ships, 1973, as modified by the Protocol of 1978 Relating Thereto: (Designation of the United States Caribbean Sea Emission Control Area and exemption of certain ships operating in the North American Emission Control Area and the United States Caribbean Sea Emission Control Area under regulations 13 and 14 and Appendix VII of MARPOL Annex VI)*, 2011, 62nd Sess, Res MEPC.202(62).

[55] See the arguments of the parties at *Pulp Mills, supra* note 5, 2009 Oral Proceedings, accessed 9 January 2015 at http://www.icj-cij.org/docket/index.php?p1=3&p2=3&case=135&code=au&p3=2.

[56] See Part II below.

In the *Aerial Spraying* case, it might have been possible to compare Colombia's regulation of aerial spraying with the stricter approach taken in other countries. The rules governing the Colombian spray programme are among the world's most lenient. The EU, by contrast, bans aerial spraying for most purposes. Those jurisdictions that do not ban spraying impose restrictions significantly stronger than those adopted by Colombia.[57]

At the same time, and unlike the EU, Colombia was not regulating the application of chemicals for agricultural purposes; it was regulating them for law enforcement purposes in an anti-narcotics campaign.[58] Comparison with operating standards set by other countries for agricultural applications is unlikely to be an appropriate test of due diligence in the very different circumstances faced by Colombia.

So what could reasonably be expected of Colombia? Although Ecuador complained that the old Vietnam War era aircraft initially used to carry out the spraying were inappropriate to the task, Colombia substituted specialist crop-spraying aircraft, so the technology in use by 2010 was arguably more appropriate. Colombia had also adopted rules and measures to control transboundary drift. From Ecuador's perspective, what was deficient in Colombia's approach to regulation was the absence of a 'certain level of vigilance in their enforcement' referred to in *Pulp Mills*,[59] coupled with a refusal to mandate a buffer zone in border areas, although as we saw earlier Colombia had in fact operated such a buffer since 2007.

The crucial problem was thus the failure to enforce the rules Colombia had itself laid down. In this respect, the case resembles those human rights cases in which a violation consists of the State's failure to enforce its own environmental regulations, resulting in pollution which exceeds health standards or compels affected families to abandon their homes.[60] A failure to enforce the law is in many respects the simplest failure of due diligence. Other failings are less easy to prove and expert technical evidence may be required. Moreover, while governments do not normally advertise their failure to control transboundary risks by having no laws

---

[57] See *Aerial Spraying,* Reply of Ecuador, *supra* note 20 at Ch 4, s III.

[58] Pursuant to the UN Convention Against Illicit Trafficking in Narcotic Drugs and Psychotropic Substances, adopted 20 December 1988, 1582 UNTS 95 (entered into force 11 November 1990), Art 14.

[59] See *Pulp Mills*, *supra* note 5 at para 197.

[60] See e.g., *Lopez Ostra v Spain* (1994), 20 EHRR 277; *Guerra v Italy* (1998), 26 EHRR 357; *Fadeyeva v Russia*, No 55723/00, [2007] IV ECHR 376, 45 EHRR 10.

on the subject, it may be far from easy to challenge the adequacy of those laws, or the choice of technology which is alleged to cause the risk. Few countries are as helpful as the US in providing damning evidence of its own misconduct,[61] and Colombia is not one of them.

B. The Precautionary Approach

Would reliance on the precautionary approach have strengthened Ecuador's case? The ICJ accepted in *Pulp Mills* that a precautionary approach may be relevant in the interpretation and application of the provisions of a treaty.[62] In its *Advisory Opinion on Seabed Activities*, ITLOS concluded that the Nodules and Sulphides Regulations required sponsoring States and the International Seabed Authority to apply the precautionary approach, but ITLOS also held that 'the precautionary approach is also an integral part of the general obligation of due diligence of sponsoring States, which is applicable even outside the scope of the Regulations'.[63] This is hardly a controversial proposition. It was accepted by the ILC,[64] and endorsed even by the late Professor Ian Brownlie, otherwise no advocate of international environmental law.[65]

Adopting a precautionary approach means that 'lack of full scientific certainty shall not be used as a reason for postponing cost-effective measures to prevent environmental degradation'.[66] That was the conclusion of ITLOS in its *Advisory Opinion on Seabed Activities*:

---

61 See Gulf Oil Disaster Report, *supra* note 38.

62 *Pulp Mills*, *supra* note 5 at para 164.

63 *Advisory Opinion on Seabed Activities*, *supra* note 6 at para 131.

64 The ILC special rapporteur agreed that the precautionary principle is already included in the principles of prevention and prior authorisation, and in environmental impact assessment, 'and could not be divorced therefrom'. 'Report of the International Law Commission on the work of its fifty-second session' (UN Doc A/55/10) in *Yearbook of the International Law Commission 2000*, vol I, Part One (UN, 1994) at 245 (UNDOCA/CN.4/SER.A/2000).

65 'The point which stands out is that some applications of the [precautionary] principle, which is based on the concept of foreseeable risk to other states, are encompassed within existing concepts of state responsibility.' Ian Brownlie, *Principles of Public International Law,* 6th edn (Oxford: Oxford University Press, 2003) at 276. The point is missing from later editions however.

66 Rio Declaration on Environment and Development, in 'Report of the United Nations Conference on Environment and Development', 12 August 1992, UN Doc A/CONF.151/26 (Vol I), 31 ILM 874 (1992 Rio Declaration), Principle 15.

> The due diligence obligation of the sponsoring States requires them to take all appropriate measures to prevent damage that might result from the activities of contractors that they sponsor. This obligation applies in situations where scientific evidence concerning the scope and potential negative impact of the activity in question is insufficient but where there are plausible indications of potential risks. A sponsoring State would not meet its obligation of due diligence if it disregarded those risks. Such disregard would amount to a failure to comply with the precautionary approach.[67]

In *Aerial Spraying*, Ecuador could not easily prove the harm it alleged. The herbicide was not persistent and traces could not be collected from soil or water samples. The people affected lived in remote areas far from medical facilities. Health effects could not be monitored as they might be in Quito or Bogota. Effects on crops and animals would have to be directly linked to the aerial spraying, but again providing such proof was not easy. There were no dead bodies, or none that could be attributed to the herbicide spray. The unwary might therefore conclude that this is a case of scientific uncertainty. But the function of the precautionary approach in international law is not to compensate for uncertainties or deficiencies in the applicant's evidence. If harm is alleged, it must be proved, to the standard required by international courts, by the party alleging it.[68] The precautionary approach affects none of these propositions, all of which are supported by the contemporary case law.

What the precautionary approach does, as ITLOS makes clear, is to reduce the standard of proof required in order to establish that an activity poses a risk of harm.[69] It cannot answer the question whether harm to people or property in Ecuador has occurred. It can help answer the question whether aerial spraying of herbicides creates a risk of harm and therefore entails an obligation of regulation and control. The evidence required for this purpose is evidence about the toxicity of the herbicide, its known effects on people, animals and plants, and the likelihood of transboundary drift when sprayed. All of this Ecuador could arguably have proved, and to a standard that would have made reliance on the precautionary approach unnecessary.

---

[67] *Advisory Opinion on Seabed Activities*, *supra* note 6 at para 131.

[68] See *Pulp Mills*, *supra* note 5 at 14, paras 162–168, and Separate Opinion of Judge Greenwood, paras 25–26; *Application of the Convention on the Prevention and Punishment of the Crime of Genocide (Bosnia and Herzegovina v Serbia and Montenegro)*, [2007] ICJ Rep 43 at 75, para 204; *Military and Paramilitary Activities in and against Nicaragua (Nicaragua v United States of America)*, [1984] ICJ Rep 392 at 437, para 101.

[69] *Advisory Opinion on Seabed Activities*, *supra* note 6 at para 131.

Another possible argument is that 'precautionary measures' should have been taken by Colombia to avoid a known risk even if the evidence was uncertain. But here too, it seems otiose to engage 'precaution' as a concept when 'prevention' and the normal obligation of due diligence will serve the same function equally well.

Ecuador did argue that taking into account 'the size of the operation; its location, special climate conditions, [and] materials used in the activity',[70] spraying should have been banned in a buffer zone adjacent to the border. Buffer zones are required by manufacturers' labelling and were recommended by Colombia's own environment ministry as a means of protecting people and crops.[71] Such a measure was well within Colombia's power, required no additional technology or expense, and would have been effective to limit aerial drift of herbicide pollution onto the territory of Ecuador. Colombia seemed to agree: the problem disappeared in 2007 when it stopped spraying in border areas.

But do we need the precautionary approach or the concept of precautionary measures to get to this conclusion? Not really. My conclusion is that the cases show that the precautionary approach contributes little of utility to general international law.[72] Indeed, it can be a double-edged sword, relied on mainly by States unable to prove their case. Advertising the weakness of their case is not usually a good litigating tactic. Like *ius cogens*, the precautionary approach is a car best left in the garage unless it really will help.[73] Too often it seems merely a rhetorical flourish, or a desperate last resort.

C. EIA

In *Pulp Mills*, the ICJ found that 'due diligence, and the duty of vigilance and prevention which it implies, would not be considered to have been exercised, if a party planning works liable to affect the régime of the river or the quality of its waters did not undertake an environmental impact

---

[70] Draft Articles on Prevention, *supra* note 22 at 394, para 11.

[71] See *Aerial Spraying*, Reply of Ecuador, *supra* note 20 at paras 4.29–4.31.

[72] The only case in which any international tribunal has ostensibly applied the precautionary approach is *Southern Bluefin Tuna Cases (New Zealand v Japan; Australia v Japan)*, Provisional Measures, Order of 27 August 1999, ITLOS Rep 1999 at paras 77–79.

[73] I am indebted to Professor Brownlie for this characterisation of *ius cogens*.

assessment on the potential effects of such works'.[74] The Court confirmed that in appropriate circumstances, an EIA must be carried out *prior to the implementation* of a project that is likely to cause significant transboundary harm.[75] Moreover, it also held that 'once operations have started and, where necessary, throughout the life of the project, continuous monitoring of its effects on the environment shall be undertaken'.[76]

In the *Aerial Spraying* case, Colombia maintained that its international legal obligation to conduct an EIA was no more stringent than its own domestic law,[77] which allegedly exempted the aerial spraying programme from EIA.[78] But this, in my view, is a misreading of *Pulp Mills*. The ICJ held in 2010 that:

> it is for each State to determine in its domestic legislation or in the authorization process for the project, the specific content of the environmental impact assessment required in each case, having regard to the nature and magnitude of the proposed development and its likely adverse impact on the environment as well as to the need to exercise due diligence in conducting such an assessment.[79]

This paragraph makes two important points. First, an EIA need not be specifically required by law, but can be required as part of the authorisation or permitting process. What matters is that some means should exist to ensure that an EIA is carried out. Thus in *Aerial Spraying*, even if Colombia were correct to say that no EIA was required by the 1993 Law on the Environment, it was still the case that before authorising the spray programme the Environment Ministry had the power to require that an assessment of possible transboundary impacts be carried out. It was debarred from using that power for reasons of state,[80] but the means to do so existed, and they would exist in most States even in the absence of a law on EIA.

---

[74] *Pulp Mills*, *supra* note 5 at para 204; see also *Advisory Opinion on Seabed Activities*, *supra* note 6 at para 147.

[75] *Ibid* at para 205.

[76] *Ibid*.

[77] Colombia, Law 99 of 1993 on the Environment; see *Aerial Spraying*, Pleadings, *supra* note 12, 'Colombia Rejoinder', at para 4.81 (*Aerial Spraying*, Colombia Rejoinder).

[78] *Aerial Spraying*, Colombia Rejoinder, *ibid* at paras 4.18–4.24.

[79] *Pulp Mills*, *supra* note 5 at para 205.

[80] See *Aerial Spraying*, Reply of Ecuador, *supra* note 20, Ch 4; *Aerial Spraying*, Colombia Rejoinder, *supra* note 77, Ch 4.

Second, while the 'specific content' of each EIA is for the State to determine, there must be an EIA and it must have regard to 'the nature and magnitude of the proposed development and its likely adverse impact on the environment'.[81] The Court is not here saying that the content of an EIA is for the State to decide in its sole discretion. On the contrary, it is reflecting the views of the ILC and the arguments of the parties based on that commentary. The ILC 2001 Commentary to Article 7 of the Draft Articles on Prevention of Transboundary Harm from Hazardous Activities contains the following explanation.

> The specifics of what ought to be the content of assessment is left to the domestic laws of the State conducting such assessment. [*fn omitted*] For the purposes of article 7, however, such an assessment *should contain an evaluation of the possible transboundary harmful impact of the activity*. In order for the States likely to be affected to evaluate the risk to which they might be exposed, they need to know what possible harmful effects that activity might have on them ... The assessment *should include the effects of the activity not only on persons and property, but also on the environment of other States*. The importance of the protection of the environment, independently of any harm to individual human beings or property is clearly recognized.[82]

It is apparent from the ILC commentary that whatever else may be required by national law, international law requires at a minimum that an EIA assess possible effects on people, property and the environment of other States likely to be affected. If national authorities do not ensure that such an assessment is carried out, for whatever reason and on whatever legal basis, there is inevitably a breach of this obligation.

In that eventuality, we can observe from *Pulp Mills*, *Land Reclamation* and *Southern Bluefin Tuna* that, in the absence of any inquiry process comparable to the 1991 Espoo Convention, provisional measures applications to international courts may be the best remedy available to a potentially affected State seeking to enforce the obligation to carry out a transboundary EIA.[83] In *Pulp Mills*, Argentina failed in its attempt to secure provisional measures from the ICJ once Uruguay showed that it had carried

---

[81] *Pulp Mills*, *supra* note 5 at para 205.

[82] Draft Articles on Prevention, *supra* note 22, Commentary to Art 7, at 402–5 (emphasis added). Article 7 provides: 'Any decision in respect of the authorization of an activity within the scope of the present articles shall, in particular, be based on an assessment of the possible transboundary harm caused by that activity, including any environmental impact assessment.'

[83] *Southern Bluefin Tuna Cases*, *supra* note 72 at para 79; *Case concerning Land Reclamation by Singapore in and around the Straits of Johor (Malaysia v*

out an EIA,[84] but in *Land Reclamation*, ITLOS ordered the parties jointly to assess the risks and effects of the proposed works, while in *Southern Bluefin Tuna*, the effect of its order was that catch quotas could only be increased by agreement after further studies of the state of the stock.

The outcome of these cases suggests that if an EIA has not been undertaken and there is some evidence of a risk of significant harm to another State – even if the risk is uncertain and the potential harm not necessarily irreparable – an order requiring the parties to cooperate in prior assessment is likely to result even at the provisional measures stage.

D. Transboundary Application of Human Rights Law

The third element of Ecuador's case was the alleged violation by Colombia of the human rights of Ecuadoreans affected by the spraying. In this context, Ecuador argued that the spraying of toxic herbicides along the border had caused or would risk causing multiple types of harm within Ecuadorian territory:

- effects on the health of the population;
- damage to agricultural crops and effects on farm animals;
- effects on forests, water supplies and the environment;
- displacement of population in certain villages; and
- effects on the cultural life of indigenous peoples.[85]

At the most basic level, Ecuador was using human rights law to help establish that harm of the kind identified by the ILC was likely to result from any resumption of herbicide spraying: '[t]he harm must lead to a real detrimental effect on matters such as, for example, human health, industry, property, environment or agriculture in other States'.[86] The ILC

Singapore), Provisional Measures, Order of 8 October 2003, ITLOS Case No 12 (*Land Reclamation)* at para 96.

[84] *Pulp Mills*, *supra* note 5, Order of 13 July 2006, [2006] ICJ Rep 113.

[85] See *Aerial Spraying*, Pleadings, *supra* note 12, 'Application of Ecuador', accessed 9 January 2015 at http://www.icj-cij.org/docket/files/138/14474.pdf (*Aerial Spraying,* Application of Ecuador). See also Human Rights Council, *Report of the Special Rapporteur on the Situation of Human Rights and Fundamental Freedoms of Indigenous People, Rodolfo Stavenhagen, Mission to Ecuador*, UN Doc A/HRC/4/32/Add.2 (2006); Human Rights Council, *Report of the Special Rapporteur on the Right of Everyone to the Enjoyment of the Highest Attainable Standard of Physical and Mental Health: Preliminary Note on Mission to Ecuador and Colombia*, UN Doc A/HRC/7/11/Add.3 (2007).

[86] Draft Articles on Prevention, *supra* note 22 at 387, para 4.

commentary notes that 'significant' harm need not be serious or substantial but must be 'more than trivial'.[87] Ecuador's argument was that there is harm to persons when there are significant effects on human health or on private life, that there is harm to property when there is significant harm to agricultural crops or domestic livestock, and that there is harm to private life when villages or homes have to be abandoned or when forest resources can no longer be utilised for subsistence lifestyles.[88]

We know from existing precedents that environmental harm includes harm to biological resources such as forests, flora and fauna, fish stocks, ecosystems, and the availability of unpolluted fresh water supplies.[89] But in no previous case has the applicant State sought to define harm to people and property by reference to human rights precedents. Yet this is an obvious analogy. The main focus of the human rights case law dealing with pollution has been the rights to life, private life, health, water and property.[90] Pollution at a level which would constitute a violation of any

---

[87] *Ibid* at paras 4–7.

[88] See *Aerial Spraying,* Application of Ecuador, *supra* note 85.

[89] UN Compensation Commission Governing Council, *Report and Recommendations Concerning 3rd Instalment of "F4" Claims*, UN Doc S/AC.26/2003/31 (2003) at paras 47–48; UN Compensation Commission Governing Council, *Report and Recommendations Concerning 5th Instalment of "F4" Claims*, UN Doc S/AC.26/2005/10 (2005) at paras 4–58, 82; Convention on Biological Diversity, adopted 5 June 1992, 1760 UNTS 79, 31 ILM 818 (entered into force 29 December 1993), Art 3; Convention on the Transboundary Effects of Industrial Accidents, adopted 17 March 1992, 2105 UNTS 457 (entered into force 19 April 2000) (1992 Convention on Industrial Accidents), Art 1(c); Convention on the Protection and Use of Transboundary Watercourses and International Lakes, adopted 17 March 1992, 1963 UNTS 269 (entered into force 6 October 1996), Art 1(2); Convention on Civil Liability for Damage Resulting from Activities Dangerous to the Environment, adopted 21 June 1993, 32 ILM 1228 (not yet entered into force), Art 2(7) and (10); United Nations Convention on the Law of the Sea, adopted 10 December 1982, 1833 UNTS 397 (entered into force 16 November 1994), Arts 145, 194(5); Convention on the Law of the Non-navigational Uses of International Watercourses, adopted 21 May 1997, 36 ILM 700 (entered into force 17 August 2014) (1997 Convention on Watercourses), Art 20.

[90] *Öçkan and others v Turkey*, No 46771/99, [2006] at paras 14, 22 and 50; *Ivan Atanasov v Bulgaria*, No 12853/03, [2010] at para 75; *Dubetska and others v Ukraine*, No 30499/0,3 [2011] at paras 106–108; *Case of López Ostra v Spain* (1995) 20 EHRR 277; *Guerra and ors v Italy* (1998) 26 EHRR 357; *Fadeyeva v Russian Federation*, No 55723/00, [2007] 45 EHRR 10; *Öneryildiz v Turkey*, No 48939/99, [2005] 41 EHRR 20; *Taşkin and ors v Turkey*, No 46117/99, [2006] 42 EHRR 50 at paras 113–119; *Tătar v Romania*, No 67021/01, [2009] at para 88; Case of *Budayeva and Others v Russia*, No 15339/02, 21166/02, 20058/02,

of these rights will necessarily constitute 'significant harm' for the purpose of establishing that an obligation of due diligence applies and that preventive measures should have been taken by the State.

In many of these cases, the 'harm' is simply the risk implicit in living with pollution at levels that reduce air or water quality below accepted standards for human health. Thus in *Dubetska v Ukraine*, the European Court of Human Rights accepted that it was unable to establish quantifiable harm to the applicants, but it nevertheless agreed that living in an area marked by pollution in excess of minimum standards meant that the applicants had been unnecessarily exposed to increased health risks.[91] The same is true at the interstate level in cases like *Pulp Mills*, where compliance with water quality standards became the main determinant of whether there was any risk of significant harm to the river.

In the *Aerial Spraying* case, Ecuador also made use of human rights law in a more ambitious and arguably more questionable way. Its key argument was that the human rights obligations of the parties applied equally to transboundary pollution and to pollution whose effects were purely domestic. This raises the question whether human rights obligations have extraterritorial effect. There are two arguments against this. The first is that the problem is properly addressed by international law at an interstate level, not at the level of human rights law.[92] On this view the relevant *lex specialis* is the customary international law on prevention and control of transboundary harm discussed above, not international human rights law.

However, a more nuanced approach to such arguments is evident in the case law, and it is far from clear that the *lex specialis* principle operates in this way.[93] A mutually exclusive relationship between human rights

---

11673/02 and 15343/02, [2008]; *Social and Economic Rights Action Center and the Center for Economic and Social Rights v Nigeria* (2001), African Commission on Human and Peoples' Rights, Communication 155/96; *Maya indigenous community of the Toledo District v Belize* (2004) Inter-American Court on Human Rights (IACHR) No 12.053, *Annual Report of the Inter-American Court of Human Rights*: 2004, OEA/Ser.L/V/II.122 Doc. 5 rev. 1.

91 *Dubetska v Ukraine*, *supra* note 90 at paras 106–108. See also *Fadeyeva v Russia*, *supra* note 90.

92 *Aerial Spraying*, Counter-Memorial of Colombia, *supra* note 19, Ch 9, para 9.7; *Aerial Spraying*, Reply of Ecuador, *supra* note 20, Ch 7, paras 7.80–7.84.

93 See *Legality of the Threat or Use of Nuclear Weapons*, Advisory Opinion, [1996] ICJ Rep 226 at paras 25–34; Ian McTaggart Sinclair, *Vienna Convention on the Law of Treaties* (Manchester: Manchester University Press, 1982) at 96; Joost Pauwelyn, *Conflict of Norms in International Law* (Cambridge: Cambridge

law and general international law on transboundary and global environmental protection is neither consistent with the evolution of international environmental law as a whole nor with contemporary developments in international human rights law, which do in some circumstances recognise extraterritorial application.

A more compelling point is that the extraterritorial application of human rights law has normally arisen in the context of occupied territory or cross-border activities by State agents.[94] The essential principle of these cases is that the State has exercised jurisdiction or control over persons or territory abroad. These precedents bear little resemblance to transboundary pollution. Although in transboundary pollution cases, the State does exercise jurisdiction and control over the polluting activity located within its territory, it does not follow that it also exercises jurisdiction or control over affected persons or territory abroad in the manner required by the case law. Here too, however, there are counter

University Press, 2003) at 385–416; International Law Commission, *Fragmentation of International Law: Difficulties Arising from the Diversification and Expansion of International Law: Report of the Study Group of the International Law Commission – Finalized by Martti Koskenniemi*, ILC Doc A/CN.4/L.682, UNGAOR, 61st Sess, Supp No 10, UN Doc A/61/10 (2006) at paras 56–122 (ILC Report).

[94] *Legal Consequences of the Construction of a Wall in the Occupied Palestinian Territory*, Advisory Opinion, [2004] ICJ Rep 136 at paras 109, 112; *Application of the International Convention on the Elimination of All Forms of Racial Discrimination (Georgia v Russian Federation)*, Provisional Measures, Order of 15 October 2008, [2008] ICJ Rep 353 at para 109; *Ecuador v Colombia* (Admissibility) (2010) IACHR, No 112/10 at paras 89–100, Annual Report of the Inter-American Commission on Human Rights: 2010, OEA/Ser.L/V/II.140; *Armando Alejandre Jr., Carlos Costa, Mario de la Pena y Pablo Morales v Republica de Cuba* (1999) IACHR, No 86/99 at para 23, Annual Report of the Inter-American Commission on Human Rights: 1999, OEA/Ser.L/V/II.106 Doc. 3 rev.; *Coard et al v United States* (1999) IACHR, No 109/99, at para 37 Annual Report of the Inter-American Commission on Human Rights: 1999, OE/Ser.L/V/II.106 Doc. 3. Rev. at 1283 (1999); *Al-Skeini and Others v United Kingdom*, No 55721/07, [2011] 53 EHRR 18 at paras 130–142; *Öcalan v Turkey*, No 46221/99, [2005] 41 EHRR 985 at para 91; *Ilaşcu and ors and Romania (Intervening) v Moldova and Russian Federation*, No 48787/99, [2005] 40 EHRR 46 at paras 310–319, 376–394; *Issa et al v Turkey*, No 31821/96, [2004] 41 EHRR 567 at para 71; *Cyprus v Turkey*, No 25781/94, [2002] 35 EHRR 30 at para 78. See also Human Rights Committee, General Comment No 31, Nature of the General Legal Obligation on States Parties to the Covenant, UN Doc CCPR/C/21/Rev.1/Add.13 (2004) at para 10.

arguments, although they cannot easily be summarised in the form required for this chapter.[95] But the main points are as follows.

First, where it is possible to take effective measures to prevent or mitigate transboundary harm to human rights, the argument that the State has no obligation to do so merely because the harm is extraterritorial is not a compelling one. On the contrary, in keeping with the principle of non-discrimination, the environmental impact of activities in one country on the right to life, private life or property in other countries should be taken into account and given due weight in the decision-making process.[96] There is no principled basis for suggesting that the outcome of cases such as *Hatton*[97] should depend on whether those affected by excessive noise or air pollution are in the same country, or in other countries.[98] It seems entirely consistent with the case law and the 'living instrument' conception of human rights treaties to conclude that a State Party must balance the rights of persons in other States against its own economic benefit, and must adopt and enforce environmental protection laws for their benefit, as well as for the protection of its own population. In this context the State is most likely to violate the human rights of those affected extraterritorially if it does not permit them equal access to

---

[95] But they are developed in Alan Boyle, 'Human Rights and the Environment: Where Next?' (2012) 23(3) *European Journal of International Law* 613.

[96] As they would have to be in transboundary environmental impact assessments, see Convention on Environmental Impact Assessment in a Transboundary Context, adopted 25 February 1991, 30 ILM 1461 (1991) (entered into force 10 September 1997) (1991 Espoo Convention), Art 3(8).

[97] *Hatton and Others v United Kingdom*, No 36022/97, [2003] 37 EHRR 611.

[98] International Law Association, Committee on Transnational Enforcement of Environmental Law, *Final Report of 72nd Conference (Toronto Conference)* at 4, Rule 2 and commentary; and the principle of non-discrimination applied to transboundary pollution in the Organisation for Economic Co-operation and Development (OECD) Council Recommendations C (74) 224 (1974); C(76) 55 (1976); C (77) 28 (1977); C (78) 77 (1978); C (79) 116 (1979), reproduced in OECD, *OECD and the Environment* (Paris, 1986); Henri Smets, 'Le principe de Non-discrimination en Matière de Protection de l'Environnement' (2000) 2 *Revue Europeen de Droit de l'Environnement* 1; Patricia Birnie, Alan Boyle and Catherine Redgwell, *International Law and the Environment* (Oxford: Oxford University Press, 2009) 304–11; John Knox, 'Myth and Reality of Transboundary Environmental Impact Assessment' (2002) 96 *American Journal of International Law* 291.

environmental information and participation in EIA permitting procedures, or if it denies access to adequate and effective remedies within its own legal system.[99]

Second, while it is less plausible to say that the polluting State 'controls' the territory of the State affected by pollution,[100] it is entirely plausible to conclude that the victims of transboundary pollution fall within the 'jurisdiction' of the polluting State – in the most straightforward sense of legal jurisdiction. The jurisdiction of national courts to hear cases involving transboundary harm to extraterritorial plaintiffs is well-established in private international law and in environmental liability conventions.[101] The principle of non-discrimination has also been

---

[99] See the ILC's draft principles on the allocation of loss in the case of transboundary harm arising out of hazardous activities. ILC Report, *supra* note 93 at paras 51–67 (Principles on Allocation of Loss). Principle 6(1) sets out the core obligation: 'States shall provide their domestic judicial and administrative bodies with the necessary jurisdiction and competence and ensure that these bodies have prompt, adequate and effective remedies available in the event of transboundary damage caused by hazardous activities located within their territory or otherwise under their jurisdiction or control.' See also UNECE Convention on Access to Information, Public Participation in Decision-making and Access to Justice in Environmental Matters, adopted 25 June 1998, 2161 UNTS 447, 38 ILM 517 (entered into force 30 October 2001), Arts 3(9) and 9(4) (1998 Aarhus Convention).

[100] But significant transboundary pollution is arguably a violation of the permanent sovereignty of a State (and its people) over their own natural resources, and in a serious case might amount to a *de facto* expropriation. See Draft Articles on Prevention, *supra* note 22, Preamble; *Social and Economic Rights Action Center (SERAC) v Nigeria* (2002), African Commission on Human and Peoples' Rights, Comm. 155/96 at para 55, Doc ACHPR/COMM/A044/1.

[101] EC, *Council Regulation (EC) No 44/2001 of 22 December 2000 on jurisdiction and the recognition and enforcement of judgments in civil and commercial matters* [2001] OJ L 12/1 at Art 5; Protocol on Civil Liability and Compensation for Damage Caused by the Transboundary Effects of Industrial Accidents on Transboundary Waters, adopted 21 May 2003, 44 ILM 258 (not yet in force), Art 13; Convention on Civil Liability for Damage to the Environment Resulting from Activities Dangerous to the Environment, *supra* note 89, Art 19; Protocol to Amend the Vienna Convention on Civil Liability for Nuclear Damage, adopted 12 September 1997, 36 ILM 1462 (entered into force 4 October 2003), Art XI; Protocol to amend the Convention on Third Party Liability in the Field of Nuclear Energy of 29 July 1960, as amended by the Additional Protocol of 28 January 1964 and by the Protocol of 16 November 1982, adopted 12 February 2004 (not yet in force), Art 13.

adopted by the ILC in its articles on transboundary harm,[102] by the United Nations Economic Commission for Europe (UNECE) in its environmental conventions,[103] and by the Southern Common Market (MERCOSUR).[104] The Inter-American Court of Human Rights (IaCtHR) has held that 'the fundamental principle of equality and non-discrimination constitute a part of general international law'.[105]

There is little point requiring that national remedies be made available to transboundary claimants if they cannot also resort to international or regional human rights law when necessary to compel the polluting State to enforce its own court orders or laws, or to assess and take adequate account of the extraterritorially harmful effects of activities which it authorises and regulates. That is exactly how domestic claimants have successfully used human rights law in environmental cases.[106]

Thus, although it may be questionable how far victims of transboundary air pollution can bring human rights proceedings against a foreign

---

[102] Article 15 prohibits discrimination based on nationality, residence, or place of injury in granting access to judicial or other procedures, or compensation, in cases of significant transboundary harm. See Draft Articles on Prevention, *supra* note 22 at 427–9. See also Principles on Allocation of Loss, *supra* note 99, Principle 8(2); 1997 Convention on Watercourses, *supra* note 89, Art 32.

[103] In addition to the Aarhus Convention, it is listed in the preamble of the 1992 Convention on the Transboundary Effects of Industrial Accidents among 'principles of international law and custom'. 1992 Convention on Industrial Accidents, *supra* note 89, Art 9; see also 1991 Espoo Convention, *supra* note 96, Art 2(6).

[104] Protocol on Jurisdictional Cooperation and Assistance in Civil, Commercial, Labor, and Administrative Matters, adopted 27 June 1992, 2145 UNTS 421 (entered into force 17 March 1996), Ch III, Art 3. The position in NAFTA is less clear. Transboundary plaintiffs appear to have equality of standing under some US environmental statutes. See Trans-Alaska Pipeline Authorization Act, USC 43 §1653(c)(1) (1973) (which allows 'any person or entity, public or private, *including those resident in Canada*' to invoke the Act's liability provisions). North American Agreement on Environmental Co-operation, adopted 14 September 1993, 32 ILM 1480 (entered into force 1 January 1994), Art 6 (which provides for 'interested persons' to have access to legal remedies for violation of environmental laws, may also apply to transboundary litigants). See generally Shi-Ling Hsu and Austen Parrish, 'Litigating Canada-U.S. Transboundary Harm: International Lawmaking and the Threat of Extraterritorial Reciprocity' (2007) 48 *Virginia Journal of International Law* 1.

[105] See *Juridical Condition and the Rights of Undocumented Migrants* (2003), Advisory Opinion OC-18/03, Inter-American Court of Human Rights (Ser A) No 18, at para 83.

[106] *Supra* note 90.

State based on substantive violation of their right to life, private life or property, it is easier to see that in cases where those rights are affected, they will be entitled to transboundary access to environmental information, participation in EIA and other authorisation procedures, and equal access to justice in national courts.

In Europe those rights for transboundary claimants are already recognised by the Aarhus Convention on Access to Information, Public Participation in Decision-Making and Access to Justice in Environmental Matters.[107] In Asia, regional human rights law is less well developed, although there is now an ASEAN Human Rights Declaration which provides for the right to a safe, clean and sustainable environment as a human right,[108] and national law in some Asian jurisdictions is very sophisticated in the application of human rights law in environmental cases.[109] In most legal systems, foreign polluters can also be sued before national courts in jurisdictions affected by the pollution, regardless of whether there has been any breach of international law or human rights.[110] Whether this may be a useful remedy will depend on the particular legal system, and it is difficult to generalise about Asia as a whole.

E. Conclusions

Whatever the merits of Ecuador's arguments, the *Aerial Spraying* case demonstrates that interstate litigation can work effectively to end transboundary pollution. The settlement also brought to an end a dispute that had upset relations between two governments who were normally friendly and cooperative. It shows the kind of arguments that can be made, and the evidence needed to make those arguments effectively.

---

107 Article 3(9) requires that 'the public shall have access to information, have the possibility to participate in decision-making and have access to justice in environmental matters without discrimination as to citizenship, nationality or domicile and, in the case of a legal person, without discrimination as to where it has its registered seat or an effective centre of its activities'. See UNECE, *The Aarhus Convention-An Implementation Guide* (UN, 2000) at 41.

108 ASEAN Human Rights Declaration, adopted 18 November 2012 at para 28.

109 See Ben Boer and Alan Boyle, *Human Rights and the Environment*, Report of the 13th Informal Asia-Europe Meeting (ASM) Seminar on Human Rights (Singapore: ASEM, 2014).

110 See generally Campbell McLachlan and Peter Nygh (eds), *Transnational Tort Litigation: Jurisdictional Principles* (Oxford: Oxford University Press, 1996), especially Chs 1, 4 and 12.

Litigation is not an easy option. It is not cheap, it takes time. It requires good lawyers, and sound expert evidence. But if it works, the expense and the effort required will be worthwhile.

The alternative is, of course, to negotiate, either bilaterally or multilaterally. That brings me more briefly to the second paradigm, multilateral regulatory regimes.

III. MULTILATERAL REGULATORY REGIMES FOR TRANSBOUNDARY AIR POLLUTION[111]

LRTAP remains the only major regional multilateral agreement devoted to the regulation and control of transboundary air pollution.[112] It treats the European air mass as a shared problem requiring co-ordination of pollution control measures and common emission standards. Its purpose is thus to prevent, reduce and control transboundary air pollution, both from new and existing sources. The treaty came into force in 1983, and as of May 2014 has over 50 northern hemisphere parties, including all the major polluter States in Europe.[113] Canada and the United States have also ratified, although the US has stayed out of the SO_2 protocols.[114]

No concrete commitments to specific reductions in air pollution are contained in the treaty itself. Thus there is only an obligation to 'endeavour to limit' and 'as far as possible, gradually reduce and prevent' air pollution.[115] Despite its evident weaknesses, the Geneva Convention's real value is that it has provided a successful framework for cooperation and the development of further measures of pollution control. Given adequate consensus among the parties, adoption of stronger and more effective measures has been possible within this framework. For this reason, the regular conference of the parties, serviced by an ongoing secretariat is, as in other treaty regimes, of particular importance.

---

111 This section is based on Birnie, Boyle and Redgwell, *supra* note 98, Ch 6.

112 LRTAP, *supra* note 2; see UNECE, *Handbook for the 1979 Convention on Long-Range Transboundary Air Pollution and its Protocols* (UN, 2004). See also Hans Christian Bugge's chapter in this book (Chapter 10).

113 *Status of the Convention on Long-Range Transboundary Air Pollution*, accessed 9 January 2015 at https://treaties.un.org/pages/ViewDetails.aspx?src=TREATY&mtdsg_no=XXVII-1&chapter=27&lang=en.

114 But see US-Canada Air Quality Agreement, *supra* note 8.

115 LRTAP, *supra* note 2, Art 2.

What has made LRTAP a success is the development of a system of protocols covering the main sources of air pollution in Europe.[116] Of the most important is the Protocol to the 1979 Convention on Long-Range Transboundary Air Pollution on Further Reduction of Sulphur Emissions (Protocol).[117] That Protocol acknowledges the need for precautionary measures to prevent transboundary air pollution from continuing to cause damage to forests, natural resources, and the sensitive Arctic environment.[118] Based on a 'critical loads' approach, its objective is to reduce sulphur deposition below the level at which there would be significant damage to the areas where deposition is likely to occur.[119]

To this end, Article 2 of the Protocol requires parties to control and reduce sulphur emissions in order to protect human health and the environment and to ensure 'as far as possible' and 'without entailing excessive cost' that they do not exceed the critical loads specified in Annex I. These loads are based on mapping of actual SO_2 deposition and sources. No date is set for reaching this ambitious objective, but each party is given minimum emissions targets to meet within timescales that vary between 2000, 2005 and 2010.[120] Instead of the single flat rate reduction for all parties used in the earlier protocol, the newer one sets differentiated emissions targets for each party, which range from an 80 per cent reduction by 2010 for Germany to a 49 per cent increase for Greece.[121] The overall SO_2 emissions reduction for all parties combined is 50.8 per cent, and the effect of this should be to reduce the amount by which deposition exceeds the critical loads by at least 60 per cent.[122]

These emissions reductions figures are not derived solely from scientific advice, although the science is important; they also represent the outcome of a politically negotiated compromise which recognises that for

---

[116] There are now eight. For details, see UNECE accessed 9 January 2015 at http://www.unece.org/env/lrtap/status/lrtap_s.html.

[117] Protocol to the 1979 Convention on Long-Range Transboundary Air Pollution on Further Reduction of Sulphur Emissions, adopted 14 June 1994, 2030 UNTS 122 (entered into force 5 August 1998) (Protocol on Further Reduction of Sulphur Emissions); see Robin Churchill, Gabriela Kütting and Lynda M Warren, 'The 1994 UN ECE Sulphur Protocol' (1995) 7 *Journal of Environmental Law* 169; Leila Jhaveri, Joyeeta Gupta and Hermann Ott, *The LRTAP Convention/2nd Sulphur Protocol: Possible Lessons for the Climate Convention* (London: FIELD, 1998).

[118] Protocol on Further Reduction of Sulphur Emissions, *ibid*, Preamble.

[119] *Ibid.*

[120] *Ibid*, Annex II.

[121] *Ibid.*

[122] *Ibid.*

some States, such as Germany and the United Kingdom, the critical loads approach is too demanding for full implementation.[123] Nevertheless, the Protocol's differentiated commitments are fairer to all parties because they are based on calculations of actual sources and effects and require reductions only to the extent that they are needed. Not only does it address the problem from a more realistic scientific perspective, it is also more effective in requiring the parties to reduce their total emissions, not merely their transboundary fluxes of SO_2.

Here in summary is what can be achieved in a regional setting, notwithstanding major political divisions. It should not be forgotten that in 1979 when LRTAP was negotiated, the Soviet Union was still very much in existence. That did not stop collaboration between Eastern and Western Europe and North America for the purpose of tackling quite serious and economically wasteful air pollution. Multilateral environmental regimes do not require shared political values or economic integration in order to work. They do require political will plus scientific and diplomatic expertise.

Transboundary air pollution in Europe has undoubtedly fallen substantially since 1979, and especially SO_2 pollution. What is less clear is how far this improvement can be attributed directly to implementation of LRTAP's protocols. Any explanation of the reduction in sulphur emissions that has undoubtedly occurred must take account of major industrial changes and the increased use of gas or nuclear power for power generation. Nevertheless, in their reports to the Executive Body, the parties have concurred in viewing LRTAP's impact on air pollution control and air quality management as a positive one.

If transboundary air pollution in Asia is a problem, and the evidence suggests that it is a growing problem,[124] here is one potential model for addressing it that has worked elsewhere. ASEAN already has a number of agreements and other instruments which cover air pollution, including the 1995 Co-operation Plan on Transboundary Pollution,[125] the 1997

---

[123] Churchill, Kütting and Warren, *supra* note 117 at 169; Jorgen Wettestad, 'Science, Politics and Institutional Design: The Case of the Northeast Atlantic Landbased Marine Pollution Regime' (1995) 4 *Journal of Environment and Development* 165.

[124] See North-East Asian Subregional Programme for Environmental Cooperation, *Transboundary Air Pollution Mitigation*, accessed 9 January 2015 at http://neaspec.org/our-work/transboundary-air-pollution-mitigation.

[125] ASEAN Cooperation Plan on Transboundary Pollution, adopted 17 June 1995.

Regional Haze Action Plan,[126] and the 2002 Agreement on Transboundary Haze Pollution.[127] The 2002 agreement has many similarities to the 1979 Geneva Convention: it sets out objectives for reducing air pollution, it creates a conference of the parties with power to take further measures and adopt protocols, it provides for further research. Here is the basis for coordinated action within ASEAN: what it lacks are the protocols and programmes that would give it some bite.[128]

---

126 1997 ASEAN Regional Haze Action Plan, adopted 23 December 1997, accessed 9 January 2015 at http://haze.asean.org/?page_id=213 (ARHAP).

127 2002 ASEAN Agreement on Transboundary Haze Pollution, signed 10 June 2002 (entered into force 25 November 2003) accessed 9 January 2015 at http://agreement. asean.org/media/download/20140119170340.pdf; see also South Asia Co-operative Environment Programme (SACEP), Malé Declaration on Control and Prevention of Air Pollution and its Likely Transboundary Effects for South Asia, adopted 22 April 1998.

128 See Laode M Syarif and Alan Khee-Jin Tan's chapters in this book (Chapters 11 and 12 respectively).

PART III

Cooperative mechanisms for addressing transboundary pollution

10. The principle and duty to cooperate: the case of conventions on transboundary pollution in Europe

Hans Christian Bugge

I. INTRODUCTION

International environmental law contains a number of general obligations for States with the purpose of limiting pollution and other environmental problems. Some obligations are today regarded as customary international law.

One of the most fundamental obligations in this regard relates to transboundary pollution: the duty of States 'to ensure that activities within their jurisdiction or control do not cause damage to the environment of other States or of areas beyond the limits of national jurisdiction'.[1] This is explicitly stated in a number of soft law instruments[2] and multilateral conventions,[3] and has also been recognised by the

[1] Rio Declaration on Environment and Development, in 'Report of the United Nations Conference on Environment and Development', UN Doc A/CONF.151/26 (Vol I), 12 August 1992, Annex I (1992 Rio Declaration), Principle 2.

[2] Most notably, see Principle 21 of the 1972 Stockholm Declaration. Stockholm Declaration on the Human Environment, in 'Report of the United Nations Conference on the Human Environment', UN Doc A/CONF.48/14, at 2 and Corr.1 (1972) (1972 Stockholm Declaration). See also Article 2 of the ILC's 2001 Draft Articles on Prevention of Transboundary Harm from Hazardous Activities. 'International Liability for Injurious Consequences Arising out of Acts Not Prohibited by International Law (Prevention of Transboundary Harm from Hazardous Activities)' in 'Report of the International Law Commission to the General Assembly covering the work of its fifty-third session, with commentaries, 2001' (UN Doc A/56/10), Ch V, in *Yearbook of the International Law Commission 2001*, vol II, Part Two (UN, 2001) (Draft Articles on Prevention).

[3] Prominent examples are the United Nations Convention on the Law of the Sea, adopted 10 December 1982, 1833 UNTS 397 (entered into force 16

International Court of Justice (ICJ).[4] In international texts, this obligation is directly linked to the statement that States have 'the sovereign right to exploit their own resources pursuant to their own environmental and developmental policies' in accordance with the Charter of the United Nations (UN) and the principles of international law.[5] Together, they express the fundamental and closely interlinked principles in international law of territorial integrity and territorial sovereignty.

The general prohibition on damaging the environment of other States or areas beyond the limits of national jurisdiction, known as the 'no-harm' rule, raises a number of questions of interpretation. These are discussed in detail in other chapters in this book and shall not be further elaborated on here. It is, however, important to note that there have been few international court cases between States on the basis of this rule. States that have been victim of transboundary pollution to a very limited extent have used judicial and adversarial means against the source State in order to stop the damage or seek compensation. There may be several reasons for this. An important reason may be found in the difficulty in selecting and deploying an appropriate basis for claim.[6] The 'no-harm' rule is in no way clear; the 'due diligence' criterion requires a broad assessment of various elements. And the burden of proof may be difficult to fulfil. Another reason may be of a political rather than a legal nature: that States simply hesitate to bring other States to courts for political reasons.

A third possible reason is that many of the transboundary pollution problems of today cannot be described simply with the picture of a 'source State' that creates environmental problems in a 'victim State'. The factual situation is more complex. Today, most transboundary pollution problems are regional or even global, with the global problem of climate change as the ultimate example. The pollution problems are the mutual and cumulative effect of numerous sources of emission in many countries, if not all. Most countries are both sources and victims at the same time, although often with great differences with regard to their

---

November 1994) (UNCLOS), Art 194(2); and the United Nations Framework Convention on Climate Change, adopted 9 May 1992, 1771 UNTS 107 (entered into force 21 March 1994) (UNFCCC), Preamble; and Convention on Biological Diversity, adopted 5 June 1992, 1760 UNTS 79 (entered into force 29 December 1993) (CBD), Art 3.

[4] See *Case concerning Pulp Mills on the River Uruguay (Argentina v Uruguay),* [2010] ICJ Rep 14, at para 101.

[5] 1992 Rio Declaration, *supra* note 1, Principle 2.

[6] See James Crawford, *Brownlie's Principles of Public International Law*, 8th edn (Oxford: Oxford University Press, 2008) at 354.

contribution to the problem and the extent to which they are affected, as we see so clearly in climate change.

In this situation, other principles and duties in international law, in particular the general principle and duty of cooperation, can play a more practical role. Regional and global pollution problems cannot be prevented and solved without cooperation and mutual obligations between all States that contribute to the problem. Cooperation is also necessary when transboundary pollution also harms or risks harming common areas and resources outside State national jurisdiction where there is no particular State being affected and the principle of territorial integrity does not work directly.

II. THE PRINCIPLE AND DUTY OF COOPERATION IN INTERNATIONAL ENVIRONMENTAL LAW

The general duty for States to cooperate to solve conflicts and common problems is closely linked to and follows from such general principles of international law as good neighbourliness, peaceful settlement of disputes,[7] reciprocity, and the general principle of good faith. It may even be said to emerge 'from the very rationale for international law'.[8] It is particularly relevant and important in the field of management of natural resources and the environment. Not surprisingly, the duty of States to cooperate is recognised as a general principle of international environmental law.[9]

As such the duty to cooperate supplements – and to some extent also overlaps with – the duty not to harm the environment of other States and areas outside national jurisdiction. The two principles or obligations should be seen in combination. To some extent they overlap, in particular with regard to the procedural obligations within the 'no-harm' rule.[10] Included in both are such obligations as the duty to consult each other

[7] See Antonio Cassese, *International Law*, 2nd edn (Oxford: Oxford University Press, 2005) at 58.

[8] Alexandre Kiss and Dina Shelton, *Guide to International Environmental Law* (Leiden: Martinus Nijhoff Publishers, 2007) at 12.

[9] See Patricia Birnie, Alan Boyle and Catherine Redgwell, *International Law & the Environment*, 3rd edn (Oxford: Oxford University Press, 2009); Ulrich Beyerlin and Thilo Marauhn, *International Environmental Law* (Oxford: Hart Publishing, 2011); Kiss and Shelton, *ibid*; Philippe Sands and Jaqueline Peel, *Principles of International Environmental Law*, 3rd edn (Cambridge: Cambridge University Press, 2012) at 203.

[10] Beyerlin and Marauhn, *ibid* at 44.

with the aim of preventing transboundary harm; the duty to exchange relevant information on possible hazardous activities and risks; and the obligation to notify in case of emergency and to carry out cross-border environmental impact assessments and cross-border participation in decision-making when projects in one country may have significant environmental effects in another country.[11] The international rules that ensure that neighbouring States or States importing hazardous wastes and substances shall receive all necessary information through information exchange, consultation and notification as well as the possibility of banning such imports can also be regarded as reflecting both principles.[12]

Furthermore, there are links also as regards *substantive* measures. A core obligation for States will be to prevent hazardous pollution with transboundary effects and to reduce existing emissions. A prerequisite for a State to take on this type of obligation may be the harmonisation of the national environmental policies of the different States and in particular with regard to principles and emission standards for various types of economic activities. This touches the issues of fairness and due diligence, which are relevant for the application of the 'no-harm' rule.

There are numerous expressions of the general cooperation principle in international environmental law, both in soft law and treaty articles. Sands and Peel state that the obligation to cooperate 'is affirmed in virtually all international environmental agreements of bilateral and regional application, and global instruments'.[13] Principle 24 of the 1972 Stockholm Declaration states that '[i]nternational matters concerning the protection and improvement of the environment should be handled in a cooperative spirit by all countries, big and small, on an equal footing' and that '[c]ooperation through multilateral or bilateral arrangements or other appropriate means is essential to effectively control, prevent, reduce and

---

[11] This topic is discussed in other chapters of this book. An important international instrument in the field is the 1991 Convention on Environmental Impact Assessment in a Transboundary Context, adopted 25 February 1991, 30 ILM 1461 (1991) (entered into force 10 September 1997).

[12] See Basel Convention on the Control of Transboundary Movements of Hazardous Wastes and Their Disposal, adopted 22 March 1989, 1673 UNTS 57 (entered into force 5 May 1992); Rotterdam Convention on the Prior Informed Consent Procedure for Certain Hazardous Chemicals and Pesticides in International Trade, adopted 10 September 1998, 2244 UNTS 337 (entered into force 24 February 2004).

[13] Sands and Peel, *supra* note 9 at 204.

eliminate adverse environmental effects resulting from activities conducted in all spheres, in such a way that due account is taken of the sovereignty and interests of all States'.

The 1992 Rio Declaration reaffirms the duty of cooperation and further provides that:

> In view of the different contributions to global environmental degradation, States have common but differentiated responsibilities. The developed countries acknowledge the responsibility that they bear in the international pursuit of sustainable development in view of the pressures their societies place on the global environment and of the technologies and financial resources they command.

The somewhat vague expression 'on an equal footing' in the 1972 Stockholm Declaration was replaced in the 1992 Rio Declaration by the more explicit principle of common but differentiated responsibilities (CBDR). This reflects the rapidly growing sensitivity to the differences in particular between rich and poor countries with regard to both the historic responsibility for global problems like climate change, and the economic and social possibilities and preconditions for combating environmental problems.[14]

In order to make some aspects of the duty to cooperate more explicit, Principle 9 of the 1992 Rio Declaration provides that '[s]tates should cooperate to strengthen endogenous capacity-building for sustainable development by improving scientific understanding through exchanges of scientific and technological knowledge, and by enhancing the development, adaptation, diffusion and transfer of technologies, including new and innovative technologies'.

In addition, Principle 14 states that '[s]tates should effectively cooperate to discourage or prevent the relocation and transfer to other States of any activities and substances that cause severe environmental degradation or are found to be harmful to human health'.

Between the Stockholm and Rio conferences, important developments took place within the international work on transboundary pollution and other international environmental and natural resource problems. Several far-reaching regional and global conventions were negotiated and adopted, which further highlights the need for cooperation and the willingness of the States to take on commitments to cooperate.

---

[14] See Part IV of this chapter for a more detailed discussion of the principle of common but differentiated responsibilities.

For example, the 1982 UN Convention on the Law of the Sea (UNCLOS) has a special section on global and regional cooperation in its Part XII (Protection and Preservation of the Marine Environment). Article 197 provides that States shall cooperate on global and regional bases, 'directly or through competent international organizations, in formulating and elaborating international rules, standards and recommended practices and procedures consistent with this Convention, for the protection and preservation of the marine environment, taking into account characteristic regional features'. The section goes on to specify the duty to cooperate with regard to notification of imminent or actual damage, contingency plans against pollution, studies, research programmes and exchange of information and data, and scientific criteria for regulation. UNCLOS also lays down the obligation for States to cooperate in the conservation and management of living resources of the high seas[15] and the 1995 Agreement on Straddling and highly migratory fish stocks explicitly implements this duty.[16]

With regard to the protection of the ozone layer, the 1985 Vienna Convention for the Protection of the Ozone Layer also has numerous provisions on the duty to cooperate. With regard to international river management, the 1997 Convention on the Law of Non-navigational Uses of International Watercourses lays down both an obligation not to cause significant harm to other watercourse States[17] and a general obligation to cooperate. Specifically, the Convention provides that '[w]atercourse States shall cooperate on the basis of sovereign equality, territorial integrity, mutual benefit and good faith in order to attain optimal utilization and adequate protection of an international watercourse'.[18]

---

15 UNCLOS, *supra* note 3, Art 118.

16 Agreement for the Implementation of the Provisions of the United Nations Convention on the Law of the Sea of 10 December 1982 relating to the Conservation and Management of Straddling Fish Stocks and Highly Migratory Fish Stocks, adopted 4 August 1995, 2167 UNTS 3 (entered into force 11 December 2001), Arts 5 and 8.

17 Convention on the Law of the Non-navigational Uses of International Watercourses, adopted 21 May 1997, 36 ILM 700 (entered into force 17 August 2014), Art 7.

18 *Ibid*, Art 8(1).

III. PRACTICAL APPLICATION OF THE PRINCIPLE OF COOPERATION: MULTILATERAL ENVIRONMENTAL AGREEMENTS

In reality, to implement the duty to cooperate, States often conclude binding multilateral agreements with common and mutual commitments.[19] In the field of transboundary pollution abatement, the need for binding multilateral agreements is twofold. On the one hand, a solution to the problem requires that measures are taken by all countries that contribute to it. The danger of 'free riders' is evident, and the risk of free riders may discourage the willing and serious State actors from taking on commitments. It is essential to get 'everybody on board'. This may require some flexibility in the obligations of the participating States, as reflected in the CBDR principle. On the other hand, to the extent pollution has to be reduced from economic activities, the issue of pollution abatement costs and market competition becomes very important. A major purpose of much of the European Union (EU) regulations and directives on pollution, for example, is to harmonise pollution policy, principles and standards-setting for economic activities in order to create a 'level playing field' for industry competing in the European (and international) market, and by that to avoid a 'race to the bottom' effect. So, in order to combat transboundary pollution, important purposes of the international cooperation through multilateral agreements are first, to avoid 'free riders' and second, to create 'a level playing field' for economic activities that are exposed to international competition.

Cooperation to combat transboundary pollution requires a 'package' – or 'cluster' – of closely interrelated measures and commitments. This often requires a multilateral agreement, and in many cases, a regional agreement with legal force but which is at the same time flexible and dynamic and adapted to the situation in the participating countries.

Concluding a multilateral agreement with broad or even universal support is, of course, not simple. First, it is necessary to come to an agreement on what the problem is, its causes and what remedial actions are required, or even whether there is a problem at all. Disagreement and difficulties may appear already at this initial stage. Then, the concerned States must agree on the objective of the agreement and the principles on which it shall build, which is not necessarily uncontroversial either. Then,

---

19 As Alexandre Kiss and Dina Shelton observe, the principle of cooperation underlines all treaty obligation. See Kiss and Shelton, *supra* note 8 at 12.

they have to agree on the concrete and specific obligations and commitments, the institutions to set up for the implementation of the agreement, issues such as monitoring, compliance and reporting mechanisms, sanctions in case of breach of commitments, and settlement of disputes rules. It is also most often necessary to agree on research activities and to exchange scientific and technological knowledge and provide mutual support in the work.

The field of international environmental law is an extremely dynamic area. New substances and activities create need for new regulations. New scientific knowledge of causes and effects of pollution problems, including possible damage to nature and hazards to health, makes it necessary to adopt more stringent measures than was first conceived. New technologies may pose new problems but may also create new possibilities in solving the problems. Also, public values and attitudes, economic conditions, and political priorities shift. These factors require a flexible legal approach and gradual development of international instruments.

Therefore, it is not surprising that the practical approach in this field has been the adoption of framework agreements that lay down the main objectives and principles and establish the formal framework and institutions for the implementation of the convention. Once the framework agreements are adopted and in force, the more specific commitments for the individual States Parties are negotiated, and often in several stages. A step-by-step development is needed in order to succeed.

Another development is noteworthy. Increasingly, the negotiated commitments differ between the parties. This may be a political necessity in order to get everybody on board. It is, as mentioned, also a result of the growing sensitivity to differences in economic and social situations and technological levels and thus is reasonable from a fairness point of view. But as we shall see, there may also be other reasons for differentiations, reasons that have to do with variations in natural conditions, and considerations of economic efficiency and cost-effectiveness.

IV. DUTY OF COOPERATION AND DIFFERENTIATION OF OBLIGATIONS

For other areas of international law, the duty of cooperation through multilateral conventions is based on the principle of equality for all parties to the convention with regard to their respective rights and obligations. In the field of multilateral environmental agreements, however, this is not so. Gradually, since the 1972 Stockholm Conference, international environmental treaty-making has changed from providing for identical treatment of all contracting States to providing for differential treatment of parties. There has been a recognition that the countries of the world are in different situations with regard to social and economic development, technological levels and natural conditions, which justify differences in their international commitments. The purpose is to bring about effective – rather than formal – equality among *de facto* unequal States and to ensure the participation of all countries in international environmental agreements.[20] As a matter of fact, differentiation is often a prerequisite for participation of all countries in an agreement, which usually is highly desirable and even necessary in order to solve the common environmental problem in question.

The CBDR principle is most clearly provided for in Principle 7 of the 1992 Rio Declaration.[21] It was built on a development that had started long before. Article 207 of UNCLOS recognises the need to take into account the 'economic capacity of developing States and their need for economic development'.[22] The preamble to the 1985 Vienna Convention for the Protection of the Ozone Layer states that 'the circumstances and particular requirements of developing countries' shall be taken into account.[23] Furthermore, Article 2 states that parties shall cooperate 'in accordance with the means at their disposal and their capabilities', and its 1987 Montreal Protocol gives a general grace period of ten years for developing countries to start activities and obligations laid down in the protocol.[24] Increasingly, we find explicit references to the CBDR principle, or to 'special needs of developing countries' and countries with economies in transition.

The CBDR represents an important qualification of the duty to cooperate, which is generally acknowledged as a just and necessary

---

[20] See Christina Voigt, 'Equity in the 2015 Climate Agreement: Lessons from Differential Treatment in Multilateral Environmental Agreements' (2014) 4 *Climate Law* 50.

[21] The literature on the CBDR principle is abundant. For general discussions, see e.g., Lavanya Rajamani, *Differential Treatment in International Environmental Law* (Oxford: Oxford University Press, 2006); Birnie, Boyle and Redgwell, *supra* note 9 at 132–6; Sands and Peel, *supra* note 9 at 233–6; Beyerlin and Marauhn, *supra* note 9 at 61–72.

[22] UNCLOS, *supra* note 3, Art 207(4).

[23] Vienna Convention for the Protection of the Ozone Layer, adopted 22 September 1987, 1513 UNTS 293 (entered into force 22 September 1988), Preamble.

[24] Montreal Protocol on Substances that Deplete the Ozone Layer, adopted 16 September 1987, 1522 UNTS 3 (entered into force 1 January 1989), Preamble.

principle. It also clearly contributes to making global negotiations even more complex. However, this does not mean that the principle is always necessary or proper under all circumstances and for all environmental problems. For example, there is no room for differentiation between developing and developed countries when it comes to international rules to combat pollution for ships, which have to be uniform due to the truly international character of the shipping industry. Similarly, for example, rules on trade in endangered species and nuclear safety have to be uniform.[25]

In multilateral environmental agreements, differentiation or differential treatment manifests itself in different ways.[26] The differentiation may be explicit or implicit. It may apply *explicitly* to substantive obligations or duties related to procedure or technical support. A treaty or related protocols or annexes can simply impose lesser substantive obligations on some countries or totally exempt them from such obligations. One of the few examples, and the most striking, is the UN Framework Convention on Climate Change (UNFCCC)[27] and its Kyoto Protocol.[28] It may grant longer timeframes ('grace periods') for implementation or less stringent reporting requirements. In many global treaties, developed countries have obligations to financial and technology transfers to developing countries to assist them in treaty implementation. The fulfilment of such obligations may even be an explicit condition for developing countries' compliance with their commitments, as is laid down in UNFCCC:

> The extent to which developing country Parties will effectively implement their commitments under the Convention will depend on the effective implementation by developed country Parties of their commitments under the Convention related to financial resources and transfer of technology and will take fully into account that economic and social development and poverty eradication are the first and overriding priorities of the developing country Parties.[29]

---

25 Birnie, Boyle and Redgwell, *supra* note 9 at 136.

26 A more detailed and elaborated analysis is carried out by Voigt, *supra* note 20.

27 See UNFCCC, *supra* note 3, Art 3(1) and Annex I.

28 See Kyoto Protocol to the United Nations Framework Convention on Climate Change, adopted 11 December 1997, 2303 UNTS 148 (entered into force 16 February 2005), Arts 3 and 10.

29 UNFCCC, *supra* note 3, Art 4(7).

Differentiation can also be *implicit* by qualifying the parties' obligations by references to 'particular conditions',[30] 'national circumstances', 'capabilities'[31] or 'means'. An important implicit method of differentiation is also the formulation of 'soft' obligations in conventions, whereby, for example, the States shall carry out their commitments 'as far as possible and as appropriate'.[32] With such rules it becomes difficult to establish the exact content of the norm and, respectively, its breach. It is not uncommon to see the combination of two or more of such elements in one and the same treaty.

The CBDR principle as stated in the 1992 Rio Declaration aims mainly at the distinction and the differences between 'developed' and 'developing' States. This distinction is, however, becoming increasingly blurred. Today, the two groups are no longer homogenous but marked by major internal differences and constant changes as international economic development and national policies change the traditional patterns.

Finding operational, flexible and dynamic criteria or means to define various different groups and allocate rights and duties accordingly has turned out to be extremely difficult and controversial by its very nature, especially in the context of climate change negotiations.[33] This has become a particularly challenging aspect of the duty of cooperation in today's international environmental law, in particular in negotiations of global instruments.

There are also other criteria and bases for differentiations, based not primarily on the countries' social and economic situation and capacities, but on differences in environmental conditions and technological options. Such differences may justify different obligations in the perspective of economic efficiency and cost-effectiveness in addition to considerations of equity. This type of differentiation is seen inter alia in two regional treaties on transboundary pollution in Europe, namely the Convention on Long-Range Transboundary Air Pollution (LRTAP)[34] and – to a lesser

---

30 CBD, *supra* note 3, Art 6(a).

31 See e.g., Protocol to the 1972 Convention on the Prevention of Marine Pollution by Dumping of Wastes and Other Matter, adopted 7 November 1996, 2006 ATS 11 (entered into force 24 March 2006), Art 2. UNCLOS notes the 'need to take account of states' economic capacity', UNCLOS, *supra* note 3, Art 207, and the Vienna Convention on the Protection of the Ozone Layer refers to 'means at their disposal and their capabilities', *supra* note 23, Art 2(2).

32 CBD, *supra* note 3, Arts 6(b), 7–10, and 11.

33 Voigt, *supra* note 20.

34 Convention on Long-range Transboundary Air Pollution, adopted 13 November 1979, 1302 UNTS 217 (entered into force 16 March 1983) (LRTAP).

extent – the Convention on the Protection of the Marine Environment in the North East Atlantic (OSPAR)[35] to which we shall now turn.

V. TRANSBOUNDARY AIR POLLUTION: LRTAP

The LRTAP, better known as the European 'acid rain convention', was initiated by Norway and Sweden in the mid-1970s, and adopted by the United Nations Economic Commission for Europe (UNECE) in 1979. Being adopted during the Cold War as a follow-up to the 1975 European Security Conference in Helsinki, the LRTAP was seen as a breakthrough for East-West cooperation. As of September 2014, 51 States and the EU are parties to LRTAP, including the United States, Canada and most of the former States of the Soviet Union.[36]

A. Background

During the 1950s and 1960s, numerous rivers and lakes in the Nordic countries, and in Norway in particular, had increasing levels of sulphur and nitrogen. Their pH level[37] fell gradually below the so-called 'critical load', reaching a level at which the existing fish stocks and other aquatic life could no longer live. The rivers and lakes in Southern Norway were particularly vulnerable and seriously affected due to their unique geological conditions with weak buffering capacity. It is estimated that in total more than 9,000 local fish stocks were lost and over 5,000 were severely depleted due to acidification. In addition, sulphuric emissions resulted in damage to forests on the European continent, and this became a political issue, particularly in Germany.[38]

Damage to fish stocks and forests is not solely an issue of biodiversity loss. When fish stocks disappear and forests die, economic losses will be

---

[35] Convention on the Protection of the Marine Environment in the North East Atlantic, adopted 22 September 1992, 32 ILM 1228 (entered into force 25 March 1998) (OSPAR).

[36] See LRTAP Status, accessed 9 January 2015 at https://treaties.un.org/pages/ViewDetails.aspx?src=TREATY&mtdsg_no=XXVII-1&chapter=27&lang=en.

[37] The pH level is an expression of acidity – the lower the level, the more acidic the solution. If the pH level of a solution is below 7, it is defined as acid. If the pH level is above 7, it is defined as basic (alkaline).

[38] See Johan Sliggers and Willem Kakebeeke (UNECE), *Clearing the Air: 25 years of the Convention on Long-range Transboundary Air Pollution* (United Nations Publication, 2004).

heavy. In addition to its effects on ecosystems, acid rain damages buildings, sculptures and other parts of our cultural heritage. Furthermore, the serious health effects of some of these pollutants have become evident.

In the early 1970s, Norway started research on the causes of the acidification of our lakes and rivers. Over time, it became clear that the main source of the problem was literally 'acid rain': sulphur and nitrogen pollution transported by air, mainly from other countries, and falling as rain and snow over Norway. At that time, more than 90 per cent of the sulphur and 80 per cent of the nitrogen deposited in Norway originated from other European countries. Power plants, industrial processes, especially metal production, and transport were the main sources of acidifying emissions. The 'acid rain' resulted from these European countries' 'high stacks' policies, which were aimed at reducing local air pollution but instead transported the pollution over long distances.

The theory of transboundary acid rain was not immediately accepted by all countries in question. A Norwegian research programme that developed into an international programme provided important research and data to the study of the 'acid rain' phenomenon. The damage to forests and buildings on the European continent due to sulphuric emissions provided additional evidence and added to the general concern and recognition of the problem. There was thus a growing understanding and agreement of the need to curb the emissions that caused these effects. States from different parts of Europe, as well as the United States and Canada, came together under the umbrella of UNECE and started discussions on how to develop cooperative measures to reduce the air pollution. This ultimately led to LRTAP, the first legally binding instrument on air pollution in the world.[39]

B. Could the 'No Harm' Rule have been Applied?

It is not clear whether Norway and Sweden ever considered legal steps against the most responsible countries on the basis of the 'no harm principle', which at the time was already provided in Principle 21 of the 1972 Stockholm Declaration. Ultimately, this could have ended in a case before the ICJ, which would definitely have been an interesting one. I

---

[39] For a detailed description and analysis of the process that lead to the adoption of the convention, see *ibid*. For a well-documented presentation of this and other international instruments to combat air pollution, see e.g., Alexander Gillespie, *Climate Change, Ozone Depletion and Air Pollution* (Leiden: Martinus Nijhoff Publishers, 2006).

presume, however, that one of several legal problems for Norway would have been the fact that emissions from national sources in Norway had also contributed to the problem, albeit to a lesser degree. Here we see one of the practical limits of the 'no-harm principle' and also the importance of the duty of cooperation.

This being said, I do not exclude that an international court would arrive at a fairly clear conclusion to the effect that the main source States at least should endeavour to reduce their emissions to a reasonable degree. It is in this connection interesting to note that LRTAP explicitly distinguishes between two groups of States. Article 5 mentions, on the one hand, 'Contracting Parties which are actually affected by or exposed to a significant risk of long-range transboundary air pollution' and, on the other hand, 'Contracting Parties within which and subject to whose jurisdiction a significant contribution to long-range transboundary air pollution originates'.[40] The two groups are to hold 'consultations' on request.[41] It is interesting, and may be legally significant, that the main source States accepted such a distinction and implicitly agreed that they had caused the transboundary harm.

But at the same time, it is clear that the LRTAP does not deal with the issue of possible State liability for damage. This is stated quite particularly for an international convention, namely a footnote to Article 8 which states that '[t]he present Convention does not contain a rule on State liability as to damage'. This question is thus left open.

C. From Doubts and Hesitation to Acceptance and Urgency

When LRTAP was adopted in 1979, there was still not unanimous agreement on the magnitude and urgency of the problem. For example, in its preamble the parties only recognised the existence of *possible* adverse effects, in the short and long term, of air pollution including transboundary air pollution, and were concerned that 'a rise in the level of emissions of air pollutants within the region as forecast *may* increase such adverse effects'.[42]

As a consequence, the obligations laid down in LRTAP itself were defined in rather timid and general wording. According to LRTAP, the parties should 'endeavour to limit and, as far as possible, gradually reduce and prevent air pollution including long-range transboundary air

---

40 LRTAP, *supra* note 34, Art 5.
41 *Ibid*.
42 *Ibid*, Preamble (emphasis added).

pollution'.[43] This was to be achieved through 'policies and strategies' based on exchanges of information, consultation, research and monitoring.[44] In particular, the scientific aspect and exchange of information and knowledge were important elements in the convention, further elaborated on in separate articles.[45] Importantly, it sets up a cooperative programme for monitoring and evaluation of the long-range transmission of air pollution in Europe (EMEP).[46] This turned out to be a key to the further development.

LRTAP is a framework convention, like most conventions in the environmental field these days. The complexities of the issues make it very difficult to agree on precise commitments at an early stage. The ambition was first simply to establish a mutual acceptance of the *possible* problems and institutions to work with them.

Very soon it was felt that more precise obligations had to be laid down through protocols on specific substances. In the Protocol to the 1979 Convention on Long-Range Transboundary Air Pollution on the Reduction of Sulphur Emissions or their Transboundary Fluxes by at least 30 per cent (1985 Helsinki Protocol),[47] the tone is significantly sharpened. The preamble talks about the urgency of the issue, about the present emissions of air pollutants that are causing 'widespread damage to natural resources of vital environmental and economic importance, such as forests, soils and waters, and to materials (including historical monuments) and, under certain circumstances, have harmful effects on human health'.

The 1985 Helsinki Protocol lays down one quite simple obligation, which is equal for all parties: to reduce national annual sulphur emissions or their transboundary fluxes by at least 30 per cent as soon as possible and at the latest by 1993, compared to 1980 levels.[48] Today, this may be seen as a rather primitive approach, but that was the thinking and understanding at the time and a really important achievement.

---

[43] *Ibid*, Art 2.

[44] *Ibid*, Arts 3 and 4.

[45] *Ibid*, Arts 7 and 8.

[46] *Ibid*, Art 9.

[47] Protocol to the 1979 Convention on Long-Range Transboundary Air Pollution on the Reduction of Sulphur Emissions or their Transboundary Fluxes by at least 30 per cent, adopted 8 July 1985, 1480 UNTS 215 (entered into force 2 September 1987).

[48] *Ibid*, Art 2.

D. The 'Critical Load' Approach

It was, however, just a beginning. Very soon the time was ripe for further steps. In 1988, the Sofia Protocol concerning the Control of Emissions of Nitrogen Oxides (1988 Sofia Protocol) to LRTAP was adopted. Again, obligations are rather general, in fact mainly not to increase the emissions of nitrogen oxides beyond the 1987 level, but a new and important concept was introduced: the concept of '*critical load*'. 'Critical load' is defined as 'a quantitative estimate of the exposure to one or more pollutants below which significant harmful effects on specified sensitive elements of the environment do not occur according to present knowledge'.[49] The idea is that emissions shall gradually be reduced to the extent necessary to keep the exposure within nature's tolerance limit. Since 1988, this has been a central element in the development of the cooperation on air pollution in Europe.

Over the years altogether eight protocols under LRTAP have been adopted,[50] covering emissions of sulphur, nitrogen oxides,[51] volatile organic compounds,[52] further reductions of sulphur emissions,[53] heavy metals,[54] persistent organic pollutants (POPs)[55] and eutrophication and

---

49 Protocol concerning the Control of Emissions of Nitrogen Oxides or their Transboundary Fluxes, adopted 31 October 1988, 1593 UNTS 287 (entered into force 14 February 1991) (1988 Sofia Protocol), Art 1(7).

50 See Sands and Peel, *supra* note 9 at 248–57, for useful summaries of the protocols.

51 1988 Sofia Protocol, *supra* note 49.

52 Protocol concerning the Control of Emissions of Volatile Organic Compounds or their Transboundary Fluxes, concluded 18 November 1991, 31 ILM 568 (1992) (entered into force in 1991).

53 Protocol to the Convention on Long-range Transboundary Air Pollution of 1979 on Further Reduction of Sulphur Emissions, adopted 14 June 1994, 33 ILM (1994) (entered into force 5 August 1998) (1994 Oslo Protocol).

54 Protocol to the 1979 Convention on Long-range Transboundary Air Pollution on Heavy Metals, adopted 24 June 1998, 2237 UNTS 4 (entered into force 29 December 2003).

55 Protocol to the 1979 Convention on Long-range Transboundary Air Pollution on Persistent Organic Pollutants (POPs), adopted 24 June 1998, 2230 UNTS 79 (entered into force 23 October 2003).

ground-level ozone.[56] And revisions have taken place at intervals, laying down steadily more stringent commitments.[57]

The cooperation has become more and more advanced both scientifically, technically, legally and as regards organisation, also with regard to compliance mechanisms. However, it is an important fact that not all Parties to LRTAP are parties to all the Protocols, so there is *de facto* differentiation in commitments between the Parties.

E. Differentiation of Commitments

A major new step was taken with the 1994 Oslo Protocol on Further Reduction of Sulphur Emissions (1994 Oslo Protocol). Unlike the original system where all the parties make the same percentage reductions, the 1994 Oslo Protocol lays down differentiated emission reduction targets for sulphur in the various countries. The extremes were the reduction obligation for 2005 compared to the base year 1980 of an 87 per cent reduction for Germany and only 3 per cent for Portugal and Greece.[58]

This differentiation is based on an advanced scientific programme where the following factors are considered:

- the effects of pollutants based on natural conditions in various countries, in particular the level of critical loads with regard to assimilative capacity of pollutants in the different regions ('critical sulphur deposition'),[59] which vary considerably due to differences in geological and soil conditions;
- meteorological conditions which define the transport and spreading of pollutants from the various sources;
- mapping of major emission sources; and
- abatement options and their costs, and economic benefits of measures.

---

[56] Protocol to the 1979 Convention on Long-range Transboundary Air Pollution to Abate Acidification, Eutrophication and Ground-level Ozone, adopted 30 November 1999, 1593 UNTS 287 (entered into force 14 February 1991) (1999 Gothenburg Protocol), Art 2.

[57] The latest revision to the Gothenburg Protocol was adopted by the Executive Body in May 2012. It lays down emission reduction commitments for the main air pollutants by 2020.

[58] See Annex II of the 1994 Oslo Protocol, *supra* note 53.

[59] Measured in centigrams of sulphur per square metre per year.

These factors are brought together in an '*integrated assessment model*', which then serves as the basis for defining necessary emission reductions for the various States Parties. The objective is to reduce pollution to within the critical loads for the various pollutants in the different parts of Europe, while achieving cross-national cost-efficiency. On this basis, the negotiations finally determine emission ceilings for each Member State.

This effects-based approach, which aims at gradually attaining critical loads, sets long-term targets for reductions in emissions. LRTAP has proven to be an effective and flexible framework for joint action, not least by fostering close contact between the scientific experts and the national authorities and policymakers. In fact, the system relies heavily on advanced science and may in that sense be vulnerable to changes in the parties' capability and willingness to support the necessary research.

The 1999 Protocol to Abate Acidification, Eutrophication and Ground-level Ozone (1999 Gothenburg Protocol) lays down emission ceilings for 2010 for four pollutants, namely sulphur (SO_2), nitrogen oxides (NO_x), volatile organic compounds (VOCs) and ammonia.[60] It applies the principle of critical load and differentiated obligations. The emission ceilings for each State Party were negotiated on the basis of scientific assessments of pollution effects and abatement options. Parties whose emissions have a more severe environmental or health impact and whose emissions are relatively inexpensive to reduce are required to make the biggest cuts. For example, Germany had to reduce sulphuric emissions by 90 per cent between 1990 and 2010, while Armenia at the other end of the spectrum had no reduction obligations.[61] The EU on average is required to reduce emissions by as much as 75 per cent by 2010 under the 1999 Gothenberg Protocol.[62]

Another important element in the cooperation under LRTAP is the adoption of tight limit values for specific emission sources, such as combustion plants, electricity production, dry cleaning, cars and lorries. This has been developed on the basis of technological discussions aimed

---

[60] 1999 Gothenburg Protocol, *supra* note 56, Art 2.

[61] *Ibid*, Annex II.

[62] There are special rules for Canada and the United States. These two countries have also agreed to special bilateral arrangements to address air pollution issues arising from emissions of SO_2 and NO_x through the 1991 Canada-US Air Quality Agreement. See Agreement between the Government of Canada and the Government of the United States of America on Air Quality, signed 13 March 1991, 1852 UNTS 79 (1991 US-Canada Air Quality Agreement).

at defining the best available techniques. Increasingly, it has also included the setting of air quality standards with a view in particular on protecting public health. The importance of air quality for public health has gradually become more and more obvious, and public health considerations play an increasingly important role in the development of new rules and obligations.

F. Compliance

LRTAP has a traditional clause for the settlement of disputes through 'negotiation or by any other method of dispute settlement acceptable to the parties to the dispute'.[63] This has, however, never been applied. Instead, implementation problems are dealt with in a proactive manner through a special committee.

An important new feature was introduced in relation to the adoption of the 1994 Oslo Protocol, namely a decision to establish an Implementation Committee.[64] The Committee, consisting of eight members, shall analyse and evaluate on a periodic basis information related to compliance with parties' obligations. In the case of non-compliance, the Committee shall work with the party in question and experts with a view to securing 'constructive solutions'. A submission may be brought before the Committee by one or more parties to a protocol that have reservations about another party's compliance, or by the secretariat if it becomes aware of possible non-compliance by a party. A submission may also be brought forward by the party itself if it finds that, despite its best endeavours, it is or will be unable to comply fully with its obligations. Such a submission shall explain the specific circumstances that the party considers to be the cause of its non-compliance.

The Committee then considers what measures to recommend, including how to best assist and support the party in question. The Committee forwards recommendations to this effect to the decision-making Executive Body of the Convention, which may decide upon measures of a non-discriminatory nature to bring about full compliance with the protocol in question, including active measures to assist a party in its efforts of compliance. The main idea is that enforcement shall be 'non-confrontational'.[65] The experience is positive. This attitude of mutual

---

[63] LRTAP, *supra* note 34, Art 13.

[64] 1994 Oslo Protocol, *supra* note 53, Art 7.

[65] It is explicitly stated that the application of these compliance procedures shall be without prejudice to the dispute settlement provisions of the protocols.

assistance and support in case of non-compliance of international obligations is quite effective. It is no longer unique for LRTAP. For example, the same idea is behind the 'facilitative branch' of the Kyoto Protocol's Compliance committee.

It is important to note that the EU, as a party to LRTAP, alongside its individual Member States, has put much effort into implementing the protocols through directives and regulations.[66] In this way, the implementation by Member States is subject to the strict formal procedures for enforcement of the EU, in addition to the LRTAP compliance system.

G. Results

The European emissions of air pollutants have been greatly reduced over the last 20–30 years – roughly 50 per cent in spite of considerable economic growth in the same period, with variations between the different pollutants.[67] This is largely due to technical improvements in industrial processes, and a transition from coal to fuels with less sulphur in energy and other production. A decisive factor in the implementation is that the EU as a party to the Convention has based its own legislation on LRTAP protocols and issued directives and regulations that are binding on the EU Member States.

From a Norwegian perspective, LRTAP has been immensely important. Since 1990, the area with damage to fish stocks has been nearly halved and it continues to decrease. Norwegian forests showed a decrease in vitality during the 1990s, but since then conditions have improved. In the last few years, the health of Norwegian forests has remained stable.

The science-based approach of LRTAP and its subsequent protocols have served as an example for other regions of the world where similar problems exist. Thus, the following regional initiatives have established regional schemes with LRTAP in mind:

---

[66] Most notably, Directive 2001/81/EC, known as the 'Emission Ceilings Directive'.

[67] The reductions vary for the different substances. For example, the total reduction of SO_2 emissions for all Convention Parties was 70 per cent between 1990 and 2008. For detailed information on reductions and on the status of the work and further work under the Convention, see UNECE, *Convention on Long-Range Transboundary Air Pollution: Strategies and Policies for Air Pollution Abatement*, UN Doc ECE/EB.AIR/123 (2013).

(i) the 1998 Malé Declaration on Control and Prevention of Air Pollution and Its Likely Transboundary Effects for South Asia (eight participating countries);[68]
(ii) the 1998 Acid Deposition Monitoring Network in East Asia (13 participating countries);[69]
(iii) the 1995 ASEAN Cooperation Plan on Transboundary Pollution, which establishes a cooperative mechanism to assess the origins, causes, nature and extent of both national and regional pollution;[70]
(iv) the 2002 ASEAN Agreement on Transboundary Haze Pollution (ten States Parties);[71] and
(v) the 1997 Air Pollution Information Network for Africa (15 countries).[72]

At the global level, the 1998 Persistent Organic Pollutants (POPs) Protocol paved the way for a global agreement, the 2001 Stockholm Convention on POPs, negotiated under the United Nations Environment Programme (UNEP).[73] Similarly, the 1998 Protocol on Heavy Metals has inspired the newest global agreement in this field, the 2013 UN Minamata Convention on Mercury, which after several years of negotiations under the auspices of UNEP was agreed in January 2013 and was opened for signature in October 2013.[74] In both cases, the assessment of risks, emission sources and mitigation options have much in common

68 1998 Malé Declaration on Control and Prevention of Air Pollution and Its Likely Transboundary Effects for South Asia, adopted 22 April 1998, accessed 12 January 2015 at http://www.sacep.org/html/docs_mos_maledec.htm.

69 See http://www.eanet.asia/eanet/brief.html (accessed 9 January 2015).

70 1995 ASEAN Cooperation Plan on Transboundary Pollution, adopted 17 June 1995. See also ASEAN Cooperation on Transboundary Haze Pollution, accessed 12 January 2015 at http://environment.asean.org/asean-cooperation-on-transboundary-haze-pollution/.

71 2002 ASEAN Agreement on Transboundary Haze Pollution, signed 10 June 2002 (entered into force 25 November 2003) accessed 12 January 2015 at http://agreement.asean.org/media/download/20140119170340.pdf. See Adella Anjani Putri, 'Indonesia Ratifies Haze Treaty after 12 Years of Stalling', *The Jakarta Globe* (17 September 2014) accessed 12 January 2015 at http://www.thejakartaglobe.com/news/indonesia-ratifies-haze-treaty-12-years-stalling/.

72 See http://www.sei-international.org/rapidc/apina.htm (accessed 12 January 2015).

73 2001 Stockholm Convention on Persistent Organic Pollutants, adopted 22 May 2001, 2256 UNTS 119 (entered into force 17 May 2004).

74 2013 Minamata Convention on Mercury, adopted 13 October 2013, accessed 12 January 2015 at https://treaties.un.org/doc/Treaties/2013/10/20131010%2011-16%20AM/CTC-XXVII-17.pdf.

with those prepared under LRTAP. These examples are testimonies to the potential of a regional initiative, such as LRTAP, to facilitate other regional and global undertakings.

H. Summing Up: A 'Package' of Interlinked Obligations

LRTAP itself is relatively simple and apparently not very ambitious. It was, however, an important step at the time as it created the formal basis and institutional framework for cooperation among all European States, the United States and Canada with the purpose of abating air pollution.

Over the years, cooperation under LRTAP has developed step by step into something far more ambitious and complex than was foreseen in the beginning. It illustrates the complexity of the task of abating transboundary pollution in a just, effective and economically efficient way based on a few important objectives and principles:

- the objective is to bring the pollution level below the *critical load* as well as the criteria on which emission reduction commitments of each State Party are based;
- the *principle of differentiated obligations* in order to reach the critical load effectively and in the most cost-effective way, based on scientific assessment of the varying natural conditions and national circumstances, emission sources, transport of pollutants, and abatement options and costs;
- close cooperation and interplay between scientific research and policymaking;
- *gradual harmonisation and tightening* of air quality standards and emission standards for various activities, based on technological developments and discussion in order to apply best available techniques; and
- the principle of *mutual information, reporting, advice and support*, and a 'constructive' system for compliance and enforcement.

VI. TRANSBOUNDARY MARINE POLLUTION: OSPAR

A. Marine Pollution in International Environmental Law

Marine pollution may be the area where the 'no-harm' rule is the most problematic from a practical point of view. Establishing legally relevant cause-effect relationships is very difficult or even impossible, unless there is one or a few point sources of emissions of easily identified substances, such as radioactive substances from nuclear power plants. Adjudication on the basis of the general duty not to harm the environment of other States seems therefore particularly difficult, and so the duty to cooperate is most relevant and even essential in this field.

International cooperation within this area is important and based on numerous treaties. It started in 1969 with several treaties related to pollution from ships as a direct result of the *Torrey Canyon* oil spill accident off the southern coast of England: the conventions on liability for oil spills[75] and on a State's right to intervention on the high seas in case of a pollution accident.[76] Since then, the international cooperation to limit and prevent pollution from shipping has constantly been expanded and strengthened, mainly under the auspices of the International Maritime Organization (IMO), with the 1973 International Convention on the Prevention of Pollution from Ships (MARPOL) as a centrepiece. Another important agreement is the 2004 IMO Convention for the Control and Management of Ship's Ballast Water. As a truly international industry, it is particularly important that international shipping is regulated by universal and harmonised environmental rules. Here, the CBDR principle has no place.

When it comes to marine environment protection, UNCLOS is the central instrument. Although it does not directly express the 'no-harm' rule as stated in Principle 21 of the 1972 Stockholm Declaration, it lays down a general obligation 'to protect and preserve the marine environment'.[77] In addition, it explicitly confirms the sovereign right of States to exploit their natural resources 'pursuant to their environmental policies and in accordance with their duty to protect and preserve the marine environment'.[78] It is interesting to note that the exercise of sovereignty here is explicitly 'limited' by the duty to protect the environment. A similar explicit link is unfortunately neither expressed in Principle 21 of

---

[75] International Convention on Civil Liability for Oil Pollution Damage, adopted 29 November 1969, 973 UNTS 3 (entered into force 19 June 1975), superseded by the 1992 Protocol to the 1969 International Convention on Civil Liability for Oil Pollution Damage, 1956 UNTS 255 (amended by the 2000 Protocol).

[76] International Convention relating to Intervention on the High Seas in Cases of Oil Pollution Casualties, concluded 29 November 1969, 970 UNTS 211 (entered into force 6 May 1975).

[77] UNCLOS, *supra* note 3, Art 192.

[78] *Ibid*, Art 193.

the 1972 Stockholm Declaration or in Principle 2 of the 1992 Rio Declaration.[79]

Prevention of marine pollution is a prominent topic in UNCLOS. Part XII provides an elaborated list of measures that States shall take to prevent, reduce and control marine pollution from various activities,[80] and lays down detailed rules on States' duties and rights with regard to enforcement in the various zones of the sea under State jurisdiction.[81] It clearly reflects the fact that cooperation is essential. Part XII, section 2 deals explicitly with global and regional cooperation and lists a number of activities to that effect.[82] Here, regional cooperation is particularly practical and relevant to which numerous activities and conventions bear witness. This includes the UNEP Regional Seas programme and various treaties on cooperation in specific regional sea areas, partly as a result of UNEP's work.[83] Altogether, there are some 16 regional seas conventions and/or action plans in force around the world.[84] One example is the East Asian Seas programme to which Singapore and all other countries in the area, including Australia, are parties, which is not a treaty, but a regularly updated action plan. In many other regions, the cooperation is based on a multilateral convention.[85]

B. OSPAR: Background, Principles and Main Content

OSPAR is the mechanism established by 15 governments of the Western coasts and catchments of Europe and the EU to protect the marine

---

[79] However, it is worth noting that the World Charter for Nature, GA Res 37/7, UNGAOR, 37th Sess, Supp No 51, UN Doc A/37/51 (1982), 22 ILM 455 (1983), also explicitly makes this link between sovereignty and environmental protection, stating in its preamble: 'Conscious of the spirit and terms of its resolutions 35/7 and 36/6, in which it solemnly invited Member States, in the exercise of their permanent sovereignty over their natural resources, to conduct their activities in recognition of the supreme importance of protecting natural systems, maintaining the balance and quality of nature and conserving natural resources, in the interests of present and future generations.'

[80] UNCLOS, *supra* note 3, Arts 194, 207–212.

[81] *Ibid*, Arts 213–233.

[82] For a more detailed discussion on UNCLOS marine protection environment provisions, see Robert Beckman's chapter in this book (Chapter 6).

[83] A broad overview is provided in Daud Hassan, *Protecting the Marine Environment from Land-Based Sources of Pollution* (Farnham: Ashgate, 2006) at 103–49.

[84] *Ibid* at 125.

[85] See http://www.unep.org/regionalseas/Programmes/unpro/eastasian/ (accessed 12 January 2015).

environment of the North-East Atlantic.[86] Work under OSPAR is managed by the OSPAR Commission, which is headquartered in London.

OSPAR has been a central instrument in the considerable work that has been done in Europe to prevent and reduce pollution of the sea from various sources. Indeed, it covers pollution from land-based sources – which means rivers running into the sea – and it has direct consequences for the pollution status of freshwater sources, including also parts of rivers and lakes, far inland, as well as estuaries.[87]

OSPAR's origins go back to the late 1960s and early 1970s. In the late 1960s, the problem of dumping of waste on the high sea became a serious concern to States around the North Sea in particular. At the initiative of the Norwegian government, a convention was negotiated which – roughly speaking – simply prohibited any dumping of hazardous substances at sea in the North-East Atlantic area.[88] This was the 1972 Convention for the Prevention of Marine Pollution by Dumping from Ships and Aircraft, also called the 'Oslo Convention', which inspired the International Convention on the Prevention of Marine Pollution by Dumping of Wastes and other Matter. The cooperation between the European coastal States thus inspired further and more significant steps. The most important source of marine pollution is run-off from land-based sources through rivers. In 1974, States Parties to the Oslo Convention came to Paris to negotiate the Convention for the Prevention of Marine Pollution from Land-Based Sources (Paris Convention). In 1992, these two conventions with their subsequent amendments, were merged, strengthened and expanded to also cover pollution from offshore activities, and consequently became OSPAR.[89]

In the OSPAR preamble, the contracting Parties recall 'the relevant provisions of customary international law' reflected in Part XII of UNCLOS and, in particular, Article 197 on 'global and regional cooperation for the protection and preservation of the marine environment'. The parties also *recognise* 'the inherent worth of the marine environment of the North-East Atlantic and the necessity for providing

[86] See http://www.ospar.org/ (accessed 12 January 2015).

[87] OSPAR, *supra* note 35, Art 3.

[88] Those parts of the Atlantic and Arctic Oceans and their dependent seas that lie north of 36 north latitude and between 42 west longitude and 51 east longitude but exclude the Baltic Sea and the Mediterranean Sea, which are subject to their proper environmental conventions.

[89] When in force in 1998, OSPAR fully replaced the Oslo and Paris Conventions but Decisions, Recommendations and all other agreements adopted under those Conventions continued to be applicable.

coordinated protection for it'.[90] This implies that it should be protected independently of its direct utility for the contracting States.[91]

The marine environment in this part of the Atlantic is subject to serious hazards and stress, including high levels of some toxic substances, considerable oil pollution from shipping and the offshore petroleum industry, ocean acidification and serious eutrophication, particularly in the southern part of the North Sea. An excessive amount of litter and plastic waste is also a problem. In addition, fish stocks are in decline, mainly due to overfishing, along with seriously declining stocks of some sea birds and marine mammals, in particular seals. Important coral reefs have also been destroyed or are under threat. Solutions to these problems will require cooperation between the coastal States.

OSPAR is another framework convention that lays down general obligations and principles, and establishes institutions and procedures for cooperation. More specific obligations are expressed in four major annexes,[92] which form 'an integral part of the Convention'[93] and are continuously developed through decisions and recommendations. It has become a very dynamic instrument.

The general obligation of the Contracting Parties is to 'take all possible steps to prevent and eliminate pollution and ... take the necessary measures to protect the maritime area against the adverse effects of human activities so as to safeguard human health and to conserve marine ecosystems and, when practicable, restore marine areas which have been adversely affected'.[94] To this end, the parties 'shall, individually and jointly, adopt programmes and measures and shall harmonize their policies and strategies'.[95]

OSPAR is in many ways quite progressive. It states explicitly that the Contracting Parties shall apply the precautionary principle:

> by virtue of which preventive measures are to be taken when there are reasonable grounds for concern that substances or energy introduced, directly or indirectly, into the marine environment may bring about hazards to human health, harm living resources and marine ecosystems, damage amenities or

---

[90] OSPAR, *supra* note 35, Preamble.

[91] Kiss and Shelton, *supra* note 8 at 198.

[92] These include: (i) Annex I on Pollution from land-based sources; (ii) Annex II on Pollution from dumping and incineration at sea; (iii) Annex III on Pollution from offshore sources; and (iv) Annex IV on regular assessments of the quality of the marine environment.

[93] OSPAR, *supra* note 35, Art 14.

[94] *Ibid*, Art 2(1)(a).

[95] *Ibid*, Art 2(1)(b).

> interfere with other legitimate uses of the sea, even when there is no conclusive evidence of a causal relationship between the inputs and the effects.[96]

It also generally lays down a duty for the contracting parties to apply the 'polluter pays' principle by virtue of which the costs of pollution prevention, control and reduction measures are to be borne by the polluter, and the principles of best available techniques and best environmental practice.[97]

OSPAR also includes an article on access to environmental information which in fact anticipated the UNECE Convention on Right to Environmental Information and Participation (Aarhus Convention) according to which parties are obliged to make available 'to any natural or legal person, in response to any reasonable request, without that person's having to prove an interest, without unreasonable charges, as soon as possible and at the latest within two months' any available information 'on the state of the maritime area, on activities or measures adversely affecting or likely to affect it and on activities or measures introduced in accordance with the Convention'.[98] No provision under OSPAR shall be interpreted as 'preventing the Contracting Parties from taking, individually or jointly, more stringent measures with respect to the prevention and elimination of pollution of the maritime area or with respect to the protection of the maritime area against the adverse effects of human activities'.[99]

Cooperation in the form of joint programmes of scientific or technical research, monitoring and exchange of knowledge and information, are important parts of the cooperation under OSPAR.[100]

C. Implementation

OSPAR implementation is taken forward through the adoption of decisions, recommendations and other agreements. Decisions are legally binding on the parties and are usually made by consensus. Recommendations are formally not legally binding and serve as guidelines. However,

---

96 *Ibid*, Art 2(2)(a).

97 *Ibid*, Art 2(2)(b) and (3)(b).

98 *Ibid*, Art 9, which also lists types of information that the State may refuse to make available.

99 *Ibid*, Art 2(5).

100 *Ibid*, Art 8.

parties strive to implement them in a faithful manner through special agreements, monitoring programmes and information collection, among others.[101]

Recommendations are the most common instrument, although there are also examples of binding decisions on important substantial issues. These may relate to the treatment of specific hazardous substances, such as the 1985 decision on Programmes and Measures on Limit Values and Quality Objectives for Cadmium Discharges, to take just one example;[102] the 1989 decision to cease all dumping of industrial waste in the North Sea by 31 December 1989 and in other Convention waters by 31 December 1995, with some exceptions;[103] the 1990 decision to phase out the dumping of sewage sludge by the end of 1998;[104] and the decision on final disposal of offshore petroleum facilities, which requires that such facilities as a rule shall be taken onshore for destruction and not be dumped at sea.[105]

Annex I on the prevention and elimination of pollution from land-based sources is particularly significant and interesting. It applies to 'discharges to the maritime area, and releases into water or air which

---

[101] Altogether, some 160 recommendations and decisions and a similar number of 'other agreements' are listed on the OSPAR Commission's website. See http://www.ospar.org/content/content.asp?menu=00040400000000_000000_000000 (accessed 12 January 2015).

[102] Paris Commission (PARCOM) established under the Convention for the Prevention of Marine Pollution from Land-based Sources (Paris Convention), PARCOM Decision 85/2: Programmes and Measures of 31 December 1985 on Limit Values and Quality Objectives for Cadmium Discharges, adopted 1985, Brussels.

[103] Oslo Commission (OSCOM) established under the Convention for the Prevention of Marine Pollution by Dumping from Ships and Aircraft (Oslo Convention), OSCOM Decision 89/1 of 14 June 1989 on the Reduction and Cessation of Dumping Industrial Wastes at Sea.

[104] OSCOM, OSCOM Decision 90/1 of 23 June 1990 on the Cessation of Dumping of Sewage Sludge at Sea.

[105] This came as a reaction to the famous and controversial *Brent Spar* case: 'Brent Spar' or 'Brent E' was a North Sea oil storage and tanker loading buoy in the Brent oilfield on the UK continental shelf, operated by Shell UK. With the completion of a pipeline connection to the oil terminal at Sullom Voe in Shetland the storage facility was considered to be of no further value as of 1991. Brent Spar became an issue of public concern in 1995, when the British government announced its support for Shell's application for disposal in deep Atlantic waters. After protests by Greenpeace and considerable public concern and international attention, the buoy was finally in 1998 brought to Norway and used as a structure in the construction of new harbour facilities near the city of Stavanger.

reach and may affect the maritime area'[106] and thus in reality lays down principles for abating both air and water pollution in the Member States. It requires that point source discharges and releases shall be strictly subject to authorisation or regulation by the parties.[107] Such authorisation or regulation shall implement relevant decisions of the Commission which bind the relevant party. In this regard, the parties shall require, either individually or jointly, the use of best available techniques for point sources, and best environmental practice for point and diffuse sources including, where appropriate, clean technology.[108] The parties shall also take preventive measures to cooperate over the risk of pollution caused by accidents.[109]

From the beginning, the central objective was reducing and gradually the complete phasing out of substances that are toxic, persistent and liable to bioaccumulate. And during the first decades of OSPAR, the work was concentrated on limiting discharges of the most hazardous substances to the sea, through various measures. This also necessitated strict control of emissions into rivers and lakes. Ambitious strategies were adopted to reduce the levels of the most toxic substances, such as heavy metals, radioactive substances and eutrophication. For example, early decisions established limit values and quality objectives for cadmium and mercury discharges by sectors.[110] This necessitated a detailed analysis of technical possibilities for emission reduction in various types of industry and other economic activities in particular. Here, similarly to the work regarding air pollution under LRTAP, the work aimed at finding solutions based on best available techniques. This required very demanding work by experts in the respective fields.

The 1998 OSPAR Ministerial Meeting laid down the 'one generation cessation target' as the key objective in its Strategy with regard to Hazardous Substances.[111] In doing so, Ministers agreed to make every endeavour to move towards the target of complete cessation of discharges, emissions and losses of hazardous substances by the year

---

[106] OSPAR, *supra* note 35, Annex I, Art 2.

[107] *Ibid.*

[108] *Ibid*, Annex I, Art 1.

[109] *Ibid.*

[110] PARCOM, PARCOM Decision 85/1: Programmes and Measures of 31 December 1985 on Limit Values and Quality Objectives for Mercury Discharges by Sectors other than the Chlor-alkali Industry, adopted 1985, Brussels.

[111] OSPAR Commission, Progress Report to the 5th North Sea Conference (Bergen 2002) at 93, accessed 12 January 2015 at http://www.ospar.org/html_documents/ospar/html/nsc_kap_5_78-118.pdf.

2020.[112] This implied an ultimate aim of concentrations in the environment near natural levels – 'background values' – for naturally occurring substances and close to zero concentrations for man-made synthetic substances. This work is well underway.

OSPAR's work on hazardous substances has greatly influenced the EU legislation in this field. Over the last decade, OSPAR's work on hazardous substances has been reduced, partly due to the fact that the EU has taken a much more active role in the field of water and marine pollution through the Water Framework Directive (Directive 2000/60/EC), the Registration, Evaluation, Authorisation and Restriction of Chemicals (REACH) Directive (Directive 2006/121/EC), and the Marine Strategy Framework Directive (Directive 2008/56/EC). There is now a certain division of functions between the EU and the OSPAR Commission. On the other hand, OSPAR has broadened into other fields of protection of the marine environment. Having adopted the ecosystem approach as an important principle, it now includes monitoring and protection of the marine biodiversity and ecosystems in a broad sense.

This brings OSPAR into the complex field of international fisheries management and regulation, which is also the responsibility of other international organisations. OSPAR nevertheless wants to play an active role in this field, as spelled out in Annex 5, Article 4:

> no programme or measure concerning a question relating to the management of fisheries shall be adopted under this Annex. However where the Commission considers that action is desirable in relation to such a question, it shall draw that question to the attention of the authority or international body competent for that question. Where action within the competence of the Commission is desirable to complement or support action by those authorities or bodies, the Commission shall endeavour to cooperate with them.

In 2010, the OSPAR Commission adopted a new Strategy for the Protection of the Marine Environment of the North-East Atlantic 2010–2020.[113] It brings together the various topics and in particular has a notable Part I called 'Implementing the ecosystem approach'. Its strategic objectives with regard to biodiversity and ecosystems are to halt and prevent by 2020 further loss of biodiversity in the OSPAR maritime area; to protect and conserve ecosystems; and to restore, where practicable,

---

112 *Ibid.*

113 Strategy of the OSPAR Commission for the Protection of the Marine Environment of the North-East Atlantic 2010–2020, accessed 12 January 2015 at http://www.ospar.org/html_documents/ospar/html/10-03e_nea_environment_strategy.pdf.

marine areas which have been adversely affected. To this end, the OSPAR Commission aims at improving the status of threatened and/or declining species and habitats, including the status of fish stocks. Recent decisions lay down quality norms for the marine environment to be reached. Seven marine protected areas (MPAs) have been established in sea areas outside national exclusive economic zones (EEZ).

In OSPAR there are some differentiations between the obligations of the parties, although not to the same extent as in LRTAP. The most important distinction is between the States bordering the North Sea and States not bordering the North Sea (Spain, Portugal and Iceland), in particular with regard to combating eutrophication. There is a general possibility to make reservations to decisions and recommendations, but apparently this is seldom used.

The strategy adopted in 2010 lays down a number of ambitious objectives towards 2020, such as:

- 'to combat eutrophication in the OSPAR maritime area, with the ultimate aim to achieve and maintain a healthy marine environment where anthropogenic eutrophication does not occur';
- 'to prevent pollution of the OSPAR maritime area by continuously reducing discharges, emissions and losses of hazardous substances … with the ultimate aim to achieve concentrations in the marine environment near background values for naturally occurring substances and close to zero for man-made synthetic substances'; and
- 'to move towards the targets of the cessation of discharges, emissions and losses of hazardous substances by the year 2020'.

D. Summing Up: A Package of Obligations

Much like LRTAP, OSPAR has provided a very useful and dynamic framework for a coherent set of measures – a package or cluster – and mutual and gradually strengthened commitments for the Member States:

- scientific monitoring of the marine environment, initially primarily with regard to hazardous substances, later also covering biodiversity and environmental conditions more broadly, based on an ecosystem approach;
- close cooperation between research and policymaking;
- ambitious long-term targets for reducing hazardous substances; and
- joint efforts to reduce emissions from various types of emission sources: developing technical requirements and emission standards based on best available technology for various activities, which are

laid down in decisions and recommendations that serve to harmonise national rules and measures.

VII. CONCLUSION: EFFECTIVE COOPERATION REQUIRES A 'CLUSTER' OF LONG-TERM MUTUAL COMMITMENTS

The duty to cooperate is a general principle in international environmental law and is highly relevant with regard to transboundary pollution. Most transboundary pollution problems cannot be prevented or reduced without a close and active cooperation between all affected parties. The problems are complex with regard to causes, effects and possible solutions, and the necessary measures may be costly in the short term.

To solve transboundary problems is, in most cases, a demanding task. The cooperation has to be broad, flexible and consistent. It must consist of 'packages' or 'clusters' of some essential and closely interlinked activities and commitments: research and close cooperation between research and policymaking, target-setting, agreement on common principles, technological development, harmonisation of standards to the extent reasonable, national differentiations to the extent necessary for the reason of equity or cost-effectiveness, exchange of information and knowledge, and supportive and 'constructive' enforcement mechanisms. It is also important to be aware of the complexities involved. As is illustrated by the two European conventions which have been presented here, a step-by-step approach is required and a long-term perspective and commitment is necessary.

11. Evaluating the (in)effectiveness of ASEAN cooperation against transboundary air pollution

Laode M Syarif

I. BACKGROUND CONTEXT AND THE SOURCES OF TRANSBOUNDARY AIR POLLUTION (TAP)

As a regional organisation with ten Member States,[1] ASEAN is one of the most important global players in the world with a total land area of 4,435,670 square kilometres (sq km), a population of 598,498 million people, a combined gross domestic product (GDP) of US$2,066 billion and an average growth rate above 7 per cent.[2] In terms of environmental richness, the ASEAN region is a paradise of flora and fauna and one of the most important biodiversity hotspots in the world. Unfortunately, the steady increase in population in most ASEAN Member States and the growing population's dependence on timber, fuel wood and other forest products, as well as the conversion of forests into agricultural and industrial lands, have taken their toll on the region's forests, with a total loss of 555,587 sq km between 1980 and 2007. In addition, the annual loss of mangrove forests is around 628 sq km, nearly a 26 per cent decline in 25 years.[3]

1 Brunei, Cambodia, Indonesia, Laos, Malaysia, Myanmar, the Philippines, Singapore, Thailand and Vietnam.

2 ASEAN, *ASEAN Community in Figures 2011* (ASEAN Secretariat, 2012). For more up-to-date information, see the ASEAN Secretariat website, accessed 12 January 2015 at http://www.asean.org/resources/publications/asean-publications.

3 For more complete information, see ASEAN Centre for Biodiversity, *ASEAN Biodiversity Outlook* (ASEAN, 2010). For up-to-date information, see http://chm.aseanbiodiversity.org/index.php?option=com_content&view=article&id=146&Itemid=232 (accessed 12 January 2015).

The high deforestation rate and biodiversity lost in the ASEAN region, especially in Indonesia, is also due to unsustainable forest management and the use of the ancient method of slash-and-burn land clearing. This unsustainable practice has put the region, especially Indonesia, Singapore, Malaysia and Brunei, in a danger zone of increasing levels of transboundary air pollution (TAP) during almost every dry season.

It is important to note that undisturbed tropical forests are normally quite fire resistant, but they will burn during severe drought, particularly after they have been logged.[4] In the last three decades, most major fire incidents in the region, especially in Indonesia, occurred on forest concessions and large plantations. Although the history of forest fires in Indonesia can be traced back 17,500 years,[5] more frequent and extensive forest fires have only occurred recently over the last three decades after the expansion of logging concessions and large plantations.

One of the most severe series of fires ever experienced in Indonesia occurred between June 1982 and May 1983. During that period, severe El Niño weather conditions and poor management of logging concessions in East Kalimantan made the forests vulnerable to fire. As a result, 3.2 million hectares were burned. One study estimated the cost of the 1982–1983 fires at US$9 billion, of which nearly US$8.3 billion constituted the loss of standing timber.[6]

In 1991, another series of fires struck East Kalimantan, burning approximately 500,000 hectares, and another series of fires in 1994 burned almost 5 million hectares. Similar to the 1982–1983 fires, most of the fires originated from logging concessions and plantations. It was reported that TAP caused by these fires disrupted air and sea transportation and sharply elevated air pollution levels in the affected countries.[7]

---

4 David Glover and Timothy Jessup (eds), *Indonesia's Fires and Haze: The Cost of Catastrophe* (Singapore: Institute of Southeast Asian Studies, 1999) at 4.

5 Scientific evidence based on radiocarbon dating of charcoal deposits in East Kalimantan indicates that lowland forest areas have repeatedly burned since at least 17,500 years ago. See Johann Georg Goldammer (ed), *Fire in the Tropical Biota: Ecosystem Processes and Global Challenges, Ecological Studies Vol 84* (Berlin: Springer-Verlag, 1990) at 84.

6 Peter Hess, 'Forest Fire in East Kalimantan 1982/1983: Effects, Damages and Technical Solutions' (Paper delivered at the Workshop on Forest Rehabilitation and Forest Protection from Fire, Samarinda, Indonesia, 31 January–2 February 1994) (unpublished).

7 Emily Matthews (ed), *The State of the Forest: Indonesia* (Washington, DC: World Resources Institute, 2000) at 54.

The episodes of forest fires reached their peak in 1997–1998 with a catastrophic series of fires, which according to the United Nations Environment Programme (UNEP) was among one of the largest environmental catastrophes since 1950 and among the most damaging in recorded history,[8] with the area affected by fires totalling 11.7 million hectares.[9] During that period, Indonesia and Malaysia experienced the worst recorded droughts in recent history. The El Niño Southern Oscillation brought long dry conditions (April 1997–May 1998) to the region and made the tropical forests more vulnerable to fires.[10] Due to the size of the area affected and the loss experienced by ASEAN Member States, these events will be remembered as the darkest days (both literally and metaphorically) by people in the ASEAN region. It was the first time in ASEAN history that TAP affected almost all States.

As indicated above, most forest fires in Indonesia and in other ASEAN Member States are anthropogenic in nature, generally for land management purposes such as land clearing for agriculture (especially for large plantations), grazing, fishing or hunting.[11] More specifically, the Indonesian forest fires in 1997–1998 mostly originated from logging concessions, large plantations, and transmigration projects and a small percentage of agriculture shifting, in Kalimantan and Sumatra, as confirmed by satellite images from the National Oceanographic and Aeronautic Administration (NOAA).[12]

Unfortunately, the Indonesian government did not learn from the lessons of the 1997–1998 catastrophes, because in the years since, the incidents of intended burning still continue to create TAP and blacken the sky of Singapore and most Malaysian States and cities. For example,

---

[8] Secretariat of the Convention on Biological Diversity (CBD), 'Impacts of Human-caused Fires on Biodiversity and Ecosystem Functioning, and their Causes in Tropical, Temperate and Boreal Forest Biomes' (2001) at 5, accessed 12 January 2015 at https://www.cbd.int/doc/publications/cbd-ts-05.pdf.

[9] The estimate of the area of forest burned in 1997–1998 in Indonesia varies, but most reports from research institutes and scientific reports conclude that the area affected by fires was more than 10 million hectares. See Luca Tacconi, 'Fires in Indonesia: Causes, Costs and Policy Implications' (Center for International Forestry Research (CIFOR) Occasional Paper No 38, 2003) at v, accessed 12 January 2015 at http://www.cifor.org/fire/pdf/pdf23.pdf.

[10] S Tahir Qadri (ed), *Fire-Smoke and Haze the ASEAN Response Strategy* (Manila: Asian Development Bank, 2001) at 36–7.

[11] Anne Guyon and Dicky Simorangkir, *The Economics of Fire Use in Agriculture and Forestry – A Preliminary Review for Indonesia* (Jakarta: Project Fire Fight South East Asia, 2002) at 3.

[12] Matthews, *supra* note 7 at 57. See also Goldammer, *supra* note 5 at 7.

forest fires in 1999 spread thick haze in Riau, Jambi, South Sumatra, Central and South Kalimantan and Singapore. In that particular period, the Ministry of Science, Technology and Environment of Singapore decided not to announce air pollution levels because the government did not want to 'drive away the tourists'.[13] In July 2000, fires burned again in parts of Sumatra's forests and the haze reportedly crept over the Malacca Straits to Singapore and Malaysia.[14] In July 2001, extensive fires were burning in some parts of Sumatra and Kalimantan and spread haze as far as Malaysia and Southern Thailand.[15]

Serious forest fires in Indonesia continued in 2002. NOAA detected at least 222 hot spots throughout Riau Province, resulting in the local residents of Riau having to wear masks.[16] Fires also occurred in Kalimantan, forcing the residents of Kalimantan to wear masks, with the haze also reported to have reached Brunei, Malaysia and Singapore.[17] In June 2003, the government of Riau Province encouraged people to wear masks due to the thick smoke from forest fires in that province.[18]

Unfortunately, similar incidents also occurred in June and July 2004, when choking haze blanketed Sumatra, Malaysia and Singapore, causing health problems for the people of Riau Province and disrupting air traffic in Sumatra and Malaysia.[19] Another set of large-scale forest fires occurred again from July to September 2005. Similar to previous fires, the 2005 fire originated from several large plantations in the Indonesian islands of Sumatra and Riau. It darkened the sky of several cities on the

---

[13] Down to Earth, 'Forest Fires Burn Again', *Down to Earth No. 43* (November 1999) accessed 12 January 2015 at http://www.downtoearth-indonesia.org/story/forest-fires-burn-again.

[14] Chris McCall, 'Legal cold water on Indonesia's forest-fire haze', *Christian Science Monitor* (25 July 2000) accessed 12 January 2015 at http://www.csmonitor.com/2000/0725/p7s2.html.

[15] 'Indon haze thickens over Pontianak, peat fires blamed', *Brunei Press* (13 July 2001) accessed 12 January 2015 at http://www.ecologyasia.com/news-archives/2001/jul-01/brunet.bn_news_bb_fri_jul13b4.htm.

[16] See 'Hundreds of hot spots found in Sumatra', *The Jakarta Post* (20 February 2002) accessed 12 January 2015 at http://www.ecologyasia.com/news-archives/2002/feb-02/thejakartapost.com_20020220.A07.htm.

[17] See 'Borneo hit by choking haze', *CNN* (20 September 2002) accessed 12 January 2015 at http://asia.cnn.com/2002/WORLD/asiapcf/southeast/09/20/indo.haze/.

[18] 'People told to don masks as haze, pollution worsen' *The Jakarta Post* (10 March 2003) accessed 12 January 2015 at http://www.thejakartapost.com/news/2003/03/10/people-told-don-masks-haze-pollution-worsen.html.

[19] See 'Indonesians Pay Heavy Price for Illegal Fires' and 'In Malaysia: Schools to Close if Haze Gets Worse', *The Straits Times* (24 June 2004).

two islands, severely affecting Kuala Lumpur and a further 32 towns on the Malaysian Peninsula. The air quality in Perak reached an 'unhealthy level'.[20] On 11 August 2005, the Malaysian government declared an emergency in two regions, closing workplaces and calling on mosques to hold special prayers for rain to rid the country of hazardous haze drifting from forest fires in Indonesia.[21]

The next catastrophic series of forest fires occurred between June–July 2013 in Riau, Sumatera and Kalimantan and forced the Malaysian government to declare a state of emergency and the government of Singapore to urge its people to stay indoors.[22] At the time of writing this chapter, some parts of the Indonesian forests are still burning and according to the Meteorological and Geophysics Agency of Indonesia, 261 hotspots have been detected in Riau Province alone[23] and have the potential of sending a 'black spell' to Singapore and Malaysia if the Indonesian government does not make serious efforts to put out the fires.

Finally, it is important to emphasise that the main cause of these fires is the use of fire in land clearing for large, commercial plantations in Riau, Sumatera and Kalimantan.

II. ASEAN'S RESPONSE TO TAP

The history of ASEAN Member States' efforts in protecting their environments only began in 1978. Environmental considerations were not in the minds of the founders when they established ASEAN in 1967. As

---

[20] See 'Forest Fires in Indonesia Blanket Malaysian Cities with Unhealthy Haze', *Associated Press* (3 August 2005) accessed 12 January 2015 at http://www.enn.com/top_stories/article/2225.

[21] See 'Malaysia haze triggers emergency', *BBC News* (11 August 2005) accessed 12 January 2015 at http://news.bbc.co.uk/2/hi/asia-pacific/4140660.stm. See also Vijay Joshi, 'Malaysia haze prompts state of emergency', *Associated Press* (11 August 2005).

[22] See Thalif Deen, 'Indonesia's Recurring Forest Fires Threaten Environment', *Inter Press Service (IPS)* (10 July 2013) accessed 12 January 2015 at http://www.ipsnews.net/2013/07/indonesias-recurring-forest-fires-threaten-environment/. See also John Vidal, 'Indonesia's forest fire smoke blows deeper into Malaysia', *The Guardian* (24 June 2013) accessed 12 January 2015 at http://www.theguardian.com/environment/2013/jun/24/indonesia-forest-fire-malaysia-singapore.

[23] See Banda Haruddin Tanjung, 'Kebakaran Hutan di Riau Semakin Memburuk (Forest Fires in Riau is Getting Worse)', *SINDO News* (7 February 2014) accessed 12 January 2015 at http://daerah.sindonews.com/read/2014/02/07/24/833693/kebakaran-hutan-di-riau-semakin-memburuk.

recorded in the 1967 ASEAN Declaration,[24] the main objectives of ASEAN were limited to the following: 'to accelerate economic growth, social progress and cultural development, to promote regional peace and stability, to promote active collaboration and mutual assistance on education, agriculture, industry, transportation, communications, and to maintain cooperation with existing international and regional organisations'.[25]

It took approximately 11 years for ASEAN leaders to realise that they needed to coordinate environmental policy to save the ASEAN region from further environmental degradation. Consequently, ASEAN leaders established the ASEAN Experts Group on Environment (AEGE) to discuss environmental protection through a series of annual meetings at the regional level. The first meeting was held in Jakarta in 1978, followed by the 1979 Penang meeting and the 1980 Manila meeting.

In 1989, the AEGE was replaced by a more formal meeting called the meeting of the ASEAN Senior Officials on the Environment (ASOEN). Different from the AEGE, the ASOEN meeting was attended by senior representatives of each ASEAN Member State. ASOEN has the power to establish subsidiary bodies in the form of Working Groups. Originally, ASOEN had six Working Groups (WGs), namely: (i) ASEAN Seas and Marine Environment; (ii) Environmental Economics; (iii) Nature Conservation; (iv) Environmental Management; (v) Transboundary Pollution and Environmental Information; and (vi) Public Awareness and Education. In 1998, the six WGs were restructured and simplified into three WGs: (i) WG on Nature Conservation and Biodiversity; (ii) WG on Coastal and Marine Environment; and (iii) WG on Multilateral Environmental Agreements. ASOEN meets once a year to discuss the reports of these WGs.[26]

The highest decision-making body on ASEAN environmental issues is the ASEAN Ministerial Meeting on Environment (AMME), which usually meets once a year. The main tasks of AMME include:[27]

---

[24] The ASEAN Declaration (Bangkok Declaration), adopted 8 August 1967, accessed 12 January 2015 at http://agreement.asean.org/media/download/20140117154159.pdf.

[25] *Ibid* at 2.

[26] Donna G Craig, Nicholas A Robinson and Koh Kheng-Lian (eds), *Capacity Building for Environmental Law in the Asian and Pacific Region: Approaches and Resources, Volume II* (Manila: Asian Development Bank, 2002) at 290.

[27] See 2009 ASEAN Socio-Cultural Community Blueprint 2009–2015 at 14–20, accessed 12 January 2015 at http://www.asean.org/archive/5187-19.pdf.

(i) addressing global environmental issues;
(ii) managing and preventing transboundary environmental pollution (transboundary haze pollution and transboundary movement of hazardous wastes);
(iii) promoting sustainable development through environmental education and public participation;
(iv) promoting environmentally sound technology;
(v) promoting quality living standards in ASEAN cities/urban areas;
(vi) harmonising environmental policies and databases;
(vii) promoting the sustainable use of coastal and marine environment;
(viii) promoting sustainable management of natural resources and biodiversity;
(ix) promoting the sustainability of freshwater resources;
(x) responding to climate change and addressing its impacts; and
(xi) promoting sustainable forest management.

The culmination of the recognition of the importance of environmental protection in ASEAN is embodied in the objectives of the 2007 ASEAN Charter,[28] which clearly states that one of the objectives of ASEAN is '[t]o promote sustainable development so as to ensure the protection of the region's environment, the sustainability of its natural resources, the preservation of its cultural heritage and the high quality of life of its peoples'.[29] This objective not only obligates ASEAN Member States to protect the ASEAN environment, but also to enhance the quality of life of the ASEAN people. It is unfortunate that at the time of writing this chapter, the annual TAP is still posing a serious threat to the ASEAN region.

A. General Environmental Soft Law Instrument Development

Similar to the development of major international environmental agreements, the legal basis for environmental management and cooperation in ASEAN was first addressed through soft law instruments and then followed by the establishment of hard law instruments. In general, ASEAN soft law instruments can be divided into two models: (i) general environmental instruments; and (ii) specific environmental instruments.

---

[28] Charter of the Association of Southeast Asian Nations, adopted 20 November 2007 (entered into force 15 December 2008) accessed 12 January 2015 at http://www.asean.org/archive/publications/ASEAN-Charter.pdf.

[29] *Ibid*, Art 1(9).

The first attempt to develop an ASEAN environmental soft law instrument culminated with the adoption of the 1981 Manila Declaration on the ASEAN Environment (1981 Manila Declaration).[30] This 1981 Manila Declaration can be classified as a 'general instrument' and its main objective called upon the ASEAN Member States 'to ensure the protection of the ASEAN environment and the sustainability of its natural resources so that it can sustain continued development with the aim of eradicating poverty and attaining the highest possible quality life for the people of the ASEAN countries'.[31] The 1981 Manila Declaration also provides policy guidelines to achieve that objective, by encouraging the ASEAN Member States to foster common awareness and enforce environmental protection measures. By the standards of 1981, the 1981 Manila Declaration can be considered quite advanced, as it incorporated basic elements of modern environmental law.

The 1981 Manila Declaration was followed by other 'general' soft law instruments in the form of declarations, resolutions, accords and action plans, which include:

(i) the 1987 Jakarta Resolution on Sustainable Development (1987 Jakarta Resolution);[32]
(ii) the 1990 Kuala Lumpur Accord on Environment and Development (1990 Kuala Lumpur Accord);[33]

[30] The 1981 Manila Declaration was signed by five ministers of the original ASEAN members (Emil Salim of Indonesia, Ong Pang Boon of Singapore, Tan Sri Ong Kee Hui of Singapore, Imelda Romualdez Marcos of Philippines and Thinakorn Bhandhugra of Thailand). Manila Declaration on the ASEAN Environment, signed 13 April 1981, accessed 12 January 2015 at http://environment.asean.org/manila-declaration-on-the-asean-environment.

[31] *Ibid* at para 1(a).

[32] Jakarta Resolution on Sustainable Development, adopted 30 October 1987, accessed 12 January 2015 at http://environment.asean.org/jakarta-resolution-on-sustainable-development.

[33] The Kuala Lumpur Accord on Environment and Development Issued by The ASEAN Ministers for the Environment at The Fourth ASEAN Ministers for the Environment Meeting, adopted 19 June 1990, accessed 12 January 2015 at http://environment.asean.org/the-kuala-lumpur-accord-on-environment-and-development-issued-by-the-asean-ministers-for-the-environment-at-the-fourth-asean-ministers-for-the-environment-meeting/.

(iii) the 1992 Singapore Resolution on Environment and Development (1992 Singapore Resolution);[34]
(iv) the 1994 Bandar Seri Begawan Resolution on Environment and Development (1994 BSB Resolution);[35]
(v) the 1997 Jakarta Declaration on Environment and Development (1997 Jakarta Declaration);[36] and
(vi) the 2000 Kota Kinabalu Resolution on the Environment (2000 Kota Kinabalu Resolution).[37]

Although these soft law instruments are general in nature, as they set out general environmental protection and management principles for the ASEAN region, they also include specific provisions dealing with TAP issues.

B. Specific Environment Soft Law Instrument Development on TAP

An examination of the comprehensive strategies and mechanisms included in the following specific soft laws will provide a comprehensive understanding of the ASEAN response to combat TAP and forest fires in the region.

1. The 1987 Jakarta Declaration

Although the 1987 Jakarta Declaration is general in nature, as it addresses several other aspects of environmental cooperation in ASEAN,[38] its primary concern is the mitigation of TAP and forest fires. This is reflected in the preamble of the 1987 Jakarta Declaration, in

---

[34] Resolution on Environment and Development, adopted 18 February 1992, accessed 12 January 2015 at http://Environment.Asean.Org/Resolution-On-Environment-And-Development-2/.

[35] Resolution on Environment and Development, adopted 26 April 1994, accessed 12 January 2015 at http://environment.asean.org/resolution-on-environment-and-development/.

[36] Jakarta Declaration on Environment and Development, adopted 18 September 1997, accessed 12 January 2015 at http://environment.asean.org/jakarta-declaration-on-environment-and-development/ (1997 Jakarta Declaration).

[37] 2000 Kota Kinabalu Resolution on the Environment, adopted 7 October 2000, accessed 12 January 2015 at http://cil.nus.edu.sg/rp/pdf/2000%20Kota%20Kinabalu%20Resolution%20on%20the%20Environment-pdf.pdf.

[38] The 1997 Jakarta Declaration concentrates on three environmental elements: the impact of economic growth on the ASEAN environment, transboundary pollution, and urging developed nations to fulfil their obligations under UNCED, Agenda 21 and the Basel Convention on the Control of Transboundary

which the ASEAN Ministers for the Environment state that they are '[d]eeply concerned on the ecological, economic, health and political impacts brought about by transboundary pollution caused by land and forest fires'.[39]

The 1987 Jakarta Declaration can be classified as the first ASEAN declaration that concentrates on the issue of TAP. It specifically calls upon ASEAN Member States to cooperate and render assistance wherever possible and to prevent and control all domestic sources of pollution and activities that could contribute towards transboundary pollution, including haze formation.[40]

It is important to note that when the 1987 Jakarta Declaration was negotiated, the sky in the ASEAN region was darkened by pollution from forest fires in Indonesia. The language of the 1987 Jakarta Declaration, however, failed to describe the actual situation of the environment in the ASEAN region at that moment and should have been expressed in stronger language. The 1987 Jakarta Declaration should have acknowledged that TAP is a serious threat to the ASEAN environment and its people and needs an immediate response both at the national and regional levels to prevent it from happening again. The idea of adopting 'strong language' in the 1987 Jakarta Declaration was proposed by Mr Datuk Law Hieng Ding, the Minister of Science, Technology and the Environment of Malaysia, but the Indonesian delegation opposed the idea on the basis that it constituted an intervention in its domestic affairs.[41]

2. The 1995 ASEAN Cooperation Plan on Transboundary Pollution

The issue of transboundary pollution was first highlighted in the 1990 Kuala Lumpur Accord and was later addressed in the 1992 Singapore Resolution and in the 1994 BSB Resolution. At the informal AMME held in Kuching, Malaysia on 21 October 1994, the ASEAN Ministers discussed the problem of transboundary pollution and agreed that

---

Movement of Hazardous Wastes and Their Disposal. 1997 Jakarta Declaration, *supra* note 36.

[39] *Ibid*, Preamble.

[40] *Ibid* at para 3.

[41] See Lim Kit Siang, 'Call on all Malaysians to give full support to the "Clean Air For Our Children – Let 1997 Haze be the last worst haze" mass signature campaign to demand compensation from Indonesia as the most effective way to ensure that the 1997 Haze is the last worst haze' (Address delivered at the Forum on 'Clean Air for Our Children – Let 1997 Haze be the last worst haze', 10 October 1997) accessed 12 January 2015 at http://www.limkitsiang.com/archive/1997/Oct97/sg658.htm.

ASEAN Member States should join forces to actively build Member States' expertise and capacity to address the problem and minimise its effects.[42]

In response to these previous commitments, the ASEAN Ministers for the Environment held a Meeting on the Management of Transboundary Pollution in Kuala Lumpur in June 1995 to develop the idea of ASEAN as one ecosystem by adopting the ASEAN Cooperation Plan on Transboundary Pollution (1995 Cooperation Plan).[43] The establishment of the 1995 Cooperation Plan was triggered by the severe forest fires in Indonesia in 1994 that had caused widespread TAP across the region.

The 1995 Cooperation Plan is one of the most significant initiatives for environmental cooperation taken amongst ASEAN Member States. It goes precisely to the heart of transboundary environmental problems in the region. The 1995 Cooperation Plan consists of three programme areas, namely: TAP, Transboundary Ship-borne Pollution, and Transboundary Movement of Hazardous Wastes. In relation to the issue of TAP, the 1995 Cooperation Plan asks ASEAN Member States to take the following actions: (i) assess the origin, main causes and nature of transboundary atmospheric pollution; (ii) prevent and control the source of haze at both national and regional level by applying environmentally sound technologies; and (iii) develop and implement national and regional emergency response plans.[44]

In the short term, the 1995 Cooperation Plan offers attainable measures, such as timely detection and prevention of forest fires through early warning systems, prohibition of biomass burning on development projects, reduction of any generation of pollution from local sources, and the promotion of investments in alternative uses of biomass. In addition, in the long-term, ASEAN Member States are encouraged to promote zero-burning methods and technologies.[45]

The 1995 Cooperation Plan also sets out certain activities that should be undertaken by the ASEAN Member States to prevent and mitigate TAP in the region, which include: (i) to establish National Focal Points; (ii) to expand the role of the ASEAN Specialised Meteorological Centre;

---

42 James Cotton, 'The "Haze" over Southeast Asia: Challenging the ASEAN Mode of Regional' (1999) 72 *Pacific Affairs* 331 at 343.

43 1995 ASEAN Cooperation Plan on Transboundary Pollution, adopted 17 June 1995 (1995 Cooperation Plan). See also ASEAN Cooperation on Transboundary Haze Pollution, accessed 12 January 2015 at http://environment.asean.org/asean-cooperation-on-transboundary-haze-pollution/.

44 *Ibid*, Objectives of the Programme Area 1.

45 *Ibid*, Strategies of the Programme Area 1.

(iii) to establish procedures for reporting/alerting of fires; (iv) to develop common air quality index; and (v) to develop a regional fire danger rating system and several other measures.

These objectives, strategies and activities should be relatively easy to achieve, because ASEAN Member States have been given clear guidance. Certain governments in ASEAN, in particular Indonesia, Malaysia, Singapore and Brunei, should be able to help each other to implement these measures, because they have experienced many bitter lessons from forest fires since the 1980s. However, the continuing failure of ASEAN Member States to implement these measures is a clear sign that the instrument is not strong enough to influence its Member States to cooperate on transboundary pollution issues.

Finally, the 1995 Cooperation Plan authorised ASOEN to undertake all measures for its implementation and evaluation, including acting as a coordinator for all issues regarding transboundary pollution and finding financial support to implement the 1995 Cooperation Plan. ASOEN is also responsible for producing a progress report on the implementation of the 1995 Cooperation Plan to the ASEAN Ministers for the Environment.[46] In implementing its responsibilities, ASOEN adopted an Appendix to the 1995 Cooperation Plan concerning the Establishment of the Haze Technical Task Force (HTTF).[47] The main responsibilities of the HTTF include demarcating critical areas of forest fires, identifying critical periods of haze, developing a system for the National Focal Point, facilitating the proper collection and dissemination of meteorological data, developing a monitoring system on the ground, and monitoring and reporting on the status of projects relating to the management and control of transboundary haze pollution involving international organisations and developed countries.[48]

3. The 1997 Regional Haze Action Plan

Following the establishment of the HTTF in 1995, the ASEAN Ministers for the Environment met in Jakarta to formulate another initiative called the ASEAN Regional Haze Action Plan (ARHAP).[49] The main objectives of ARHAP are (i) to prevent land and forest fires through better

---

46 *Ibid*, Implementation and Evaluation.

47 The HTTF was established by the 6th Meeting of ASEAN Senior Officials on the Environment (ASEON) in Bali, Indonesia, on 20–22 September 1995.

48 See 1995 Cooperation Plan, *supra* note 43.

49 1997 ASEAN Regional Haze Action Plan, adopted 23 December 1997, accessed 12 January 2015 at http://haze.asean.org/?page_id=213 (ARHAP). The

management policies and enforcement; (ii) to establish operational mechanisms to monitor land and forest fires; and (iii) to strengthen regional land and forest firefighting capabilities and other mitigating measures.[50] Florano summarised these objectives in three words: prevention, monitoring and mitigation.[51]

In addition, the main characteristics of ARHAP were based on three pillars: the spirit of volunteerism, the no-fault-finding rule, and the offer of assistance based on expertise and/or capability. Although these three pillars are not written into ARHAP or other ASEAN environmental agreements, according to Florano, they have been used as the basic model of environmental cooperation on haze pollution in ASEAN. The implementation of the first pillar for instance, can be witnessed in 1997, when Malaysia and Singapore assisted Indonesia in extinguishing the forest fires, while the second pillar was reflected in September 1997, when Soeharto apologised to his fellow ASEAN leaders about the failure of Indonesia to stop forest fires. In addition, the implementation of the third pillar was demonstrated in the distribution of tasks for the prevention of forest fires, where Indonesia, Singapore and Malaysia were assigned specific responsibilities based on their capability.[52] Unfortunately, this mode of cooperation suffered from poor implementation and supervision.

In preventive measures, ARHAP recognises the need to strengthen national policies and strategies to mitigate land and forest fires and the importance of curbing activities that may cause land and forest fires, including the prohibition of open burning and strict control of slash-and-burn practices. In addition, ASEAN Member States are also urged to enforce legislation on forest fires and to implement the air quality monitoring system both for mobile and stationary sources. The establishment of a national task force/committee and the utilisation of information technology to prevent the spread of fires are also encouraged.[53] ARHAP also established a Regional Monitoring Mechanism to strengthen the

---

ARHAP was established by the ASEAN Ministerial Meeting on Haze on 23 December 1997.

50 *Ibid*, Objectives.

51 Ebinezer R Florano, 'Regional Environmental Cooperation without Tears or Fear: The Case of the ASEAN Regional Haze Action Plan' (Paper delivered at the International Environmental Governance Conference, Paris, 15–16 March 2004) at 4, accessed 12 January 2015 at http://cleanairinitiative.org/portal/sites/default/files/articles-69345_paper.pdf.

52 *Ibid* at 5–8.

53 ARHAP, *supra* note 49 at paras 4–5.

region's early warning and monitoring system to provide an alert of the first outbreak of land and forest fires, an assessment of meteorological conditions, a prediction of the spread of smoke haze, a systematic tracking of the control and spread of fires and haze, and the necessary data to support enforcement action.[54] In addition, ARHAP reiterates the need to strengthen national and regional land and forest firefighting capability to combat future forest fires.[55]

Moreover, certain mechanisms introduced by ARHAP seem to be insufficient in dealing with multidimensional problems faced by ASEAN. The ASEAN policymakers need to adopt a radical approach and a new strategy to manage its TAP problem. Future cooperation on TAP should be replaced with a more legally oriented agreement that reflects the obligations of all ASEAN Member States and the consequences of non-compliance. This kind of regulation may be difficult to attain, but it would bring greater benefit for the region rather than repeating the soft law approach.

4. ASEAN Working Group on the Environment

Apart from previous mechanisms, ASEAN has also established Working Groups to be responsible for environmental protection. Under the current ASEAN structure, Working Groups are located under the ASOEN. So far, ASEAN has established three Working Groups, which include: (i) the Working Group on Nature Conservation and Biodiversity; (ii) the Working Group on Marine and Coastal Environment; and (iii) the Working Group on Multilateral Environmental Agreements. In addition, there are two other bodies that operate on the same level as Working Groups: the HTTF and another body responsible for Other Environmental Activities.[56]

In the area of TAP, instead of a Working Group, the ASOEN created the HTTF to coordinate the implementation of all agreements and policies on haze-related issues. However, the HTTF has similar functions and hierarchy to a Working Group. In order to carry out its functions, the HTTF established the Coordination and Support Unit (CSU) within the ASEAN Secretariat. It has three main functions: (i) coordination, planning and programme development; (ii) information management; and (iii) ARHAP management and support. To some degree, the CSU can implement certain projects as the fulfilment of existing agreements at a

---

54 *Ibid*, Regional Monitoring Mechanisms.

55 *Ibid*.

56 See ASEAN, *Second ASEAN State of the Environment Report 2000* (ASEAN Secretariat, 2000) at 165.

national level, but ideally, it should be done through cooperation with the relevant national authorities.

Certain programmes that have been implemented are focused on capacity-building to help local government units and local communities in developing comprehensive action plans for land and forest fire haze management, such as in West Kalimantan (Indonesia), and a programme that helped the local government of the Riau Province in drafting and disseminating the Governor's Decree on *Pusat Pengendalian Kebakaran Hutan dan Lahan* (Centre for Land and Forest Fires Management).[57] However, being a small operation with only five staff, the CSU seems inadequate to carry out the massive task of managing, implementing and supervising all transboundary haze problems.

Despite their limited capacity, according to Adelina Kamal, the ASEAN Senior Officer at the CSU, they can mitigate transboundary haze pollution in the region, if they have sufficient support from national authorities.[58] Some projects initiated by the CSU have been implemented, such as coordinating sub-regional firefighting arrangements; publishing 'Guidelines for the Implementation of Controlled Burning Practices'[59] and the 'Guidelines for the Implementation of the ASEAN Policy on Zero Burning';[60] and involving NGOs, local government agencies, local communities, and a plantation company in West Kalimantan (Indonesia) to discuss the implementation of the zero burning policy.

These programmes are a good sign that ASEAN has been involved directly at the national, as well as local, government levels to implement its programmes. However, the question still remains as to how much control ASEAN actually has if these programmes are not fully implemented by national and local governments.

C. ASEAN Hard Law Instruments

As compared to other 'mature' regional organisations, such as the European Union (EU), that have produced a number of hard law

---

57 See ASEAN's response to the forest fires and transboundary haze, accessed 12 January 2015 at http://haze.asean.org/?page_id=222.

58 Interview with Adelina Kamal at ASEAN Secretariat, Jakarta, 8 July 2004.

59 ASEAN, *Guidelines for the Implementation of Controlled Burning Practices* (ASEAN Secretariat, 2004).

60 ASEAN, *Guidelines for the Implementation of the ASEAN Policy on Zero Burning* (ASEAN Secretariat, 2003).

instruments, ASEAN only has a few hard law instruments on environmental management and protection. These instruments include: (i) the 1985 ASEAN Agreement on the Conservation of Nature and Natural Resources (1985 ASEAN Conservation Agreement);[61] (ii) the 1995 Agreement on the Cooperation for the Sustainable Development of the Mekong River Basin (1995 Mekong Basin Agreement);[62] and (iii) the 2002 ASEAN Agreement on Transboundary Haze Pollution (2002 ASEAN Haze Agreement).[63] It is important to note that the 1995 Mekong Basin Agreement cannot be fully considered an ASEAN agreement, because it was established before Cambodia, Laos and Vietnam had become ASEAN Member States.[64]

1. The 1985 ASEAN Conservation Agreement

As the first hard law instrument of ASEAN in environmental protection, the 1985 ASEAN Conservation Agreement has established a foundation for environmental protection cooperation in ASEAN. Since its conclusion in 1985, it has been regarded as one of the most progressive agreements for the preservation of nature.[65] It adopts a holistic approach rather than a sectoral one, because it integrates not only conservation issues but also all elements of an ecosystem within its provisions. The 1985 ASEAN Conservation Agreement has also been praised as a far-sighted document as it integrates several principles of the 1972 Stockholm Declaration,

---

[61] ASEAN Agreement on the Conservation of Nature and Natural Resources, adopted 9 July 1985 (entered into force 16 November 1997) accessed 12 January 2015 at http://agreement.asean.org/media/download/20140119153602.pdf (1985 ASEAN Conservation Agreement); Iwona Rummel-Bulska and Seth Osalo (eds), *Selected Multilateral Treaties in the Field of the Environment: Volume 2* (Cambridge: Grotius, 1991) at 343–51.

[62] Agreement on the Cooperation for the Sustainable Development of the Mekong River Basin, adopted 5 April 1995, accessed 12 January 2015 at http://www.mrcmekong.org/assets/Publications/policies/agreement-Apr95.pdf (1995 Mekong Basin Agreement).

[63] ASEAN Agreement on Transboundary Haze Pollution, signed 10 June 2002 (entered into force 25 November 2003) accessed 12 January 2015 at http://agreement.asean.org/media/download/20140119170340.pdf (2002 ASEAN Haze Agreement).

[64] At the time the 1995 Mekong Basin Agreement was established, Thailand was the only ASEAN Member State party to it. 1995 Mekong Basin Agreement, *supra* note 62.

[65] Koh Kheng Lian and Nicholas A Robinson, 'Strengthening Sustainable Development in Regional Inter-Governmental Governance: Lessons from the "ASEAN Way"' (2002) 6 *Singapore Journal of International and Comparative Law* 640 at 675.

those developed by the 1987 Brundtland Commission, and the many concerns, policies and institutional mechanisms discussed at the 1992 United Nations Conference on Environment and Development.[66]

The main strength of the 1985 ASEAN Conservation Agreement is in the content of its provisions, because it addresses important concepts and principles that serve as prerequisites in achieving sustainable development. Instead of simply protecting individual animals or plants, as adopted by the 1973 Convention on International Trade in Endangered Species of Wild Fauna and Flora,[67] the 1985 ASEAN Conservation Agreement aims to preserve the entire ecosystem and emphasises the importance of genetic diversity.[68] Its carefully crafted preamble integrates all the ingredients of a good environmental agreement such as intergenerational equity, balance between conservation and socio-economic development, and the importance of international cooperation.[69]

Although the 1985 ASEAN Conservation Agreement was not designed to manage TAP issues, it imposes a specific responsibility on the Parties to not pollute the air, soil, fresh water and the marine environment.[70] More specifically, it calls upon Parties to take all appropriate measures towards air quality management.[71] In addition, in the area of forest management, the 1985 ASEAN Conservation Agreement has addressed the importance of preventing forest fires and developing plans for reforestation and afforestation.[72]

---

66 Koh Kheng Lian, 'Ecosystem Management Approach to Biodiversity Conservation: The Role of ASEAN' in Craig, Robinson and Koh, *supra* note 26 at 310–11.

67 Convention on International Trade in Endangered Species of Wild Fauna and Flora, adopted 3 March 1973 (entered into force 1 July 1975) accessed 12 January 2015 at http://www.cites.org/sites/default/files/eng/disc/E-Text.pdf (CITES).

68 Dianne M Kueck, 'Using International Political Agreements to Protect Endangered Species: A Proposed Model' (1995) 2 *University of Chicago Law School Roundtable* 345 at 350.

69 See 1985 ASEAN Conservation Agreement, *supra* note 61 at 1.

70 *Ibid*, Art 11.

71 *Ibid*, Art 9.

72 Article 6(2) states in part that, '[t]hey shall, in particular, endeavour to: (a) control clearance of vegetation; prevent bush and forest fires; prevent overgrazing by, inter alia, limiting grazing activities to periods and intensities that will not prevent regeneration of the vegetation' and '(d) in reforestation and afforestation planning avoid as far as possible monoculture causing ecological imbalance'. *Ibid*.

However, it is also important to recognise that the language used in some of the provisions of the 1985 ASEAN Conservation Agreement is weak. For example, the wording of certain provisions uses the terms 'shall endeavour' and 'wherever possible' when referring to State responsibility, such as 'they *shall endeavour* to control clearance of vegetation; prevent bush and forest fires'.[73]

One of the main challenges in implementing the 1985 ASEAN Conservation Agreement is the lack of ratification. At the time of writing, seven countries have ratified or acceded to the 1985 ASEAN Conservation Agreement. It is also interesting to observe that the countries ratifying the 1985 ASEAN Conservation Agreement include, among others, Indonesia, the Philippines and Thailand, which are also known as having among the worst environmental records in the region.[74] On the other hand, the countries with better environmental records, especially Singapore, Malaysia and Brunei are cautious in ratifying it, without giving any clear explanation from their governments as to why that might be so.

Non-ratification by these three countries is also difficult to understand, as the 1985 ASEAN Conservation Agreement does not impose strict responsibilities on its Parties and there are no legal sanctions on those who do not comply. The only possible explanation for the reluctance of Singapore in ratifying the 1985 ASEAN Conservation Agreement is the fact that it requires ASEAN Member States to carry out environmental impact assessments (EIA) for every major and important project. This provision may be difficult for the government of Singapore to fulfil, because to date, the main environmental legislation of Singapore, the Environment Pollution Control Act, does not recognise the importance of EIA. The non-ratification of the 1985 ASEAN Conservation Agreement has created negative sentiment among ASEAN Member States. Indonesia, for example, used to refer to the non-ratification of the 1985 ASEAN Conservation Agreement as an excuse for not ratifying the 2002 ASEAN Haze Agreement.

---

[73] *Ibid.* See also Art 4, which states that 'Parties … *shall endeavour* to develop, adopt and implement management plans …' (emphasis added).

[74] The 1985 ASEAN Conservation Agreement, *supra* note 61 has been ratified or acceded to by Thailand, Indonesia, the Philippines, Vietnam, Laos, Myanmar and Cambodia, accessed 12 January 2015 at http://agreement.asean.org/agreement/detail/207.html. For Lao accession to the Agreement, http://cil.nus.edu.sg/1997/1997-second-protocol-for-the-accession-of-the-lao-people%E2%80%99s-democratic-republic-to-asean-agreements-signed-on-16-october-1997-in-subang-jaya-malaysia/ (accessed 12 January 2015).

2. The 2002 ASEAN Haze Agreement

The idea of creating a regional hard law agreement to manage TAP in ASEAN began in the early 1990s and reached its culmination after the catastrophic 1997–1998 forest fires in Indonesia that blanketed five ASEAN Member States. As mentioned previously, the issue of TAP was highlighted by the 1990 Kuala Lumpur Accord, the 1992 Singapore Resolution and the 1994 BSB Resolution. Although these documents do not specifically call for the establishment of a hard law instrument, they clearly suggest that ASEAN should engage in regional cooperation to mitigate the problems of TAP.

Following these previous commitments, the AMME held a specific session on the Management of Transboundary Pollution in Kuala Lumpur in June 1995 to develop the idea of the ASEAN region as 'one ecosystem' by adopting the 1995 Cooperation Plan.[75] One important programme area of the 1995 Cooperation Plan is TAP, under which the ASEAN Member States were urged to assess the main causes of the fire and to prevent the fires.

Unfortunately, the 1995 Cooperation Plan failed to mitigate the acute problems of TAP in the region. As a result, ASEAN Member States initiated several other programmes such as the establishment of the HTTF and the conclusion of ARHAP. Since these initiatives brought very little change in the practice of ASEAN Member States, in particular Indonesia, the ASEAN Secretariat also introduced the possibility of establishing a hard law agreement on transboundary haze. This idea began in 1998 through cooperation between the ASEAN Secretariat, the Asian Development Bank and UNEP. At the first stage, the ASEAN Secretariat was assisted by the Indonesian Centre for Environmental Law (ICEL) in conducting a background study.[76] At the same time, UNEP also conducted its own study and this was used as a comparison by the ASEAN Secretariat. These studies were subsequently discussed at the ASEAN ministerial level meeting and it was concluded that the idea of establishing a hard law instrument was viable.[77]

According to Ms Kamal of the ASEAN Secretariat, the Malaysian government had acted as the main driver for the 2002 ASEAN Haze

---

[75] See 1995 Cooperation Plan, *supra* note 43.

[76] See Indonesian Centre for Environmental Law, *Study on Possible Legal Instruments Dealing with Transboundary Haze Pollution* (1999) (unpublished); Indonesian Centre for Environmental Law, *Synthesis of Approaches to Addressing Transboundary Air Pollution Drawn from Global, Regional and Bilateral Agreements and Arrangements* (1999) (unpublished).

[77] See Interview with Adelina Kamal, *supra* note 58.

Agreement, supported by Singapore and Brunei. It came as a surprise that the Indonesian government was cooperative in the process and responded positively to the idea of establishing a hard law agreement. It is also interesting to note that the governments of victim States (Malaysia, Brunei and Singapore) have never insisted on applying stricter State responsibility standards to the polluter States. In fact, all the ASEAN countries agreed to adopt a typically 'ASEAN model' agreement, which avoids legal responsibility and relies heavily on national institutions.[78] These negotiations culminated in the signing of the 2002 ASEAN Haze Agreement during the World Conference on Land and Forest Fires Hazards in Kuala Lumpur on 10 June 2002.

a. Strengths of the 2002 ASEAN Haze Agreement The willingness of ASEAN Member States to embark on a hard law instrument can be considered as one of the main strengths of the 2002 ASEAN Haze Agreement, because it is in stark contrast to the 'ASEAN way' that usually prefers a soft law approach. Similarly, the fact that the Agreement has entered into force, as all ten ASEAN Member States have ratified it,[79] should be considered a significant development in comparison with the 1995 ASEAN Conservation Agreement, which has also entered into force but is still waiting for three instruments of ratification from the original ASEAN Member States of Malaysia, Singapore and Brunei.[80]

Another strength of the 2002 ASEAN Haze Agreement is the existence of a secretariat that is responsible for arranging meetings of Conference of the Parties, transmitting notifications and reports, and ensuring coordination with other international agencies.[81] In the reporting provision, the Agreement obliges its Parties to transmit reports on its implementation at a national level to the Secretariat in an agreed time

---

78 *Ibid.*

79 See ASEAN Instruments Ratification Status, accessed 12 January 2015 at http://agreement.asean.org/agreement/detail/238.html. Indonesia ratified the Agreement on 16 September 2014. Adella Anjani Putri, 'Indonesia Ratifies Haze Treaty after 12 Years of Stalling', *The Jakarta Globe* (17 September 2014) accessed 12 January 2015 at http://www.thejakartaglobe.com/news/indonesia-ratifies-haze-treaty-12-years-stalling.

80 The 1985 ASEAN Conservation Agreement, *supra* note 61 entered into force on 16 November 1997.

81 2002 ASEAN Haze Agreement, *supra* note 63, Art 19(2).

frame.[82] In addition, the 2002 ASEAN Haze Agreement has no reservation clause, therefore the Parties are not allowed to make any reservations[83] and it provides possibilities to adopt new protocols, amendments and annexes.[84] Furthermore, another strength of the 2002 ASEAN Haze Agreement is the existence of the ASEAN Transboundary Haze Pollution Control Fund, which is intended to be utilised to prevent and mitigate the reoccurrence of transboundary haze pollution.[85] However, with only US$500,000 in the fund,[86] it is far too small an amount compared to the magnitude of the problem.

Another aspect of the 2002 ASEAN Haze Agreement that can be considered a strength is the adoption of the mandatory language 'shall' in its provisions, such as '[the] parties shall cooperate to prevent and monitor transboundary haze pollution and to control source of fires'.[87] In addition, it states that '[the] parties shall take legislative, administrative and/or other measures'.[88] According to Tan, such a general provision adopted by the AHA is 'wide enough to embrace the enactment of strong anti-burning laws, the prosecution of offenders and the imposition of adequate penalties to deter the use of fires effectively'.[89]

The last important factor of the 2002 ASEAN Haze Agreement that can be considered a strength is the establishment of the ASEAN Coordinating Centre for TAP (ASEAN Centre). The functions of the ASEAN Centre are to facilitate cooperation and coordination among the Parties in managing the impact of forest fires, including establishing and maintaining regular contact with National Monitoring Centres,[90] donor

---

82 *Ibid*, Art 25.

83 *Ibid*, Art 30.

84 *Ibid*, Arts 21–23.

85 *Ibid*, Art 20(2).

86 See ASEAN, Media Release, '14th Informal ASEAN Ministerial Meeting on the Environment and 9th Meeting of the Conference of the Parties to the ASEAN Agreement on Transboundary Haze Pollution' (25 September 2013) accessed 12 January 2015 at http://www.asean.org/news/asean-secretariat-news/item/14th-informal-asean-ministerial-meeting-on-the-environment-and-9th-meeting-of-the-conference-of-the-parties-to-the-asean-agreement-on-transboundary-haze-pollution.

87 2002 ASEAN Haze Agreement, *supra* note 63, Art 4(1).

88 *Ibid*, Art 4(3).

89 Alan Khee-Jin Tan, 'The ASEAN Agreement on Transboundary Haze Pollution: Prospects for Compliance and Effectiveness in Post-Suharto Indonesia' (2005) 13 *New York University Environmental Law Journal* 647 at 662–3.

90 Under the 2002 ASEAN Haze Agreement, *supra* note 63, Art 7, each Party shall designate one or more bodies to function as National Monitoring

States, and disseminating practical information on the implementation of the Agreement.[91] The ASEAN Centre, to some extent, will play a greater role in the future, as it has more functions compared to the Secretariat. The existing functions of the ASEAN Centre, as described in the Annex, can be extended in the future upon consensus of the Parties.[92]

b. Weaknesses of the 2002 ASEAN Haze Agreement One of the weaknesses of the 2002 ASEAN Haze Agreement is the 'absence' of a special secretariat, as it uses the existing ASEAN Secretariat as its secretariat. In addition, the Secretariat under the Agreement has no power to monitor ASEAN Member States' compliance as adopted by many international treaties. The power of the Agreement's Secretariat is limited to administrative work, such as arranging meetings, transmitting information, reporting, and notification, and arranging coordination with related international agencies to enhance the functions of the Secretariat.[93] This constitutes the principal weakness of the 2002 ASEAN Haze Agreement and needs to be improved in the future. It is ineffective to have a legally binding regional agreement that has no institution to ensure its enforcement.

Another serious weakness of the 2002 ASEAN Haze Agreement is the absence of measures that can be taken by the ASEAN Secretariat if one or more Member States fail to comply with the Agreement. Likewise, the 2002 ASEAN Haze Agreement is not equipped with a special body to conduct observation and inspection. The Agreement merely states that when transboundary pollution occurs and originates from within their jurisdictions, they should respond promptly to a request sought by the affected State or States.[94] This provision does not even require the polluter State to inform its neighbouring States that forest fires have occurred within their national jurisdictions, if victim States are silent about it. In addition, a State's obligation under the 2002 ASEAN Haze Agreement is too general and weak as it only contains the 'suggestion' to the Parties to cooperate in developing and implementing measures to prevent and monitor transboundary haze pollution. Meanwhile, the Parties are also expected to take legislative, administrative and other

---

Centres to monitor all fire prone areas, land and forest fires, the environmental conditions conducive to such fires, and haze pollution arising from such fires.

[91] For more detail, see 2002 ASEAN Haze Agreement, *supra* note 63, Art 5 and Annex.

[92] *Ibid*, Art 23.

[93] *Ibid*, Art 19(2).

[94] *Ibid*, Art 4(2).

measures to implement their obligations under the Agreement.[95] This demonstrates that the implementation of the Agreement depends heavily on the political will of individual States.

In relation to standards of performance, the 2002 ASEAN Haze Agreement only depends on procedural matters rather than measurable ones. Procedural performance standards adopted by the Agreement are monitoring, assessment and prevention. Under monitoring mechanisms, the Parties are required to monitor all fire-prone areas and all land and/or forest fires.[96] Meanwhile, the ASEAN Centre is expected to assess consolidated and analysed data sent by the respective National Monitoring Centres to determine the risk to human health or the environment and then distribute them, where possible, to the National Focal Points.[97]

In the prevention mechanism, the 2002 ASEAN Haze Agreement requires the Parties to develop and implement legislative and policy measures, strengthen the local firefighting capabilities, and to halt open burning and the use of fire for land clearing.[98] These procedural standards clearly indicate that the Agreement imposes no measurable standards on the Parties. Such measurable standards as a timetable in which to eradicate TAP from the ASEAN region or to establish emission reduction targets are absent from the text. In addition, the Agreement imposes no legal sanctions on polluters.

As a result of the absence of explicit standards and liability provisions, the 2002 ASEAN Haze Agreement limits itself to '*consultation and negotiation*' in its settlement dispute provision.[99] This is another sign that the ASEAN Member States are still hesitant in pursuing a legal solution for their transboundary environmental problem. Unlike similar agreements in other regions, the Agreement provides no provision on the possibility of using international courts or arbitration tribunals to settle TAP disputes among Parties. This characteristic pre-empts the enforcement of compliance through the legal principles of State responsibility and international liability. Therefore, it would be difficult for the victim States or their citizens to claim compensation for losses incurred as a result of TAP from other States. However, it is important to note that in theory, any dispute may still be referred to a judicial or arbitral resolution as long as the parties involved agree to do so. According to Tan, by not providing for what happens should consultation or negotiation fail, the

---

95 *Ibid*, Art 4(1) and (3).
96 *Ibid*, Art 7.
97 *Ibid*, Art 8.
98 *Ibid*, Art 9.
99 *Ibid*, Art 27 (emphasis added).

Agreement actually promotes even more uncertainty for ASEAN's legal and political order.[100]

In addition to these weaknesses, by its name, the 2002 ASEAN Haze Agreement concentrates on 'haze' rather than on 'pollution' issues. The use of the term 'haze' rather than 'pollution', according to Tay, understates the risk to human health. In fact, the emissions from forest fires have a greater negative impact than normal city pollution. Use of the term 'haze' also suggests that the presence of haze in Malaysia, Singapore and Brunei depends on wind direction.[101] Therefore, the emphasis and use of the term 'haze' should be discouraged, because the main problem here is 'atmospheric pollution' with all its elements.

According to Florano, a possible explanation for these weaknesses is that the 2002 ASEAN Haze Agreement was drafted within the framework of the 'ASEAN Way', 'that is a decision-making style in which decisions are made through small groups and committees, and follow the non-confrontational consensus model'.[102] The cultural roots of this practice come from the Indonesian words *musyawarah* (consultation) and *mufakat* (agreement or unanimous decision).[103] This decision-making style is widely practiced in the region, especially in Indonesia, Malaysia, Brunei, and to a lesser degree in Singapore and the southern part of Thailand. The influence of the 'ASEAN Way' is reflected in the absence of a liability provision in the 2002 ASEAN Haze Agreement.[104]

---

[100] Tan, *supra* note 89 at 664–5.

[101] Simon S C Tay, 'Southeast Asian Fires: The Challenge for International Law and Development' (1999) 11 *Georgetown International Environmental Law Review* 241 at 272.

[102] Florano, *supra* note 51 at 7. See also Pushpa Thambipillai, 'Negotiation Styles' in Kernial Singh Sandhu, Sharon Siddique, Chandran Jeshurun, Ananda Rajah, Joseph LH Tan and Pushpa Thambipillai (eds), *ASEAN Reader* (Singapore: Institute of Southeast Asian Studies, 1992) at 72–5.

[103] The origin of word the '*musyawarah*' actually comes from the Arabic word '*shawarah*', which means 'consultation'. These words can be found in the *Holy Qur'an*, Chapter 42 (As Shura), verse 38, which states that '[they] who conduct their affairs by mutual consultation'. Ali beautifully comments on this Chapter by saying: 'their conduct in life is open and determined by mutual consultation between those who are entitled to a voice …'. See Abdullah Yusuf Ali, *The Holy Qur'an: Text, Translation and Commentary* (Brentwood, MD: Amana Corporation, 1989) at 1257, nn 4578 and 4579.

[104] Florano, *supra* note 51 at 7; see also Thambipillai, *supra* note 102 at 72–5.

However, it is also important to underscore that '*musyawarah*' and '*mufakat*' are merely methods of reaching an agreement, and not specifically designed to avoid legal responsibility. As a tool, *musyawarah* can also give birth to a more legalistic document that contains stricter liability provisions with a wider range of dispute settlement provisions. Several ASEAN trade agreements recognise more measurable provisions, which set clear and strict timeframes in their dispute settlement mechanisms.[105] The result of *musyawarah* (that is, *mufakat*) depends on the actors involved in that process. As such, to blame *musyawarah* and *mufakat* as a cause of the weak liability provisions in the Agreement is not accurate.

Another obvious weakness of the 2002 ASEAN Haze Agreement is the absence of a regional body that can ensure its implementation. The Agreement, for instance, will establish the ASEAN Coordinating Centre for Transboundary Haze Pollution Control (ASEAN Centre) to coordinate and facilitate cooperation among parties in managing the impacts of TAP. However, the ASEAN Centre only works on the basis that the national authority will act first to stop the spread of the fire. It is the role of the national authority to declare a state emergency and it depends on the discretion of the national authority to ask for assistance from the ASEAN Centre. According to Tan, such arrangements seem at odds with the whole spirit of the 2002 ASEAN Haze Agreement, which requires timely cooperation and consultation. Ideally, there should have been a procedure for the ASEAN Centre to prescribe urgent, remedial measures to be undertaken by the Parties, accompanied by possible sanctions for non-compliance. In addition, since there is no clear provision on the amount and nature of information to be provided by the Parties, the ASEAN Centre does not appear to have the authority to require further data and information should the initial data provided by the Parties be inaccurate or inadequate.[106]

Apart from the identified weaknesses, the 2002 ASEAN Haze Agreement needs to be improved in several ways to achieve its goals. For instance, the Agreement has no provision to distinguish atmospheric pollution from other types and sources of pollution. This discouraged Indonesia from proceeding faster with its ratification process, since it

---

105 Since 1996, the AMME meetings have established the *Protocol on Dispute Mechanism Settlement*, which was revised in 2004 with the *ASEAN Protocol on Enhanced Dispute Settlement Mechanism*, adopted 29 November 2004, accessed 12 January 2015 at http://www.asean.org/news/item/asean-protocol-on-enhanced-dispute-settlement-mechanism.

106 Tan, *supra* note 89 at 666–7.

perceives that the Agreement was designed to target Indonesia as the main atmospheric polluter in ASEAN.

Finally, although by definition, the 2002 ASEAN Haze Agreement is a hard law document, it possesses soft law characteristics. In Florano's language, with the absence of all necessary elements of a 'good agreement', the Agreement is a 'blind and toothless tiger'.[107] This characterisation has some merit, as forest fires and consequent TAP still occurs on a regular basis. The prospect of the 2002 ASEAN Haze Agreement helping ASEAN to prevent and mitigate its TAP is still far from realisation.

c. Implementation challenges The main barrier to the full implementation of the 2002 ASEAN Haze Agreement in its first 12 years was the delay of the Indonesian Parliament in ratifying the Agreement. The Agreement was only ratified on 16 September 2014, after 12 years of waiting and uncertainty. Based on the author's personal discussion with Ms Liana Bratasida from the Ministry of Environment, the Indonesian Parliament might have tried to use the Agreement's ratification process as a bargaining chip with Singapore[108] to put pressure on Singapore to agree to sign an agreement on extradition with Indonesia.[109] After Indonesia decided to impose an export ban on sand and gravel exports from several Indonesian small islands in Riau Province to Singapore in 2007,[110] the Singaporean government showed some willingness to sign an extradition

---

[107] Florano, *supra* note 51 at 16.

[108] The interview took place at the Ministry of Environment during workshop preparation on Environmental Law Enforcement between the Ministry of Environment and the Supreme Court of Indonesia (Jakarta, 13 June 2008).

[109] Natasha Hamilton-Hart, 'Indonesia-Singapore Relation', in N Ganesan and Ramses Amer (eds), *International Relation in Southeast Asia between Bilateralism and Multilateralism* (Singapore: Institute of Southeast Asian Studies, 2010) at 206–8. Indonesia has been trying to persuade Singapore to sign an extradition agreement, because it believes that most Indonesian corruptors use Singapore as their safe haven, including those who are involved in illegal logging and forest fires.

[110] Indonesian sand was said to have helped Singapore expand its size from only 581.5 square kilometres in 1960 to the current 650 square kilometres. See 'Singapore's Shifting Sands', *Asia Sentinel* (30 January 2007) accessed 12 January 2015 at http://www.asiasentinel.com/econ-business/singapores-shifting-sands/. The Indonesian public and Parliament supported the ban due to concerns over the environmental damages that the sand exploitation activities caused. See 'Sand Mining Destroys Community Resources', *Down to Earth, No 51* (November 2001) accessed 12 January 2015 at http://www.downtoearth-indonesia.org/story/sand-mining-destroys-community-resources.

treaty.[111] In the end, the two governments agreed to conclude a treaty on extradition and a treaty on defence cooperation as a package.[112] However, the package deal could not get through the Indonesian Parliament because it was deemed as 'not favourable to Indonesia'.[113] This issue slowed down progress in the Indonesian Parliament to ratify the 2002 ASEAN Haze Agreement, which became a major barrier to the implementation of the Agreement in the first 12 years.

Weak environmental law enforcement in Indonesia will also create a significant barrier for the implementation of the 2002 ASEAN Haze Agreement, because until now, the Indonesian government has been struggling to improve the capacity of its judicial system, particularly in the environmental field. The continuation of forest fires in Indonesia is an obvious example of the failure of the judiciary and law enforcement in Indonesia to combat forest fires.

Without Indonesia's active implementation, the 2002 ASEAN Haze Agreement will only have minimal impact in mitigating TAP in the region. Indonesia, as the main source of TAP in ASEAN, should take a leading role in combating forest fires within its jurisdiction, while other ASEAN Member States, especially Brunei, Malaysia and Singapore, as victim States, should help Indonesia to implement its responsibilities based upon mutual cooperation as stated in the Agreement.

Apart from these factors, there are some other aspects that may hinder the implementation of the 2002 ASEAN Haze Agreement. The

---

[111] See 'Sand in Singapore's Gears', *Asia Sentinel* (5 March 2007) accessed 12 January 2015 at http://www.asiasentinel.com/politics/sand-in-singapores-gears/.

[112] Donald Greenlees, 'Indonesia and Singapore Sign Two Landmark Treaties', *The International Herald Tribune* (27 April 2007) accessed 12 January 2015 at http://www.nytimes.com/2007/04/27/world/asia/27iht-indo.3.5474698.html?_r=0.

[113] 'Indonesia's House wants extradition treaty with S'pore', *The Jakarta Post* (25 October 2013) accessed 12 January 2015 at http://www.thejakartapost.com/news/2013/10/25/house-wants-extradition-treaty-with-s-pore.html. See also Parliamentary Debates Singapore: Official Report, vol 83 at col 988 (16 July 2007) (Minister George Yeo) accessed 12 January 2015 at http://sprs.parl.gov.sg/search/topic.jsp?currentTopicID=00002537-WA&currentPubID=00004741-WA&topicKey=00004741-WA.00002537-WA_1%2B%2B. Controversial provisions include, for example, an article in the Extradition Agreement providing that Indonesia cannot ask Singapore to transfer assets of Indonesian citizens that take refuge in Singapore, and an article under the Defence Agreement that allows Singaporean armed forces to conduct military exercises in Indonesian water and airspace.

inadequate capacity of the CSU within the ASEAN Secretariat to deal with TAP in the region is one factor, because it is only supported by very limited resources.[114] In addition, the fact that the CSU has no direct power to implement their programmes at the national and local levels might become another barrier that needs to be addressed. The enforcement mechanisms within ASEAN are also very weak and remain heavily dependent on national institutions. This makes it even more difficult for ASEAN to interfere if its Member States fail to comply with the Agreement. However, the drafters of the 2002 ASEAN Haze Agreement understood that the implementation of the Agreement would only work on a voluntary basis and very much relied on the willingness of each individual Member State to comply.

d. The enforcement of the 2002 ASEAN Haze Agreement at the domestic level The regional legal framework differs significantly from that at the national level, as most ASEAN Member States have national laws that deal with environment and forestry-related matters. The constitutions of Indonesia and the Philippines, for example, recognise the constitutional rights of their citizens to a clean and healthy environment. These constitutional rights then translate into their national environmental legislation. In the case of Indonesia, where most ASEAN TAP originates, they at least have a complete set of legislation that can be utilised for the prevention and prosecution of environmental crimes, including forest fires.

The Indonesian Law No 32/2009 on Environmental Protection and Management Act (EPMA) provides a complete set of provisions that deal with almost all the elements of modern environmental law.[115] The EPMA recognises all international principles on environmental management, including:

(i) sustainable development;
(ii) intergenerational equity principle;
(iii) pre-cautionary principle;
(iv) strict liability principles;

---

[114] For example, the core staff of CSU consists of only one Senior Officer, two Technical Officers, one Technical Assistant and a secretary, accessed 12 January 2015 at http://www.cornerstone-msc.net/peat-portal/newsmaster.cfm?&menuid=55&action=view&retrieveid=457.

[115] Republic of Indonesia, Environmental Protection and Management Act, Law No 32/2009 (3 October 2009) accessed 12 January 2015 at http://faolex.fao.org/docs/pdf/ins97643.pdf (EPMA).

(v) polluter pay principle;
(vi) access to environmental information;
(vii) public participation in environmental decision; and
(viii) environmental impact assessment (EIA).

The EPMA also recognises the legal standing of NGOs and has a special provision on class action. In addition, it also provides all necessary measures for its implementation, as it has complete rules on administrative, civil and criminal law enforcement. In addition to these common features of modern environmental law, the EPMA has even created some breakthroughs in environmental law development by mandating the government to conduct strategic environmental studies, spatial planning, environmental audits and risk assessments, and environmental-based budgeting.[116]

In relation to forest fires, the EPMA clearly states that it is a criminal offence for a private company or one of its representatives to clear forest land by using fire.[117] In addition, the Forestry Act also clearly states that the licence holder is responsible for any fire that occurs on its concession.[118] The EPMA and the Forestry Act complement Indonesia's Government Regulation (GR) No 4/2001 on Forest Fires[119] and GR No 45/2004 on Forest Protection.[120] The GR No 4/2001 has existed for more than a decade, but only a few cases have been investigated and prosecuted under this law. It is also important to mention the newly enacted law, Law No 18/2013 on Prevention and Eradication of Forest Destruction,[121] which imposes significant penalties for forestry-related crimes.

---

116 For more comprehensive information, see Laode M Syarif, 'Current Development of Indonesia Environmental Law' (2010) *IUCN Academy Environmental Law e-Journal, Vol 1*, accessed 12 January 2015 at http://gnhre.org/literature/linked-journals/the-iucn-academy-of-environmental-law-e-journal.

117 See EPMA, *supra* note 115, Art 69(1).

118 Republic of Indonesia, Forestry Act (Kehutanan), Law 41/1999 (30 September 1999) accessed 23 January 2014 at http://www.setneg.go.id/components/com_perundangan/docviewer.php?id=34&filename=UU_no_41_th_1999.pdf, Art 47.

119 Republic of Indonesia, Government Regulation (GR) No 4/2001 on Forest Fires, accessed 23 January 2015 at http://www.setneg.go.id//components/com_perundangan/docviewer.php?id=585&filename=PP_No_4_th_2001.pdf.

120 Republic of Indonesia, Government Regulation (GR) No 45/2004 on Forest Protection, accessed 23 January 2015 at http://www.dephut.go.id/uploads/files/45_04.pdf.

121 Republic of Indonesia, Prevention and Eradication of Forest Destruction (Pencegahan Dan Pemberantasan Perusakan Hutan) Law No 18/2013, accessed

Unfortunately, the above laws and regulations are not subject to serious enforcement. As a result, only a few forest fire cases have been brought to court, with most of the cases involving 'small players', such as labourers, field operators and traditional farmers who live near large plantations.

Serious enforcement of the above laws was only conducted in one case against a large plantation, PT Adei Plantation, where the manager Mr Gobi, a Malaysian citizen, and the corporation were prosecuted under the old environmental act, Law No 23/1997 on Environmental Management[122] (*PT Adei Plantation Case*).[123] The court sentenced Mr Gobi to eight months in jail and imposed a fine of Rp 100 million (about US$10,000). In addition to criminal enforcement, the Ministry of Environment sued PT Adei Plantation and Industry in a civil case and demanded the amount of Rp 273 billion (about US$30.3 million) to compensate for the economic and environmental damages. The result of the civil settlement of this case was not announced because both parties agreed to settle out of court.

After more than ten years since the *PT Adei Plantation Case*, the first significant judgment on environmental liability in Indonesia was handed down in Meulaboh-Aceh in the *Rawa Tripa Case*,[124] where the District Court of Meulaboh ordered PT Kallista Alam to pay damages in the amount of IDR 114,333,419,000 (US$1,335,670) and environmental rehabilitation in the amount of IDR 251,765,250,000 (US$2,941,170).[125] In addition, the judge also ordered the defendant to replant 1,000 hectares of destroyed peat land.[126] It is important to note that the chief judge of this landmark decision was Judge Rachmawati, who recently completed her environmental law training and was appointed by the Supreme Court of Indonesia as a Certified Environmental Judge. This decision has been

---

23 January 2015 at http://ppid.dephut.go.id/files/info_kemenhut/UU_Nomor_18_tahun_2013_(P3H)2.pdf.

[122] Republic of Indonesia, Environmental Management Act, Law No 23/1997, accessed 12 January 2015 at http://faolex.fao.org/docs/html/ins13056.htm.

[123] *State v PT Adei Plantation/Bangkinang Forest Fires* Criminal Case (2001).

[124] Republic of Indonesia Ministry of Environment, Fact Sheet, at 1–2, accessed 12 January 2015 at http://www.menlh.go.id/DATA/kebakaran_hutan.PDF.

[125] *Ibid.*

[126] *Ibid.*

appealed to the High Court, but the government and many environmental activists are optimistic that this ruling will stand.[127]

It is also important to note that after the *Rawa Tripa Case*, the Ministry of Environment in collaboration with other law enforcement agencies (police, prosecutors and the Ministry of Forestry) are investigating several more large plantations that were allegedly involved in the burning of forests in 2013 and 2014. These cases are still under investigation and according to the latest information from the UKP4 (Presidential Working Unit for Supervision and Management of Development) are likely to be brought to court shortly.

It is expected that after the current Indonesian Parliament ratification of the 2002 ASEAN Haze Agreement, the Indonesian government will enhance its commitment to the enforcement of its domestic laws and at the same time, other ASEAN Member States, especially Singapore and Malaysia, are also expected to contribute more substantial funds to mitigate forest fires in Indonesia.

III. CONCLUSION

Based on the above analysis, it can be concluded that the 2002 ASEAN Haze Agreement is ineffective in preventing and mitigating transboundary atmospheric pollution in the ASEAN region, due to the following reasons: (i) most of the Agreement's provisions include very weak language; (ii) most of the implementing mechanisms in the Agreement are given to the national authorities; (iii) the Agreement does not provide any clear sanctions for the non-compliance of ASEAN Member States; and (iv) the absence of a 'specific regional agency' that is responsible for the coordination and implementation of the Agreement at the national level also contributes to its slow implementation.

However, it is important to underscore that the 2002 ASEAN Haze Agreement has successfully 'influenced' ASEAN Member States to develop their national legislation in regard to the prohibition of fire for land clearing, because most ASEAN Member States have now enacted such laws. In addition, the current developments regarding the serious enforcement of environment and forestry-related laws against large plantations and corporations that were involved in the forest fires of 2013

---

[127] Republic of Indonesia Ministry of Environment, Media Release, 'MenLH Menangkan Gugatan Kasus Kebakaran Lahan di Rawa Tripa-Aceh' (13 January 2014) accessed 12 January 2015 at http://www.menlh.go.id/menlh-menangkan-gugatan-kasus-kebakaran-lahan-di-rawa-tripa-aceh/.

and 2014, together with the enactment of Presidential Instruction No 6/2013 on A Moratorium for the New Concession on Peat Land and Primary Forest, have brought new hope that Indonesia is working to enforce its national laws.

12. 'Can't we even share our maps?': Cooperative and unilateral mechanisms to combat forest fires and transboundary 'haze' in Southeast Asia

Alan Khee-Jin Tan

This chapter assesses the transboundary smoke pollution problem in Southeast Asia arising from recurring forest and land fires in Indonesia. It analyses the problem from the perspective of the 2002 ASEAN Agreement on Transboundary Haze Pollution[1] (2002 ASEAN Haze Agreement). The initial focus in this chapter is on the 2002 ASEAN Haze Agreement's systemic weaknesses as well as relevant political and socio-economic factors within Indonesia. These factors conspire to ensure that, even with Indonesian ratification of the 2002 ASEAN Haze Agreement, the problem will still not be effectively resolved. The chapter then analyses the promises and pitfalls of cooperative mechanisms that have been proposed to deal with the fires and haze problem. The main issue here is the sharing of plantation concession maps that will supposedly enable the perpetrators to be identified and then prosecuted. This chapter then concludes by highlighting Singapore's adoption of unilateral, extraterritorial legislation to deter companies from engaging in land and forest burnings.

---

[1] ASEAN Agreement on Transboundary Haze Pollution, signed 10 June 2002 (entered into force 25 November 2003) accessed 14 January 2015 at http://agreement.asean.org/media/download/20140119170340.pdf (2002 ASEAN Haze Agreement).

I. OVERVIEW: OF FIRES AND MAPS

In recent years, large-scale forest and land fires in Indonesia, together with the resulting transboundary smoke or 'haze' pollution (as the smoke pall is euphemistically called in the region), have recurred so frequently that the governments and people in affected States in Southeast Asia have come close to exhausting their diplomatic and political means to deal with the problem. Until the latest series of forest fires and resulting haze in 2013, the most serious occurrence in recent memory had been in 1997, the year Indonesia was engulfed by the Asian financial crisis and the ensuing political upheavals that led to the ouster of President Suharto. The fires and haze of 1997 had prompted the present author to make a case then for Indonesian State responsibility for failure to control the fires.[2] Since that time, the fires and transboundary haze have continued to be an almost annual occurrence affecting Malaysia, Singapore and Brunei, with serious episodes taking place in 1999, 2002, 2004, 2006 and 2010.

In short, the problem keeps recurring and shows no sign of long-term resolution. Indeed, even as the memories of 1997 began to recede, the haze came back with a vengeance in June 2013 when massive fires in Indonesia caused unprecedented smoke pollution in Singapore and Malaysia. The air pollution measurement index in Singapore (the PSI – Pollutant Standard Index) hit a historical high of 401 on 21 June 2013, well over the 'hazardous' level of 300 and far exceeding the worst level of 226 in 1997.[3] In fact, the PSI reading breached the 300 level in Singapore on four successive days in June 2013 before abating gradually toward the end of that month.

In the intervening years since the last serious haze of 1997, the much-vaunted ASEAN Haze Agreement was adopted in 2002 by the ten-member Association of Southeast Asian Nations (ASEAN) to tackle the problem. The one Member State that matters most to the resolution of

---

[2] Alan Khee-Jin Tan, 'Forest Fires of Indonesia: State Responsibility and International Liability' (1999) 48 *International and Comparative Law Quarterly* 826.

[3] See 'Singapore haze hits record high from Indonesia fires', *BBC News* (21 June 2013) accessed 14 January 2015 at http://www.bbc.com/news/world-asia-22998592.

the problem – Indonesia – finally agreed in September 2014 to ratify that Agreement, a full 12 years after its adoption.[4]

For the affected States and their populations, the haze of June 2013 was a case of déjà vu with a wincing 'oh no, not again' familiarity. In Singapore, the smoke pall rankled the public greatly and became the biggest health crisis since the SARS flu epidemic of a decade earlier. After all, the 2002 ASEAN Haze Agreement had been crafted precisely to lay down a framework for cooperative mechanisms to deal effectively with the problem. Why did it take Indonesia so long to ratify it? Who was responsible for the fires and why could they not be controlled? This chapter seeks to offer some answers to these vexing questions and to assert that Indonesian ratification of the Agreement offers no panacea to what is a deeply complex socio-economic, legal and political problem.

Amidst the recriminations over the 2002 ASEAN Haze Agreement's ineffectiveness, a related question involving intergovernmental co-operation arose to complicate the issue. This has to do with maps showing where agricultural plantation concessions are located in Indonesia. It is well-documented that most of the serious fires over the years have arisen from intentional and systematic burning of land to clear undergrowth or old plantations for new plantings.[5] In this regard, the oil palm and timber sectors (the latter supplying feedstock for the pulp and paper industry) have long been fingered as the prime villains for using fire as a cheap and fast means to clear land.

Indeed, the environmental NGOs have consistently maintained that the crux of the issue lies in the illegal burnings by agribusiness concerns motivated by profits and emboldened by poor law enforcement and local government collusion in Indonesia.[6] The problem is exacerbated in many areas by the presence of large deposits of combustible peat in the land. This makes the fires – once ignited – that much more difficult to control

---

[4] Adella Anjani Putri, 'Indonesia Ratifies Haze Treaty after 12 Years of Stalling', *The Jakarta Globe* (17 September 2014) accessed 14 January 2015 at http://www.thejakartaglobe.com/news/indonesia-ratifies-haze-treaty-12-years-stalling.

[5] See Centre for International Forestry Research, 'Q&A on fires and haze in Southeast Asia' (4 July 2013) accessed 14 January 2015 at http://blog.cifor.org/17591/qa-on-fires-and-haze-in-southeast-asia#.UtzHtCP2PmQ.

[6] Ben Bland, 'Indonesian fires highlight weak governance and corruption', *Financial Times* (23 June 2013) accessed 14 January 2015 at http://www.ft.com/cms/s/0/a6d8c050-dbf5-11e2-a861-00144feab7de.html.

as they may continue to burn underground.[7] Indeed, some experts believe that the plantation companies' illegal clearing of peat-rich lands is the main source of the fires and smoke pollution.[8]

As far as the plantation interests are concerned, what marked a difference in 2013 was their more organised response in stoutly denying responsibility for the haze and asserting that they actually practiced a 'zero burning' policy. Indeed, the plantations have insisted that they only employ mechanical means (that is, bulldozers) to clear land and have extensive fire management and firefighting programmes in place.[9] The blame was shifted to the local communities and small-scale farmers (themselves users of fire for generations to clear land) whom the plantation interests accused of starting fires in or near their concessions.[10] The fires that these farmers started, it seemed, raged out of control and spread into the plantations, thus explaining why 'hot spots' were regularly picked up by satellite images showing raging fires within concession boundaries.

Even so, the NGOs familiar with the problem maintained that it was impossible to believe that the plantations had no role in the fires. These NGOs have analysed the fires by overlaying publicly available satellite-generated 'heat maps' with Indonesian government forestry concession maps. What this showed was that many of the fires were burning within concessions owned by and supplying some of the world's biggest oil palm and pulp and paper conglomerates, including companies based in Indonesia, Singapore and Malaysia. The NGOs' claims make the issue of the concession maps all the more compelling – if made available, these maps could be the key to identifying who the actual transgressors are.

Yet, the matter is hardly that straightforward. If only the maps and boundaries were clear, well understood and respected by all interests on the ground. It became evident that concession maps in Indonesia do not and cannot reliably pinpoint the land owners, occupiers or rights holders.

---

[7] Peat lands are attractive to planters because they are usually unoccupied and therefore easier to acquire. However, drained peat, even if not deliberately ignited, is susceptible to fire outbreaks. The high carbon content and depth of peat on Sumatra Island, in particular, make the fires there difficult to extinguish. *Ibid.*

[8] *Ibid.*

[9] Neo Chai Chin, 'Haze finger-pointing: Time for companies to show and tell', *Today* (28 June 2013) accessed 14 January 2015 at http://www.todayonline.com/commentary/haze-finger-pointing-time-companies-show-and-tell.

[10] See Bland, *supra* note 6, detailing how the companies have dismissed or downplayed accusations that they were responsible, saying they practiced 'no burning' policies and laying the blame on local communities instead.

The NGOs that tracked the fires have been using whatever maps they could procure, but it is unclear if these are even accurate or updated. Indeed, confusion abounds in Indonesia as to why different agencies issue and rely on contradictory maps, and whether the fires started by some individuals could have occurred within plantation boundaries after all.

To make matters worse, the Indonesian government has come up with the official position that it cannot make its maps public due to legal obstacles. Hence, it seems that the very basic evidential enquiry needed to resolve the problem – that is, to find out who was setting fires to whose lands – cannot even be conducted, let alone the infinitely more complex tasks of actually prosecuting perpetrators or managing land use conflicts for the longer term.

II. THE ASEAN AGREEMENT IN THE INDONESIAN DOMESTIC CONTEXT

A. On Treaty Compliance and Effectiveness

As a result of the fires of 1997 and the resulting smoke pall's severe consequences, the ten Member States of ASEAN had in June 2002 adopted a landmark regional agreement on Transboundary Haze Pollution. The 2002 ASEAN Haze Agreement came into force relatively swiftly on 25 November 2003[11] and has today been accepted by all the ASEAN Member States, including Indonesia only recently in September 2014.

In 2005, the present author published a study on whether the 2002 ASEAN Haze Agreement – *even assuming* Indonesia ratified it – would make any difference.[12] That study entailed a comprehensive analysis of the substantive provisions in the Agreement, particularly those relating to interstate cooperative mechanisms and domestic legal action that could be taken against perpetrators. My conclusions then bordered on the pessimistic, and I have little reason to change my assessment today. In essence, the Agreement contains significant pathologies that work to

---

[11] The 2002 ASEAN Haze Agreement *supra* note 1, Art 29 required the acceptance of six States as a condition for entry into force. This was fulfilled at the end of 2003.

[12] Alan Khee-Jin Tan, 'The ASEAN Agreement on Transboundary Haze Pollution: Prospects for Compliance and Effectiveness in Post-Suharto Indonesia' (2005) 13 *New York University Environmental Law Journal* 647.

compromise its effectiveness. Major deficiencies had been implanted – from weak obligations relating to requesting and receiving assistance, monitoring, reporting, exchanging information and conducting research to the complete absence of enforcement and liability provisions and compulsory dispute resolution mechanisms.

With legal niceties relating to 'cooperation', 'coordination' and 'consultation', the 2002 ASEAN Haze Agreement reiterates the traditional ASEAN platitudes of partnership and regional solidarity while omitting any mention of legal consequences for non-compliance.[13] I posited then that this was largely the result of ASEAN Member States' reluctance to deal firmly with Indonesia – the regional giant – which was then in the throes of harsh economic and political upheavals. Above all, there was the geopolitical context: the regional governments' belief that in view of ASEAN's history and culture, rules-based methods premised on assertions of culpability were unlikely to work, or worse, could prove counter-productive.[14] This often-derided 'ASEAN Way' of doing things,[15] rightly or wrongly, came to influence the drafting of the 2002 ASEAN Haze Agreement's key provisions.

The established theories of treaty compliance and effectiveness posit that the effectiveness of any given treaty in addressing an identified concern depends in large part on whether the treaty secures the requisite compliance of 'target actors' within the States Parties.[16] To such end, an international regime can be said to be 'effective' only when the rules which it prescribes are adequately implemented and enforced, eliciting a high degree of compliance by the target actors at whom they are directed *and* resulting in an overall resolution or amelioration of the problem at hand.[17]

---

[13] *Ibid* at 660.

[14] *Ibid.*

[15] On forest fires and ASEAN generally, see Nicholas A Robinson, 'Forest Fires as a Common International Concern: Precedents for the Progressive Development of International Environmental Law' (2001) 18 *Pace Environmental Law Review* 459.

[16] See e.g., Harold K Jacobson and Edith Brown Weiss, 'A Framework for Analysis' in Edith Brown Weiss and Harold K Jacobson (eds), *Engaging Countries: Strengthening Compliance with International Environmental Accords* (Cambridge, MA: The MIT Press, 1998) 1 at 1–4.

[17] In this regard, the notion of 'implementation' refers to the measures which States Parties adopt at the national level to make treaty rules effective in domestic law. Looked upon broadly, 'implementation' relates to a State's overall effort to accomplish the objectives of a body of rules. On its part, the related but narrower concept of 'enforcement' entails the State directly bringing into force

Hence, the cumulative extent and impact of implementation, enforcement and compliance go toward determining the overall level of 'effectiveness' of that treaty. In this respect, the concept of 'effectiveness' is concerned with whether the stated objectives of a body of rules have been met and whether the problems leading to the adoption of the rules have been meaningfully resolved.[18] Here, 'effectiveness' is connected to, but not identical with compliance. States and target actors may well be in compliance with treaty rules, but as will be argued in relation to the 2002 ASEAN Haze Agreement below, the rules themselves, as influenced and enshrined by the relevant political interests, may nevertheless contain inherent pathologies or weaknesses and become ineffective in attaining the relevant treaty objectives.

Where a treaty regime encounters deficiencies, particularly in relation to inadequate compliance and effectiveness, four determinative factors may be identified as being responsible.[19] The nature of the accord in question is critical, as is the precise character of the activity being regulated. In addition, the international environment within which regulation takes place forms a contextual background that may throw up forces affecting compliance and effectiveness. Finally, the characteristics peculiar to individual States Parties to the treaties and the extent of their control (or lack thereof) over target actors may affect the effectiveness of any treaty regime.

In this regard, the 2002 ASEAN Haze Agreement cannot be meaningfully understood without locating it within the specific geopolitical context that characterises relations among ASEAN Member States. As will be seen, significant pathologies have been implanted into the Agreement's provisions – deficiencies that severely compromise the Agreement's potential effectiveness. At the same time, whatever prospects the Agreement has for effectiveness are largely influenced by domestic conditions within the one State that matters most – Indonesia. Hence, the whole political economy of forest resource exploitation as

---

the relevant rules (often, on pain of penalties or sanctions) to compel conformity and 'compliance' on the part of specific target actors. Compliance thus goes beyond implementation and entails an inquiry into whether States are putting into effect the measures which they have instituted and whether target actors are in fact adhering or conforming to the relevant rules. See e.g., *ibid* and Abram Chayes and Antonia Handler Chayes, *The New Sovereignty: Compliance with International Regulatory Agreements* (Cambridge, MA: Harvard University Press, 1995).

[18] Jacobson and Weiss, *supra* note 16.

[19] *Ibid* at 4–12.

well as domestic issues relating to environmental governance and regional autonomy in Indonesia must be appreciated in order to assess the Agreement's effectiveness.

Where the surrounding political context is defined less by a *demand* by States Parties for legally enforceable commitments, but more an *expectation* for moral force and action, prospects for compliance and effectiveness, as defined above, are affected considerably. The dynamics of political intercourse among States Parties are further impacted when the treaty regime depends solely or mainly on one or a handful of States Parties for compliance and effectiveness, as happens when such parties account for a disproportionately large share of the problem as defined.[20]

It is precisely in such situations that the final two components of the quartet of forces determining compliance and effectiveness – that is, contextual politics and dominant actors – assume an even more potent relevance. Efforts will thus have to be directed toward ameliorating the impact of these specific factors if greater compliance and effectiveness are desired. Otherwise, if these overriding factors cannot be 'manipulated'[21] to a sufficient degree, compliance and effectiveness will remain stultified.

Returning to the 2002 ASEAN Haze Agreement, the twin factors of contextual politics and dominant actors impinge heavily upon effectiveness. Thus, ASEAN Member States' distaste for what they perceive to be 'confrontational' methods of dispute resolution and their prevailing instinct to uphold the principle of non-interference or non-intervention in Member States' internal affairs have led to an Agreement containing the best of intentions but severely lacking in effective sanctions and enforceability.[22] This is particularly worrying given that the Agreement is probably one of the few multilateral treaties in existence that relies, almost exclusively, on *one* State Party for meaningful compliance and effectiveness.[23] In this sense, the treaty's effectiveness – defined as its capacity to fulfil the objective of ameliorating or resolving the fires and haze problem – is substantially undermined when Indonesia's capacities are considered.

---

20 Tan, *supra* note 12 at 651–2.

21 Jacobson and Weiss, *supra* note 16 at 10–11.

22 Tan, *supra* note 12 at 652.

23 *Ibid.* The classic examples of such instruments would be the nuclear proliferation and space exploration treaties.

B. On Illegal Burnings and Enforcement Failure

As argued above, the pessimism over the 2002 ASEAN Haze Agreement's effectiveness can be traced to several factors linked to contextual politics and dominant actors – namely (i) the political forces dictating the drafting of the Agreement's obligations; (ii) the resulting pathologies implanted in the Agreement which affect treaty compliance and effectiveness; and (iii) relevant internal conditions in politically relevant States, that is, the contemporary deficiencies of Indonesia. As assessed below, all these factors collude to create a treaty regime that is highly lacking from the perspectives of compliance and effectiveness.

To be fair, the 2002 ASEAN Haze Agreement itself is not without substance. On paper, there are several notable provisions that imbue the Agreement with some force. Articles 4 and 9, in particular, outline general obligations in seemingly mandatory language. Thus, Parties *shall* cooperate to prevent and monitor transboundary haze pollution and to control the sources of fires.[24] When fires originate from within their territories, States Parties are obliged to respond promptly to requests for information and consultations by affected States.[25]

Most significantly, States Parties '*shall* take legislative, administrative and/or other measures' to implement their obligations under the Agreement.[26] In this regard, States *shall* undertake measures to prevent and control activities related to land and/or forest fires that may lead to transboundary haze pollution, including developing and implementing legislative and other regulatory measures, as well as programmes and strategies to promote zero burning policy to deal with fires and haze, and ensuring that legislative, administrative and/or other relevant measures are taken to control open burning and to prevent land clearing using fire.[27]

These provisions clearly impose obligations on Indonesia to develop the requisite legal or administrative mechanisms in order to prevent and combat transboundary haze pollution arising from forest and/or land fires. In any event, some or all of these obligations are established under

24 2002 ASEAN Haze Agreement, *supra* note 1, Art 4(1) (emphasis added).

25 *Ibid*, Art 4(2).

26 *Ibid*, Art 4(3) (emphasis added).

27 *Ibid*, Art 9 (emphasis added). Article 9 also prescribes measures to curb activities that may lead to fires, to identify and monitor fire-prone areas, to strengthen fire management and firefighting capabilities and to promote public education, community participation and the use of indigenous knowledge to prevent and manage fires.

customary international law, particularly the international obligation to take effective measures to control the fires. Hence, Indonesia would bear such international obligations under both the Agreement and customary international law.

Even if the 2002 ASEAN Haze Agreement does not elaborate on law enforcement and penalties, the general provisions seem wide enough to embrace the enactment of domestic anti-burning laws, the prosecution of offenders and the imposition of adequate penalties to deter the use of fires. Thus, while it is unfortunate that the Agreement does not spell out a positive obligation to impose penalties that are adequate in severity to deter open burning, the necessary legal, administrative and other measures needed to curb transboundary haze pollution arising from forest and/or land fires *can* conceivably be taken pursuant to the Agreement's broad provisions.[28]

On the whole, therefore, there is no lack of legal authority under the Agreement to provide for strong legislation as well as effective enforcement, prosecution and adjudication. The overriding obstacle here is the lack of political will in Indonesia to implement the Agreement's obligations (whether as a State Party or under customary international law) and to ensure effective compliance with the obligations in the ways outlined above. As far as the Agreement is concerned, one worrying feature is the complete absence of sanctions in the event that States fail to comply with their obligations. Given Indonesia's consistent past record in failing to deal with the fires effectively[29] and the absence of evidence to suggest that its future actions will be any better, the Agreement seems headed toward crippling ineffectiveness in meeting its outlined objectives.

In my 2005 study, I examined the internal political forces in Indonesia that I believed would undermine compliance with anti-burning laws and the prosecution of such transgressions. Here, it should be emphasised that 'zero burning' legislation does exist in Indonesia, although there are explicit allowances for local communities to use fire to clear land. There are also prohibitions against starting fires in or developing peat areas that are more than three metres in depth. As for penalties, any company or person found guilty of setting an illegal forest fire could be imprisoned for a period ranging from 5 to 15 years and fined up to 5 billion Rupiah (about US$420,000).[30] These are not inconsequential penalties. However,

---

28 Tan, *supra* note 12 at 663.

29 *Ibid* at 676–85.

30 Under the 1999 Forestry Law, *ibid* at 683–4. There is at least one other legal instrument that prescribes penalties for forest and land burning. This is the 2009 Law for the Protection and Management of the Environment that provides

the complexities of political governance in Indonesia result in ineffective enforcement of and compliance with these laws. There are several factors behind this state of affairs.

1. Institutional deficiencies

The major problem relates to weaknesses in the institutional structure governing forestry, agriculture, land use and environmental protection in Indonesia. For one thing, serious jurisdictional conflicts exist between central governmental agencies, between central and regional authorities, and among local authorities themselves. During the previous regime (that is, the pre-1998 political order of President Suharto), the exploitation of the nation's natural resources was synonymous with the regime's brand of corrupt patronage politics. In a system run by a centralised bureaucracy and yet effectively controlled through power networks extending all the way into the provinces, regencies and villages, natural resource exploitation in Indonesia became dominated by a clique of businessmen with political connections to the powerful Suharto family.[31]

These vested interests were able to exert inordinate influence over – and indeed, override – governmental agencies and policies, and remained largely free from regulatory supervision. At the same time, the institutions that existed for environmental and natural resource governance were fractured and weak (often by design). For instance, the environmental management effort was complicated at the time by the existence of two separate and seemingly competent agencies – the Office of the State Minister for the Environment and the Environmental Impact Management Agency (known by its Indonesian language acronym as BAPEDAL).

The Office of the State Minister has since been merged with BAPEDAL, and an integrated Ministry of Environment came into existence in 2008. However, in contrast with the more powerful ministries in charge of forestry and agriculture, the Ministry of Environment is still relatively weak in both sectoral and geographical competences. It has little authority over the forestry, agriculture and mining sectors, for instance, and has no direct competence over forest and land fires. At the same time, it possesses a weak geographical presence in the regions to

for a minimum punishment of 3 years' imprisonment and a fine of 3 billion Rupiah (US$250,000) and a maximum punishment of 10 years' imprisonment and a fine of 10 billion Rupiah (US$840,000) for intentional starting of fires. Republic of Indonesia, Law No 32/2009 on Environmental Protection and Management, Art 98(1).

[31] *Ibid* at 671–2.

enforce the relevant laws (including its own). Enforcement competence resides instead with the respective provincial (*provinsi)* and regency (*kabupaten*)[32] authorities, including the local police and prosecutors, as well as the local offices of the central ministries in charge of agriculture, mining and forestry.

Such anomalies in the structure of environmental institutions accounted for much of Indonesia's failure to deal with the fires in 1997, and again in 2013. Inter-agency coordination at the national and provincial levels remains weak, and each agency operates largely independently of the others. At its core, the problem is that the Ministry of Environment has little power over the revenue-generating decisions and activities of the ministries in charge of forestry and agriculture, along with their local-level bureaus. For one thing, the Ministry of Forestry continues to grant lucrative timber concessions that are the major sources of land-clearing by fires.[33] In addition, this Ministry claims direct competence over the fires, given that it runs provincial and regency units/directorates responsible for forest fire control. On its part, the Ministry of Agriculture has direct control over the oil palm and pulp and paper plantations, the other major sources of land-clearing by fires.

The forestry and agriculture ministries in Indonesia have traditionally been more concerned with exploiting resources and generating revenue, often in association with private interests. In particular, direct collusion between the Ministry of Forestry and the timber and plantation industries has long been documented.[34] True to form, despite overwhelming evidence to the contrary, the Ministry has downplayed the roles of the timber interests in the 1997 and 2013 fires. Similarly, the Ministry of Agriculture has done little to control the activities of the oil palm and pulp and paper plantations.

---

32 Indonesia is administratively divided into *provinsi* (provinces), *kabupaten* (regencies), *kota* (towns) and various smaller units. The *kabupaten* are usually referred to in English as districts or regencies – the term used here is 'regency'. The *kabupaten* and *kota* are at the same level within a *provinsi*. The *gubernor* (governor) heads the *provinsi*, while the *bupati* (regent) and *walikota* (mayor) head the *kabupaten* and *kota* respectively. The *kabupaten* are further divided into units such as the *kecamatan*, *keluharan* (sub-districts) and *desa* (villages).

33 Tan, *supra* note 12 at 673.

34 *Ibid*.

Postscript In October 2014, Joko Widodo was sworn in as the new President of Indonesia following his election victory. Among the President's first initiatives was to merge the environment and forestry ministries. The creation of the Ministry of Environment and Forestry has met with some skepticism. NGOs worry that the Ministry might prioritise exploitation over conservation and that its environmental 'watchdog' functions might become compromised. Indeed, the NGOs would prefer to have a separate but strengthened environment ministry to monitor and check the resource exploitation sectors. It is probably too early to dismiss the benefits of a merged ministry or to underestimate its environmental powers, but the concerns are legitimate given the traditional emphasis on exploitation in Indonesia. No doubt, there will be high expectations for the Ministry's environmental arm to display independent and effective supervision over the forestry sector.

2. The perils of regional autonomy

Some observers assert that little has changed in the years since the 1997 fires.[35] In fact, in the post-Suharto era, the institutional and enforcement problems have actually become exacerbated with the advent of the 'regional autonomy' movement. Since Suharto's ouster, substantial degrees of administrative, legislative and regulatory authority have been transferred from the central to the regional (that is, provincial and regency) governments, including competence over forestry, agriculture, mining and other land use and natural resources issues.

Indeed, one of the revolutionary features of regional autonomy in Indonesia is the transfer of substantive powers directly to the 'second-level' regencies (*kabupaten*) and towns (*kota*), as opposed to the 'first-level' provinces (*provinsi*). Hence, the regencies and towns are now recognised as 'autonomous regions' (*daerah otonom*) that possess significant competences independent of the provinces of which they are part. The regency/town has effectively become the new locus of devolved authority in present-day Indonesia, largely bypassing the provincial authorities whose limited role is to act as facilitators in issues such as forestry management. Naturally, this has not gone down well with the provincial governments, let alone the central government agencies. In the meantime, the Ministry of Forestry at the central level has continued to

---

[35] Bland, *supra* note 6.

insist that no forest policy can be set without its initiative or concurrence.[36] Moves to 'backtrack' on regional autonomy have also arisen from within the provincial governments, which are uneasy over the enhanced powers of their constituent regencies and towns. This makes for increasing tension between them.

Meanwhile, a bewildering patchwork of laws and regulations continues to be promulgated by the different levels of government, be these at the central, provincial or local levels. Often, the instruments are mutually contradictory, with the intent being simply to override or neutralise another body's actions. Even laws issued by the same body cause problems when they typically assert that previous inconsistent laws become henceforth abrogated, without specifying exactly what these laws are. It has also become a habit for new laws and regulations to be enacted every time a problem arises or a new Minister is appointed.[37] Often, little effort is made to reconcile the conflicting instruments, and varying interpretations are typically canvassed by different interest groups to suit their own purposes. The problem becomes particularly acute in relation to land tenure and natural resource matters which fall within the overlapping purview of several different governmental agencies.

One glaring example was the Ministry of Forestry's enactment of its landmark Forestry Law in 1999 (known as *Undang-Undang* or UU 41 of 1999), an avowedly pro-exploitation instrument that has never been properly reconciled with the regional autonomy laws. In the meantime, the reality on the ground became wholly different: the regional governments did not treat such laws and decrees by the central agencies as binding upon them.[38] In many situations, the regional governments began invoking the spirit of regional autonomy to institute reforms that went well beyond the authority granted to them under the central government's decentralisation laws. In large part, this reflected suspicion and resentment over the central government's half-hearted attempts at regional autonomy.[39] In fact, the regencies have enacted all forms of *Peraturan*

---

[36] Ahmad Dermawan and Ida Aju Pradnja Resosudarmo, 'Forests and Regional Autonomy: The Challenge of Sharing the Profits and Pains' in Carol J Pierce Colfer and Ida Aju Pradnja Resosudarmo (eds), *Which Way Forward?: People, Forests, and Policymaking in Indonesia* (Washington, DC: RFF Press, 2002) 325–57.

[37] Tan, *supra* note 12 at 675.

[38] A treatment of the perplexing array of forestry laws and regulations enacted in the past few decades is outside the scope of this work.

[39] For an intriguing glimpse into the politics of forest exploitation at the regency level following the advent of regional autonomy, see Ida Aju Pradnja

Daerah or Regional Regulations (whose status is elevated under the regional autonomy laws), giving the local governments more authority and direct control over natural resources and revenue streams.[40]

In relation to the lucrative timber industry, for instance, the regencies have sought to assert greater control by issuing their own district logging licences, establishing greater control over existing large-scale concessions and restructuring the provincial forestry service so that this reports directly to the regency head, as opposed to the provincial governor or central ministry. Such regulations are squarely inconsistent with the provisions in the centrist Forestry Law. Differing interpretations have also arisen on the right of governors and regents to issue concessions of varying sizes and types without the approval of the central government. Efforts by the Ministry of Forestry to halt the issuance of regency concessions have typically fallen on deaf ears, and some regional governments have even threatened not to honour pre-existing concessions issued by the central government.[41]

In such circumstances, private parties with vested interests have unsurprisingly emerged to exploit the uncertainties surrounding the decentralisation process and the accompanying weaknesses in law enforcement. In many cases, the greater autonomy devolved to local governments has created conditions even more ripe for corruption and ineffective control than in the past. Local elites have become emboldened, and there are frequent reports of timber concessionaires actively cultivating the '*putera daerah*' or '*raja kecil*' (regional princelings or little kings – typically the regency head, the *Bupati*) whose positions have been strengthened by regional autonomy.

In many cases, even the leaders of *adat* or customary communities (that are typically governed by special exceptional laws that provide them with greater autonomy) can be as corrupt as government officials, aided by opportunistic individuals who fashion themselves as community spokespersons.[42] These 'local elites' often position themselves as intermediaries between their communities and concession firms, negotiating

---

Resosudarmo, 'Closer to People and Trees: Will Decentralization Work for the People and the Forests of Indonesia?' (2004) 16 *European Journal of Development Research* 110.

40 Tan, *supra* note 12 at 701.

41 Dermawan and Resosudarmo, *supra* note 36 at 348.

42 See e.g., John F. McCarthy, 'Power and Interest on Sumatra's Rainforest Frontier: Clientelist Coalitions, Illegal Logging and Conservation in the Alas Valley' (2002) 33 *Journal of Southeast Asian Studies* 77 at 90–6, describing the actions of local leaders in the regency of Southeast Aceh in Aceh province.

settlements (such as the release of limited tracts of land) from the companies. Such settlements are frequently short-term fixes or 'buy-outs' that provide minimal long-term livelihood benefits to local communities.[43]

Democratisation in Indonesia has thus meant not only greater decentralisation of political power but increased access by local elites to the profits of natural resource exploitation. As decentralisation has gathered momentum, the reduction in supervision and budgetary allocations from the centre has clearly encouraged regional actors to pursue extralegal sources of income. In the absence of functioning state institutions and law enforcement, direct personal ties based on reciprocity, inducements and favours have come to govern relations between patrons and clients.[44] In this regard, the politics of 'clientelism' – the use of informal networks based on personal influence to secure one's interests – becomes especially strong where the rules imposed by central agencies are inappropriate, lack local legitimacy or contradict long-standing local traditions.[45]

In almost all cases, the replacement of Suharto-era 'outsider' cronies by native elites in natural resource exploitation has not led to greater distributive justice for local communities. On the ground, the benefits of logging operations have largely evaded such communities – in many cases, local peoples whose lands have been encroached are employed in illegal logging and clearing activities, often for low pay and using cheap methods such as fires to clear land.[46] In other situations, entire communities have been bought off by small compensatory payments or limited concessions given by commercial interests.

Meanwhile, wary of the legacy of harsh Suharto-era controls, the central government has been politically unwilling or unable to resolve conflicts in a robust manner. Such systemic weaknesses accompanying the process of decentralisation translate into specific problems on the ground. In particular, illegal logging, timber smuggling and land clearing (using fire, of course) have become especially rampant and appear poised to accelerate with decreasing control being exerted from the centre. The situation has hardly improved in recent years; one example is the adoption of the Law on Prevention and Eradication of Forest Destruction in 2013 that amends the Forestry Law of 1999. The new Law omits any mention of forest fires – surely the issue of the day at this time – and

---

43 Tan, *supra* note 12 at 703.

44 McCarthy, *supra* note 42 at 80.

45 *Ibid.*

46 Dermawan and Resosudarmo, *supra* note 36 at 343, citing various field reports.

even remarkably reduces the maximum prescribed sentence for illegal logging from ten years (under the 1999 law) to five years!

3. Inadequate enforcement and prosecution

Amidst these uncertainties, it is unsurprising that there has been such weak enforcement of anti-burning laws and inadequate prosecution of offenders. The few cases of reported action being taken came well after the fires had raged out of control and transboundary injury caused. During the 1997 fires, for instance, the Minister of Forestry released the identities of 176 plantation, timber and transmigration land-clearing companies suspected of illegal large-scale burning in contravention of anti-burning laws.[47] Subsequently, it was reported that 29 of these firms had had 154 of their operating licences suspended or revoked.[48] However, these licences were mostly reinstated in later months without any apparent justification.[49] In many cases, the environmental agencies' efforts to prosecute companies were thwarted by the provincial police, indicating possible collusion between law enforcers and plantation interests.[50]

The courts have not performed much better with the handful of cases coming before them. In the first place, there is a deep-seated belief in Indonesia that the judiciary is frequently paid off to prevent conviction of high-profile figures behind illegal logging and forest fires.[51] In the aftermath of the 1997 fires, a handful of prosecutions were undertaken, but few gave rise to satisfactory outcomes. Of the 176 companies identified publicly as violators, only five were ever brought to court, and only one was found guilty.[52] In most of the cases, the courts were unwilling to impose heavy penalties that would effectively deter the plantation interests. The penalties imposed were extremely lenient, such

---

47 Charles Victor Barber and James Schweithelm, 'Trial by Fire: Forest Fires and Forestry Policy in Indonesia's Era of Crisis and Reform' (2000) *World Resources Institute* at 8.

48 Derwin Pereira, 'Indonesia Gathering Evidence against 29 Forestry Firms', *The Straits Times* (8 October 1997).

49 Barber and Schweithelm, *supra* note 47.

50 *Ibid* at 23.

51 In November 2003, the then-State Minister for the Environment, Nabiel Makarim, was quoted as saying that the prosecution of illegal logging was made difficult by a corrupt Indonesian judiciary, see 'Jakarta Promises to Tackle Illegal Loggers', *Financial Times* (7 November 2004).

52 'Legal Action on Forest Fires', *Down to Earth Newsletter Volumes 53–54* (August 2002), citing then-State Minister for the Environment, Sonny Keraf, as reported in *Kompas* (Indonesia), 28 March 2000.

as the US$8,400 fine in the sole successful prosecution in the case of PT Adei Plantation, an Indonesian-Malaysian joint venture oil palm company.[53]

As noted by Indonesia's National Development Planning Agency (BAPPENAS) and the Asian Development Bank (ADB), the cases that have been decided may actually 'seriously undermine other attempts at further prosecutions'.[54] It was believed that the corruption of high-level officials had affected the outcome of many of these court cases, thus denting the value of potential sanctions against illegal logging and burning. For the recent June 2013 fires, the Indonesian Environment Minister has said that investigations are still ongoing, and has vowed tough action against culprits.[55] As recently as May 2014, the Indonesian President pledged to adopt a 'zero tolerance' attitude toward illegal land burning.[56] It appears that about a dozen companies and more than 100 individuals are currently facing trials for the 2013 fires, but there is as yet little firm or consistent indication that the prosecutors and courts are taking a more stringent view than they have done in the past.

---

[53] PT Adei was a joint venture between the well-known Malaysian company, Kuala Lumpur Kepong Sdn. Bhd. and Al Hakim Hanafiah of Indonesia; see Dicky Simorangkir, 'Convicting Forest and Land Fire Offences: A Case Study of the Legal Process in Riau, Indonesia' (2003) *Project Fire Fight*. The company was charged for starting fires in 1999 and 2000, and the Bangkinang district court handed down a two-year imprisonment term for the company's Malaysian manager and a 250 million Rupiah (US$21,000) fine. On appeal by the company, the Riau High Court reduced the penalty to eight months' imprisonment and 100 million Rupiah (US$8,400) fine. This was then upheld by the nation's highest court – the Supreme Court – at the end of 2002. A decade later, PT Adei is in the news again, this time for the 2013 fires. In September 2014, the firm's manager was sentenced to one year's imprisonment and fined 2 billion Rupiah (US$170,000) for his role in the June 2013 fires. The company itself was fined 1.5 billion Rupiah (US$125,000) and ordered to pay 15.1 billion Rupiah (US$1.25 million) to repair the environmental damage. Both verdicts are being appealed. See Wahyudi Soeriaatmadja, 'Riau court jails and fines plantation firm manager for role in forest fires', *The Straits Times* (12 September 2014).

[54] BAPPENAS, 'Planning for Fire Prevention and Drought Management Project Final Report, Annex I: Causes, Extent, Impact and Costs of the 1997/98 Fires and Drought' (1999) *Asian Development Bank Technical Assistance Grant TA 2999-INO* at 91.

[55] Zakir Hussain and Wahyudi Soeriaatmadja, 'Jakarta minister vows action against forest-fire culprits', *The Straits Times* (25 July 2013).

[56] Zubaidah Nazeer, 'Zero tolerance for illegal land burning, vows Yudhoyono', *The Straits Times* (6 May 2014).

That being said, a handful of recent cases have seen penalties being imposed on oil palm companies for using fires illegally to clear land. For instance, in January 2014, a court in Aceh province (in the north of Sumatra island) ruled in a civil case that an oil palm company, PT Kallista Alam, was in breach of the 2009 Law for the Protection and Management of the Environment for using fires illegally. It was ordered to pay 114.3 billion Rupiah (US$9.5 million) in compensation and 251.7 billion Rupiah (US$21 million) to restore the affected forests. In July 2014, another court found in a criminal case that the same company's director was guilty of illegally clearing peat forests and imposed an imprisonment term of eight months and a fine of 150 million Rupiah (US$12,500).[57] The company is currently appealing both sets of rulings, and it is too early to celebrate the case as a victory for strict enforcement given the fact that superior courts have tended in the past to reduce penalties dramatically.[58]

4. Land use conflicts

Most land use and tenure conflicts arising in Indonesia today stem from the absence of proper demarcation of land for use by different stakeholders. This is the larger context in which the land and forest fires are taking place. To begin with, notions of land ownership remain uncertain in Indonesia, with the State often claiming ultimate rights over land use in priority over local communities' *adat* (customary) rights. In the process, the land use patterns of these communities are often disturbed and their rights sacrificed in the interest of commercial exploitation. In this regard, the process of spatial planning, as well as its reconciliation with land use by local communities, has changed little since the Suharto days.

Problems such as overlaps between concession and small-scale land rights continue to arise, as do disputes among different ethnic groups and between resentful local communities and oil palm plantations, pulp and paper concessions and logging companies. In many of these conflicts, repressed and alienated groups seeking vengeance often resort to arson as

---

57 Annette Garland, 'Palm oil company director sentenced to jail for illegal forest clearance in Indonesia', *Changing Times*, accessed 14 January 2015 at http://www.4apes.com/pt/news/item/1238-http-time2transcend-wordpress-com-2014-07-16-palm-oil-company-director-sentenced-to-jail-for-illegal-forest-clearance-in-indonesia.

58 As happened with the PT Adei case in 2002, *supra* note 53. As noted above, PT Adei has been prosecuted and convicted for its role in the 2013 fires and is appealing the verdicts, *supra* note 53.

a weapon against other actors.[59] Indeed, there is evidence of a higher risk of fires in areas of unclear boundaries between community territories.[60] This aspect of forest and land fires has not been extensively investigated but is believed to be a root cause of many fires set off throughout the Indonesian archipelago. This also provides a relevant backdrop to the 'map controversy' that arose in the 2013 fires, as assessed below.

III. THE 2002 ASEAN HAZE AGREEMENT AND COOPERATIVE MECHANISMS

With the domestic context for land use and environmental governance in Indonesia set out above, it is useful at this juncture to return to the 2002 ASEAN Haze Agreement. Since this chapter's focus is on cooperative mechanisms, it is apposite to highlight what the Agreement contributes in this regard. Part II of the Agreement (Articles 5 to 15) lays down provisions relating to monitoring, assessment, prevention and response – all hallmarks of a robust interstate cooperative regime. Indeed, the Agreement envisages the setting up of an ASEAN Coordinating Centre for Transboundary Haze Pollution Control (ASEAN Centre) to facilitate cooperation and coordination among States Parties in managing the impact of transboundary haze.

To date, the ASEAN Centre has not yet been established, even though the Contracting States to the 2002 ASEAN Haze Agreement have envisaged that the ASEAN Centre shall be set up and based in Indonesia.[61] Meanwhile, coordination activities during haze emergencies are still being carried out by the respective States' national environmental

[59] Fires are also started by commercial interests such as oil palm plantations to intimidate local communities into submission or to decrease the value of their lands. In turn, peasants use fire as a defensive weapon against the takeover of their lands. See Barber and Schweithelm, *supra* note 47 at 28.

[60] See Grahame Applegate, Ross Smith, James J Fox, David Packham, Nigel J Tapper and Graham Baines, 'Forest Fires in Indonesia: Impacts and Solutions' in *Which Way Forward?: People, Forests and Policymaking in Indonesia*, *supra* note 36 at 293–308.

[61] 2002 ASEAN Haze Agreement, *supra* note 1, Art 5. In his remarks to the 15th Meeting of the Sub-Regional Ministerial Steering Committee (MSC) on Transboundary Haze Pollution, 17 July 2013, Kuala Lumpur, Malaysia, ASEAN Secretary-General Le Luong Minh mentioned that Indonesia indicated interest in hosting the ASEAN Centre upon ratification of the 2002 ASEAN Haze Agreement, accessed 14 January 2015 at http://www.asean.org/images/2013/resources/speech/SG%20remarks_15th%20MSC_final.pdf.

agencies. Overall, there is the inescapable expectation among Member States that before the ASEAN Centre can be fully functional, Indonesia must first ratify the Agreement.[62] This is the reason why the ASEAN Centre had not featured in any coordination or cooperation efforts during the 2013 fires.[63]

In any event, concessions to State sovereignty abound in the 2002 ASEAN Haze Agreement's provisions on cooperative mechanisms. For instance, the ASEAN Centre shall work on the basis that the national authority[64] designated by States Parties will act *first* to put out any fires. The national authority has the discretion to declare a state of emergency, and when it does so, *may* then make a request to the ASEAN Centre to provide assistance.[65] Hence, the ASEAN Centre is powerless to prescribe or take any action should States Parties decide to handle the problem unilaterally.

Overall, the State where the fires occur retains complete sovereignty over the admission of assistance onto its territory, even when the harm being caused to other States is egregious. Hence, there is nothing victim States can do to compel the acceptance of assistance should the State causing harm refuse offers of assistance. This was the precise challenge faced in 1997 when the Indonesian government initially refused to acknowledge that the haze was a deliberate and large-scale man-made problem. Such official intransigence reappeared during the 2013 fires when an Indonesian government minister accused Singapore of acting like a child in complaining about the problem.[66]

In practical terms, there has been little improvement since the 2002 ASEAN Haze Agreement was adopted. In 1997, Indonesia took months before finally agreeing to assistance in the form of admitting a small team of firefighters from Malaysia. Unlike that time when Malaysian firefighters were actually allowed to go into Indonesia, no such assistance was sought and provided in 2013. Instead, the Indonesian military had scrambled helicopters and aeroplanes to water bomb the fires and to

---

62 See *ibid.*

63 See Henriette Litta, *Regimes in Southeast Asia: An Analysis of Environmental Cooperation* (Berlin: Springer, 2012).

64 Pursuant to the 2002 ASEAN Haze Agreement, States are to designate Competent Authorities and a Focal Point to act on their behalf in performing their administrative functions. 2002 ASEAN Haze Agreement, *supra* note 1, Art 6.

65 *Ibid*, Art 5(2).

66 Zakir Hussain, 'Indonesian minister says Singapore "behaving like a small child" over haze', *The Straits Times* (20 June 2013).

perform 'cloud seeding' by dropping salt into the air to encourage rain. These ineffectual efforts have been described by NGOs and experts as 'playing around at the margins'.[67]

A related deficiency in the 2002 ASEAN Haze Agreement is the absence of a strong dispute resolution mechanism. Consistent with the ASEAN preference for consensus and non-confrontation, disputes as to the interpretation or application of, or *compliance* with the Agreement or its protocols 'shall be settled amicably by consultation or negotiation'.[68] Unlike other relevant treaty regimes, no provision whatsoever is made for disputes to be settled by recourse to international courts or arbitration tribunals. This wholly pre-empts the enforcement of compliance through legal principles of State responsibility and international liability. Moreover, unless consultation or negotiation is able to procure an undertaking to compensate (in itself an unlikely outcome), no compensation is due to victim States or their citizens for losses incurred as a result of transboundary haze.

Given ASEAN's operating dynamics, it is not difficult to imagine what 'consultation or negotiation' will entail – a great deal of persuasion and 'behind-the-scenes' diplomatic cajoling and discussions, all of which were readily employed during the 1997 and 2013 fires but with precious little effect. In this regard, the 2002 ASEAN Haze Agreement effectively adds little to ASEAN's traditional modes of regional engagement. To such extent, it is highly questionable if the existence of the Agreement provides anything more than a formal encapsulation of 'tried and tired' methods.

In any event, the 2002 ASEAN Haze Agreement is silent as to the recourse of victim States should 'consultation or negotiation' fail to produce satisfactory results. In this regard, the Agreement appears to have foreclosed all forms of compulsory dispute resolution. Here, one may contend that it would be unrealistic to expect a more stringent treaty regime to work within ASEAN's immense geopolitical constraints. Indeed, a more strongly worded Agreement with compulsory dispute settlement provisions would in likelihood *not* have secured the minimum number of ratifications required for entry into force. Certainly, a State like Indonesia would have had great difficulty in accepting such a treaty, and its ratification would have been completely foreclosed at the outset.[69] This is an outcome no State desires.

---

[67] Bland, *supra* note 6.

[68] 2002 ASEAN Haze Agreement, *supra* note 1, Art 27 (emphasis added).

[69] Tan, *supra* note 12 at 665.

In theory, one positive cooperative feature of the 2002 ASEAN Haze Agreement is its provision for an ASEAN Transboundary Haze Pollution Control Fund (ASEAN Fund).[70] To be administered by the ASEAN Secretariat under the guidance of the Conference of Parties, the ASEAN Fund is to receive voluntary contributions made by States Parties in accordance with decisions of the Conference of Parties. The ASEAN Fund shall also be open to contributions from external sources, including regional financial institutions, the international donor community and foreign governments and organisations. The bulk of contributions is likely to come from the States Parties themselves, presumably in accordance with the relative capacities of the various States.

However, without the assurance of immediate and effective measures to curb the fires, it is unlikely that the richer States such as Singapore, Malaysia and Brunei will be willing to throw vast amounts of money at the problem. This is despite the fact that these three States are the most hard-hit victims of any fires and haze incident. Any amount of financing offered is thus likely to be symbolic in nature. Indeed, the ASEAN Fund has attracted largely symbolic contributions of US$50,000 from each of the following States – Brunei, Malaysia, Singapore, Thailand and Vietnam.[71] Overall, Indonesia is still regarded as lacking the necessary political will to overcome the problem and this will invariably affect the amount of financing that other relevant actors are likely to contribute.

True to form, there was hardly any mention of the ASEAN Fund during the 2013 fires. In fact, it did not appear that the ASEAN Fund was even in operation, or had received any more financial injections. This was all due to Indonesia not being a party to the ASEAN Agreement at the time. Now that Indonesia has ratified the Agreement, the establishment of the ASEAN Fund will likely become a key plank in creating incentives for Indonesia to act. As such, the affected States should have to signal clearly their readiness to inject significant amounts into the ASEAN Fund for substantial fire prevention and remediation activities as well as other long-term assistance programmes. Whether the monies will be put to good use is a separate matter; the reality of politics dictates that the ASEAN Fund should be used to demonstrate the affected States' 'goodwill' to assist in resolving the problem.

---

70 2002 ASEAN Haze Agreement, *supra* note 1, Art 20.

71 ASEAN, Media Release of the Seventh Meeting of the Sub-Regional Ministerial Steering Committee (MSC) on Transboundary Haze Pollution (29 April 2009), accessed 14 January 2015 at http://www.asean.org/archive/PR-7thMSC.pdf.

One form of interstate cooperation that has actually taken place – but with limited success – has been the collaboration that the Singapore authorities have put in place with the provincial government in Jambi. This is a province in the fire-prone southern part of the island of Sumatra from where the haze hitting Singapore often comes. For a few years since the 1997 fires, financial resources and expertise had been provided to Jambi officials to set up monitoring mechanisms and fire prevention programmes and practices. There were even plans to develop alternative livelihood programmes like fish farming for the affected villagers. It was thought that such direct engagement with local-level authorities could yield positive results.

However, it is now clear that the impact of such collaboration has been limited to only small areas in Jambi, and it certainly did not spare Singapore and the region from the large-scale fires of 2013. In any event, press reports indicate that the Jambi programme has faltered due to insufficient funding and interest on the ground.[72] There has also been reluctance among officials in Jakarta to continue endorsing the programme. The programme appears to have officially lapsed in 2011 and little has been heard of it since.[73] Instead, much of the States' attention in 2013 was directed at the issue of the concession maps, to which we now turn.

IV. COOPERATIVE AND UNILATERAL MECHANISMS

A. Concession Maps and the Haze Monitoring System

As mentioned above, the 2002 ASEAN Haze Agreement's provisions on interstate cooperation and sharing of information were severely tested during the 2013 fires by the controversy over concession maps. Satellite images of burning tracts of land in Indonesia had indicated that the fires

---

[72] See e.g., 'Getting together against the haze', *The Straits Times* (4 July 2013).

[73] The Jambi programme is now defunct and Indonesia has not taken up Singapore's offer to discuss re-starting such programmes, see Transcript of Minister for Foreign Affairs and Minister for Law K Shanmugam's reply to Parliamentary questions on haze (8 July 2013) accessed 14 January 2015 at http://www.straitstimes.com/sites/straitstimes.com/files/TranscriptofK.Shanmugams replytoparliamentaryquestionsonhaze.pdf.

were occurring within large-scale agribusiness plantations.[74] Yet, the plantation interests vehemently denied that they had set the fires, blaming small-scale farmers living near or within the plantations instead. Amidst the wrangling, the Malaysian and Singapore governments pressed for an emergency meeting of regional environment ministers to be convened. The meeting, which was originally scheduled for August 2013, was brought forward to July 2013 and held in the Malaysian capital, Kuala Lumpur.[75]

Ahead of the meeting, the Singapore authorities publicly raised the question of whether Indonesia could provide and share its concession maps with the precise geographical coordinates of oil palm and timber plantations. The objective was to pinpoint the locations of the fires accurately and to identify the companies that had control over the relevant tracts of land. Crucially, when overlaid with satellite images of 'hot spots', it was thought that the maps could provide the Indonesian authorities with the necessary evidence to facilitate enforcement action as well as to anticipate fires during future burning seasons.[76]

The Singapore position has to be understood in the context of allegations – arising out of Indonesia to deflect criticisms of inaction – that the plantations were actually foreign-owned, with several being listed on the Singapore and Malaysian stock exchanges.[77] Indeed, at least seven companies with oil palm operations in Indonesia are based or listed in Singapore, including major players like Wilmar, APRIL and Golden Agri-Resources. Singapore thus took the stand that it needed the maps to verify who the concession holders were so that it could contemplate action against any Singapore-linked entity. Such action by Singapore would have to entail the adoption of extraterritorial legislation, something that Singapore government ministers signaled they were prepared to study (see below on the Transboundary Haze Pollution Act).

In meetings that were described as 'tense', the Indonesian government resisted and raised legal obstacles to the sharing of map information. The

---

74 According to the World Resources Institute (WRI), over half of Riau province's fire alerts occurring between 12 and 23 June 2013 were within concession areas. See Neo, *supra* note 9.

75 The meeting was the Fifteenth Meeting of the Sub-Regional Ministerial Steering Committee (MSC) on Transboundary Haze Pollution.

76 Neo, *supra* note 9.

77 Bland, *supra* note 6.

Indonesian Environment Minister even insisted that the country's freedom of information law prohibited the public disclosure of such information.[78] It was further claimed that data that could reveal the country's natural resources wealth such as forests could not be made public under Indonesian law. The Minister was also quoted as saying that the plantation companies would 'feel compromised in terms of their competitiveness by releasing these data'.[79] At the same time, he denied speculation that the Indonesian government was worried that the companies might sue the government if the maps were inaccurate.

Ironically, Indonesia's reluctance was shared by Malaysia. The Malaysian Natural Resources and Environment Minister pointed to the fact that land matters fell within the authority of the state governments in Malaysia and went on to express doubts as to whether the Malaysian federal government could publish details of concession maps. Under international law, States cannot raise obstacles or shortcomings in their domestic laws to excuse performing an international obligation. Therefore, the Malaysian position (as a party to the 2002 ASEAN Haze Agreement that explicitly requires sharing of information) is doubtful.

For its part, Indonesia should fare no better – the duty to cooperate (including to share information) in preventing transboundary pollution is provided in the 2002 ASEAN Haze Agreement and is also well-established under customary international law. The Indonesian government's position is also curious in that its 2008 Law on Freedom of Information is actually intended to make a range of official data publicly available, with several narrow exceptions meant to cover national security and diplomatic ties. In any event, the Law provides that the government can issue regulations providing for exemptions and allowing disclosure. The reluctance to disclose the maps has inevitably drawn criticism from international observers and NGOs for it suggests that there is an effort to keep the maps secret and to avoid scrutiny.[80]

Arising from Indonesia's and Malaysia's reluctance, the ensuing discussions could only produce a compromise for the maps to be shared in a limited manner and only between governments on a 'case-by-case basis'. To date, it is unclear if the Indonesian official maps have actually

---

[78] Zakir Hussain, 'Jakarta's information law forbids sharing of maps', *The Straits Times* (19 July 2013); Elgin Toh, 'Legal obstacle to sharing of maps to fight haze', *The Straits Times* (17 July 2013); Associated Press, 'Singapore tells Indonesia to share forest fire maps', *The Guardian* (17 July 2013).

[79] Teo Cheng Wee and Lester Kong, 'Asean finds compromise on tackling haze issue', *The Straits Times* (18 July 2013).

[80] Hussain, *supra* note 78.

been shared with the Singapore and Malaysian governments. Meanwhile, the Indonesian authorities are pursuing a major programme called the 'One Map Initiative' that aims to produce a single all-encompassing national map that will rationalise information on forest licences, agricultural concessions and other land use claims. If successful, the single-reference map is expected to resolve overlapping and contradictory claims and provide much-needed certainty on land use rights.

The problem, however, is the project's slow rate of implementation. The initiative is intended to be rolled out province by province but it appears that the fire-prone provinces like Riau and Jambi have yet to come up with their uniform maps. Indonesian government officials have estimated that the national map project may take another two years to complete.[81] Until that time, Indonesia has signalled that it is unable to share any maps with the neighbouring countries.

Given the realities of governance in Indonesia, the problem remains that whatever maps are shared might not even be accurate in terms of according with realities on the ground. To begin with, plantation concessions and their boundaries are extremely ill-defined in Indonesia. Often, the companies themselves do not know for certain where their boundaries lie.[82] There are typically overlaps with other competing rights such as community lands, state forests, protected areas, other concession areas and even indigenous peoples' ancestral claims, all of which arise from the land use uncertainties analysed above.

In the confusion, local governments and their leaders tend to award concessions to large companies without regard to the existing occupants or rights holders. Consequently, it is common to find different and mutually contradictory versions of concession maps issued by a variety of agencies.[83] On their part, the environmental NGOs have long called for the disclosure of official maps showing where the forestry, timber, mining and oil palm concessions are, how they overlay with forest and peat-rich areas and the distribution of hot spots when they occur, as well as information on fire risk and weather conditions.[84] However, such calls are premised on the often-mistaken belief that whatever maps are in existence are authoritative, accurate and adhered to by all the competing interests.

---

[81] Feng Zengkun, 'S'pore frustrated at slow pace in tackling haze', *Straits Times* (5 April 2014) accessed 23 January 2015 at http://news.asiaone.com/news/singapore/spore-frustrated-slow-pace-tackling-haze.

[82] Neo, *supra* note 9.

[83] *Ibid.*

[84] *Ibid.*

In any event, it is unwise to assume that all companies with 'concessions' own or occupy their lands and exercise effective control over them. Quite apart from the lack of reliable maps or boundaries, the loose term 'concession' may mean little more than an exclusive right to purchase the harvest (of the oil palm crop, for instance) from local farmers at a predetermined price. In addition, the companies or concession holders typically subcontract the task of planting, harvesting and clearing land (using fires, of course) to local villagers, many of whom belong to farmers' cooperatives controlled by regency or village leaders. In many cases, the companies are mandated by local laws and requirements to subcontract such activities to villagers in order to create jobs.

As a result, the companies have little actual control over the activities or modes of land clearance that take place, given the large size of their 'concessions' and their indeterminate boundaries. In turn, this allows the companies to claim that they have no role in the burnings and that the fires have been started by subcontractors, farmers, encroachers or other parties. At the same time, of course, the involvement or acquiescence of corrupt local leaders results in little or zero enforcement of anti-burning laws on the ground.

There is then the tension between local communities and the companies. Resentful villagers dispossessed of their lands are known to use fire to burn tracts of land claimed by the concessions, with the companies retaliating in similar fashion on the villagers' crops or staked lands.[85] Even in the absence of outright hostility, encroachment by villagers practicing slash-and-burn agriculture is all too common. The fires that they set to clear land can spread quickly and widely in peat-rich areas, and frequently into the companies' plantations. Again, the fundamental problem here is that there is no respect for supposed boundaries.

In such circumstances, it becomes doubtful whether the concession maps – even if freely produced and shared by the Indonesian side – can accurately identify the actual transgressors. On the part of affected States like Singapore, any legislation that allows the authorities to target and prosecute Singapore-based companies must grapple with important questions of evidence and burden of proof. As discussed below, it is conceivable that a 'strict liability' regime premised on legal presumptions that throw the burden of disproving certain facts onto the companies themselves can be employed under the victim States' own laws.

However, the robustness (and fairness) of such a system must again depend on the accuracy of the maps. Such accuracy, as it is now clear,

---

[85] Barber and Schweithelm, *supra* note 47 at 28.

cannot be assumed. It would seem that even the highly valued satellite identification technology is not without its issues – while generally regarded as reliable, the accuracy of satellite tracking can be compromised by thick cloud cover and other weather-related factors.[86]

Despite the setback with the sharing of concession maps, the meeting in Kuala Lumpur in July 2013 agreed to establish a cooperative mechanism relating to a joint haze monitoring system (HMS). This was subsequently endorsed by the ASEAN heads of governments at their summit meeting at the end of 2013. To be developed by Singapore using satellite technology (and presumably the shared official maps), the HMS is fully consistent with the 2002 ASEAN Haze Agreement's provisions on cooperative monitoring and sharing of information. It is premised on effective monitoring and early detection of fires and rapid response, with the aim being to identify responsible parties and the causes of the haze.

In reality, the HMS idea is not wholly novel, given that the 2002 ASEAN Haze Agreement had already provided for the ASEAN Centre (discussed above) and for national monitoring centres to be designated. These instruments were all conceived to provide cooperative monitoring and rapid response. And yet, nothing concrete has materialised since the 2002 ASEAN Haze Agreement's adoption. Even with a robust HMS in place, the ultimate responsibility still lies with the Indonesian authorities to employ the collected evidence against the identified perpetrators and to take the full measure of legal action against them.

So how has the HMS developed? In a word, 'slowly'. The Singapore Foreign Minister alluded to this in January 2014,[87] decrying the lack of progress since the HMS was agreed upon by the ASEAN leaders at the end of 2013. In diplomatic terms, he averred to 'some degree of, I won't say resistance, but a lack of movement'.[88] To the extent that the HMS relies on the sharing of the concession maps, Indonesia's slow progress with its 'One Map' initiative must inevitably hamper the HMS's implementation.

Without providing details, the Singapore Foreign Minister further stated that if ASEAN continued to do nothing, individual countries would be forced to take unilateral steps to handle the matter within the confines

---

[86] Lydia Lim, 'Trust needed for govt role as info provider and regulator', *The Straits Times* (9 July 2013).

[87] May Wong, 'ASEAN slow in adopting haze monitoring system: Shanmugam', *Today* (18 January 2014) accessed 14 January 2015 at http://www.todayonline.com/singapore/asean-slow-adopting-haze-monitoring-system-shanmugam.

[88] *Ibid.*

of international law.[89] This brings us squarely to the next issue thrown up by the 2013 fires, which is the intriguing prospect of affected States legislating laws with extraterritorial effect and taking court action against firms found to be using fires to clear land.

B. Affected States and Unilateral Legal Action

It is, of course, somewhat ironic to speak of unilateral action in a discussion about cooperative mechanisms. Yet, this is exactly what NGOs and a concerned citizenry have pressed the Singapore government to consider in the wake of the 2013 fires and haze. Among other measures, there were calls to legislate the disclosure of concession areas as well as planting and clearing practices among Singapore-linked plantation companies. If the companies are listed on the Stock Exchange of Singapore, it is conceivable that such disclosure need not even be mandated by law – they could simply be part of stock exchange listing requirements.[90]

The trickier question relates to what kinds of legal action the Singapore authorities can pursue against these companies for their conduct outside Singapore but which has effects in Singapore. When it comes to actually prosecuting such companies, the problem lies in proving that the fires had started in areas under their control and that they were at fault. No satellite technology can provide such proof. Indeed, it would require evidence-gathering and enforcement action by the Indonesian authorities on the ground, which are clearly lacking. In sum, even if Singapore were prepared to prosecute Singapore-based companies in Singapore courts for their or their related companies' actions outside of Singapore, there would be difficulties with securing the necessary evidence.

One way to get around this (or to supplement whatever evidence-gathering effort does take place in Indonesia) is to assume liability on the part of a Singapore-based plantation company the moment fires are detected in their concessions. The law could thus presume and impose *strict* liability based on the fires being detected in concession areas that are within the companies' control. The evidential burden would thus fall on the companies to produce proof that they had not used fire to clear land, or at the very least, to satisfy the 'due diligence' defence that they had done all that was reasonably possible to avoid the use of fire. This strategy will not fully resolve the problems relating to unclear boundaries, overlapping claims or encroachment by other parties, but it would

89 *Ibid.*

90 Neo, *supra* note 9.

go some way toward signifying a zero-tolerance attitude toward plantation interests and their activities.

In fact, Singapore can go further and even target companies with no links to or presence in Singapore. On the basis of the harm suffered by Singapore nationals and residents and the deleterious effects on public health and the economy felt in Singapore, there could be jurisdiction to prosecute *any* company that can be shown to use fires to clear land in Indonesia. Given Singapore's position as Southeast Asia's financial centre, it would be conceivable that many of the plantation interests and the entities that own or operate them would have assets in Singapore, or that their top management might visit Singapore and come within its jurisdiction. A judgment of the Singapore courts could then be enforced against these assets or individuals.

In line with such ideas, the Singapore Parliament passed a landmark law on transboundary haze pollution on 5 August 2014 after several months of discussions and public consultations. Consistent with the points above, the new Transboundary Haze Pollution Act[91] (Act) lays out an ambitious extraterritorial attempt to attach criminal and civil liability to agribusiness companies involved in using fires outside Singapore, whether these be Singapore-linked or otherwise.

The Act's centrepiece is a regime of multiple legal presumptions. First, it provides that if there are maps which show that any land is owned or occupied by a company, it shall be presumed that that company owns or occupies that land.[92] The maps can be procured from a variety of sources – the Act contemplates these to include any foreign government, any department or instrumentality of the government of a foreign state or any person who can be legally compelled to furnish any maps.[93] This would include a company or entity operating a concession being compelled to furnish its own maps.

Second, if there is serious haze pollution in Singapore and satellite and other meteorological evidence shows that at or about that time, there is a land or forest fire on any land causing smoke that is moving in the direction of Singapore, it shall be presumed that there is haze pollution in Singapore involving smoke resulting from that land or forest fire.[94] This

91 *Transboundary Haze Pollution Act 2014*, accessed 14 January 2015 at http://www.parliament.gov.sg/sites/default/files/Transboundary%20Haze%20Pollution%20Bill%2018-2014.pdf. The Act came into operation on 25 September 2014.

92 *Ibid*, Section 8(4).

93 *Ibid*.

94 *Ibid*, Section 8(1).

is so even if there may be fires in other or adjacent areas at or about the same time.[95] Third, it shall be presumed that the company that owns or occupies the land in question has engaged in conduct, or engaged in conduct that condones any conduct by another, which caused or contributed to that haze pollution in Singapore.[96]

The entity or company concerned can deny each of these presumptions but in each case, it will bear the heavy burden of proving the contrary. Further, the Act extends liability to any entity that participates in the management or operational affairs of another (second) entity, exercises decision-making control over the latter's business decision pertaining to land that it (the second entity) owns or occupies outside Singapore, or exercises control over the second entity at a level comparable to that exercised by a manager of that entity.[97] This is designed to target parent or holding companies that have subsidiaries or related entities that are the owners or occupiers of land and that engage in offending conduct on the ground.

The convicted entity that engages in conduct, or engages in conduct that condones any conduct by another entity or individual which causes or contributes to any haze pollution in Singapore (or the entity that participates in the management of a second entity that owns or occupies land and engages in the relevant conduct) can face a fine not exceeding S$100,000 (about US$80,000) for every day or part thereof that there is haze pollution in Singapore.[98] If the entity has failed to comply with any preventive measures notice, there can be an additional fine not exceeding S$50,000 (US$40,000) for every day or part thereof that the entity fails to comply with the notice.[99] Overall, the court must not impose an aggregate fine exceeding S$2 million (US$1.6 million). In addition, a civil liability regime is prescribed. Affected parties may thus bring civil suits against entities causing or contributing to haze pollution in Singapore.[100] The civil damages recoverable are theoretically unlimited and will be determined by the court based on evidence of personal injury, physical damage to property or economic loss (including a loss of profits).

As regards the series of presumptions set up by the Act, what is likely to be challenging is the first presumption. Even if maps can be obtained

---

95 *Ibid.*
96 *Ibid*, Section 8(2).
97 *Ibid*, Sections 3 and 8(3).
98 *Ibid*, Sections 5(1) to 5(4).
99 *Ibid*, Sections 5(2)(b) and 5(4)(b).
100 *Ibid*, Section 6.

from any of the relevant sources, the assumption here would be that these maps are authoritative, updated and accurate. However, as has been assessed earlier, different 'official' maps or versions may exist, depending on their source. Since the different agencies frequently issue overlapping land use rights, their respective maps might be mutually contradictory in identifying the boundaries and the respective owners or occupiers. In short, there would be uncertainty if the prosecution relies on maps that are contested.

Of course, the prosecution could avoid these problems by compelling the companies to furnish the maps that they themselves rely upon. In this manner, any area stated as within a company's map boundaries must be presumed to be owned or occupied by that company. This could compel the companies to obfuscate or qualify their own maps further by marking vast areas as being 'contested' or indeterminate. This situation could arise if the companies are operating alongside local communities who dispute the boundaries and thus encroach on the companies' land.

In such situations, the companies could still be held responsible under the Act unless they prove that the offending conduct was by another person acting without their knowledge or consent, or contrary to their wishes or instructions.[101] This will be a hugely difficult burden to discharge given the massive size of some of the concessions. Also, the Act stipulates that this defence is unavailable if the other person is an employee or agent of the companies, or is engaged, directly or indirectly, by the companies to carry out any work on land that they own or occupy, or is any person who has a customary right as regards the land and who has an agreement or arrangement with the companies relating to farming or forestry operations in respect of that land.

On the prosecution's part, there is the difficult task of proving that one entity participates in the management of a second entity to use fires on the ground (for example, by controlling the latter's decisions). For both the prosecutors and the companies alike, it will boil down to procuring reliable evidence to bolster their respective cases, and this will be true for the prosecution even if they are aided by the full weight of the presumptions.

Two further points need to be made. First, the Act seeks to impose liability on owners and occupiers of lands where fires are found. However, ownership and occupation are not straightforward legal concepts in Indonesia. Rights to the land may include the right to farm or operate plantations, to extract minerals found under the surface or to

101 *Ibid*, Section 7(2).

harvest the produce. Such rights may co-exist alongside the local communities' customary rights to the land.

To take into account such realities, the Act extends the definition of 'owner' to include not just a person who holds a valid lease, licence, permit, concession or other similar authorisation, but also one who has an agreement or arrangement with another person who has a customary right under local law as regards the land, and the agreement or arrangement relates to any farming or forestry operations to be carried out by any person in respect of that land.[102] Consequently, a company that operates on customary lands in such fashion will still fall within the Act's ambit, even if the actual farming or forestry operations are carried out by other parties. This will cover the common situation where a concession holder typically buys the harvest from adjoining customary lands that have been set aside for local communities to conduct farming.[103]

Second, it seems inevitable that the Act will be used against an entity the next time a serious haze hits Singapore. This is because the Singapore public will not accept anything less. For a start, that entity to be charged is likely to be a Singapore-based operation that is either listed on the local stock exchange or, even if not publicly listed, is operated and run out of Singapore (hence, having a clear link to or presence in Singapore).

A successful prosecution may go some way toward mollifying an angry public, but there is no guarantee that it will resolve the problem. This is simply because there are too many other entities or plantation interests using fires to clear land in Sumatra, Kalimantan and other parts of the sprawling Indonesian archipelago. Ultimately, Indonesian enforcement and prosecution action on the ground will matter most in resolving the problem at its core, together with cooperative mechanisms among regional States (including the provision of financial and technical assistance). The true value of Singapore's unilateral action lies more in bringing pressure on Indonesia to take greater action. In other words, it will take the wind out of the sails of the Indonesian argument that the victim States should look at their own entities first.

In a nutshell, the Act cannot be a long-term solution. Any higher expectation by the Singapore public would be naïve. At the same time, one cannot help but wonder if the Act's tough sanctions might drive agribusiness concerns out of Singapore altogether. Indeed, the publicly listed companies could de-list and easily relocate to places like Hong Kong and Malaysia, *and* still carry on with the *status quo* on the ground.

---

102 *Ibid*, Section 2(1).

103 Such adjoining lands may or may not be part of the concession.

If this happens, Singapore would have lost precious leverage over these players. *And it is not as if they will take the haze problem with them.*

V. CONCLUSION

It is somewhat ironic that this chapter should end with Singapore's adoption of unilateral measures against agribusiness companies operating in Indonesia. This must surely be the antithesis of interstate cooperative mechanisms. Yet, as this chapter has shown, the traditional forms of engagement have proven futile because they depend wholly on Indonesian political will to enforce laws effectively against offending companies. The reality is that such enforcement is unforthcoming when one considers the huge tracts of land involved, the magnitude of the fires that rage, the entrenched corruption and mismanagement involving vested interests, and indeed, the entire deep-seated sociopolitical nature of the problem.

As for the 2002 ASEAN Haze Agreement, the emphasis placed on depoliticised measures such as technical and managerial responses may be counterproductive in the long term, in that they detract from the meaningful restructuring of the underlying framework of forest exploitation and management in Indonesia. Moreover, the Agreement's dependence on wholly *technical* responses seems out of touch with the contemporary social landscape in Indonesia, one in which a wary public is demanding greater *political* accountability of government and corporate interests.[104] Ultimately, a treaty regime (particularly one operating within the peculiar ASEAN context and depending on one sole actor for effectiveness) cannot dictate the reordering of an entire national system of governance, particularly one with an entrenched ethos of mismanagement and corruption.

As for cooperative mechanisms, now that Indonesia has become a State Party to the 2002 ASEAN Haze Agreement, it has the obligation to enforce the Agreement in good faith. In addition, the ASEAN Centre and the ASEAN Fund will need to be properly established, and concrete cooperative mechanisms (particularly financial and technical assistance incentives provided by regional States) must be developed. In the meantime, the region and the victim States will have to contend with existing bilateral cooperative mechanisms and possibly unilateral extraterritorial action against companies that use fires.

---

[104] Tan, *supra* note 12 at 670.

13. Legal redress of transboundary air pollution through environmental cooperation

Nicholas A Robinson

I. INTRODUCTION

Traditional rules governing State responsibility for transnational environmental injury are fast becoming ineffective. The efficacy of rules of sovereignty that arose in a world with less than three billion people, modest levels of international trade, and relatively stable technologies, has eroded. World trade, direct overseas foreign investment, and technological innovation expand faster than governments can manage. To complicate affairs further, the planet Earth has left the Holocene Epoch and entered the Anthropocene Epoch, in which rapid, human-induced change appears throughout the planet's natural and geological systems.[1]

Transboundary air pollution, known as the haze in Southeast Asia, illustrates this new reality. Direct foreign investment in Indonesia, which converts forests to plantations for producing palm oil, has expanded exponentially in the past 20 years, driven by growing market demands. Although unlawful under Indonesian law, forests are burned to establish palm plantation monocultures. Smoke from the fires pollutes the air of Malaysia and Singapore, producing unhealthy conditions and impairing economic life. To compound matters, these conversions destroy peat wetlands and peatland forests, releasing greenhouse gases (GHGs), impairing GHG sinks, eliminating photosynthesis, and devastating the biological diversity of the world-renowned flora and fauna of Sumatra and Kalimantan.

For the past four decades, demands to curb transboundary air pollution from these forest fires in Indonesia, in accord with international law, have

[1] See Will Steffen, Paul J Crutzen and John R McNeill, 'The Anthropocene: Are Humans Now Overwhelming the Great Forces of Nature?' (2007) 36(8) *Ambio* 614.

been unavailing. The fires persist, producing more extreme pollution incidents in neighbouring nations. Although general principles of international law[2] provide that no nation may allow its territory to cause harm to another nation or the commons, this transboundary pollution has grown. The Association of Southeast Asian Nations (ASEAN) has negotiated a treaty, the ASEAN Agreement on Transboundary Haze Pollution (2002 ASEAN Haze Agreement), to abate the pollution, which has been ratified by Indonesia after 12 years and entered into force.[3] Repeated requests that Indonesia prevent the activity causing the pollution had been often met with the defence that, since this was an internal affair, it affronted Indonesia's sovereignty for other nations to press for action about the palm plantations. As the Indonesian fires increasingly harmed Indonesia internally, the administration of President Susilo Bambang Yudhoyono pressed for the ratification of the 2002 ASEAN Haze Agreement, which was accomplished in September 2014. This ratification permits acceleration of international cooperation to abate the fires and transboundary air pollution.

Without the acceptance of a treaty such as the ASEAN Haze Agreement, international organisation similarly offers little recourse. No international courts or tribunals have compulsory jurisdiction to hear the claims of States being polluted, save possibly the compulsory dispute settlement procedures under the United Nations (UN) Convention on the Law of the Sea (UNCLOS).[4] Resort to compulsory dispute settlement mechanisms under

---

[2] The rule is restated in Principle 21 of the 1972 United Nations Stockholm Declaration on the Human Environment and Principle 2 of the Declaration of Rio de Janeiro on Environment and Development. Stockholm Declaration on the Human Environment, in 'Report of the United Nations Conference on the Human Environment', 16 June 1972, UN Doc A/CONF.48/14, at 2 and Corr.1 (1972), 11 ILM 1416 (1972 Stockholm Declaration); Rio Declaration on Environment and Development, in 'Report of the United Nations Conference on Environment and Development', 12 August 1992, UN Doc A/CONF.151/26 (Vol I), 31 ILM 874 (1992 Rio Declaration).

[3] ASEAN Agreement on Transboundary Haze Pollution, signed 10 June 2002 (entered into force 25 November 2003) accessed 15 January 2015 at http://agreement.asean.org/media/download/20140119170340.pdf (2002 ASEAN Haze Agreement). Adella Anjani Putri, 'Indonesia Ratifies Haze Treaty after 12 Years of Stalling', *The Jakarta Globe* (17 September 2014) accessed 16 January 2015 at http://www.thejakartaglobe.com/news/indonesia-ratifies-haze-treaty-12-years-stalling.

[4] See United Nations Convention on the Law of the Sea, adopted 10 December 1982, 1833 UNTS 397 (entered into force 16 November 1994) (UNCLOS), Part XV, Section 2.

UNCLOS, however, is unlikely to end the air pollution. Other fora offer little benefit. The UN Security Council has thus far declined to entertain propositions that ongoing environmental harms are a threat to the peace, and so even the collective security of the UN Charter is illusory.

Nonetheless, environmental legal 'remedies' do exist beyond those traditionally considered for transboundary pollution. Environmental law can abate pollution while curbing the underlying activities that drive the emissions. Remedies available under several international environmental legal regimes might be deployed in ways to abate air pollution in Southeast Asia. There are four dimensions of how this might work, focusing on the use of biological 'remedies': (i) air pollution as a common concern of mankind; (ii) the reasons for neglect of extant environmental law remedies; (iii) interstate cooperation for sustainability; and (iv) the various causes of the air pollution emanating from Indonesia. However, there also may be remedies under human rights regimes,[5] rights of indigenous peoples,[6]

---

[5] The right to the environment, the human right to health and water, and procedural rights under international law merit further study. The emergence of a right to the environment is a worldwide phenomenon. Most nations embody this right in their national constitutions. It is quite advanced in several ASEAN Member States, such as the Philippines, where citizens are authorised to bring actions to enforce the environmental laws. See Republic of the Philippines, Supreme Court, Rules of Procedure for Environmental Cases, Resolution of 13 April 2010 (AM No 09-6-8-SC) at Section 5, accessed 15 January 2015 at http://www.lawphil.net/courts/supreme/am/am_09-6-8-sc_2010.html. The wanton destruction of biodiversity arguably violated human rights and the rights to the environment in Indonesia. Forest clearing by fires is often hard for a distant central government to know about, much less to prevent. Local communities and non-governmental organisations (NGOs) can act locally. The land conversions in Sumatra appear to violate many Indonesian national laws. What is needed, as Principle 10 of the 1992 Rio Declaration and now the Sustainable Development Goals emphasise, is public participation and ensuring access to justice. See 1992 Rio Declaration, *supra* note 2, Principle 10; UN Sustainable Development Goals, accessed 15 January 2015 at http://sustainabledevelopment.un.org/index.php?menu=1300 (SDGs). The right to water is affected by the pollutions that emanate from palm oil plantations, as well as by changes in the aquifers and groundwater table. The human right to health and the right to the environment are compromised by the acute air pollution within Sumatra, and as local communities experience this pollution, they are entitled to access to justice within Indonesia.

[6] Rights under the UN Declaration on the Rights of Indigenous Peoples are violated by the destruction of the forests. *Declaration on the Rights of Indigenous Peoples*, GA Res 61/295, UN GAOR, 61st Sess, Supp No 49, UN Doc A/RES/61/295 (2007). The Indigenous People's Alliance of the Archipelago, an indigenous NGO in Indonesia, represents one-third of the overall Indonesian population. Their interests can be examined at the annual UN Permanent Forum

the law of the sea,[7] trade agreements,[8] energy law,[9] labour law,[10] and other sectors. Analysis of the immediately relevant biological 'remedies' illustrates how cooperation might abate the haze.

---

on Indigenous Issues. Indigenous and traditional communities deserve access to justice; they experience human rights violations, environmental degradation, economic marginalisation and suffer loss from clearing of forests. Debating how to recognise and accommodate indigenous rights draws attention to the fires and makes them more than just an economic issue of the haze. Converting peatlands to plantations is a practice that entails denials of human rights. Eventually, this is the sort of issue that may be addressed by the UN Human Rights Council.

[7] Chapters 1 and 6 in this book address the role of UNCLOS. One of the indirect impacts of deforestation is erosion and land-based marine pollution, along with incremental deposition of air contaminants into seawater. The need to help mangroves and coastal wetlands migrate and adapt to increasing sea levels offers ways to combine marine law discussions with the objectives of the Ramsar Convention. Convention on Wetlands of International Importance, especially as Waterfowl Habitat, adopted 2 February 1971, 996 UNTS 245 (No 14583), ILM 963 (entered into force 21 December 1975) (Ramsar Convention). Cooperation measures could also address the obligations of UNCLOS, *supra* note 4 at Article 195(5), and its duties toward protecting and conserving rare or fragile ecosystems and endangered marine life. UNCLOS duties can be synergistically addressed also through the Convention on Biological Diversity, adopted 5 June 1992, 1760 UNTS 79 (entered into force 29 December 1993) (Convention on Biological Diversity). Marine protection has been emphasised by the Thematic Programme on the Conservation and Sustainable Use of Marine and Coastal Biological Diversity CBD, Jakarta Mandate on the Conservation and Sustainable Use of Marine and Coastal Biological Diversity, decided at COP-2 (Jakarta, Indonesia, 6–17 November 1995) accessed 15 January 2015 at http://www.cbd.int/decision/cop/default.shtml?id=7083, subsequently adopted at the COP-4 (Bratislava, Slovakia, 4–15 May 1998) accessed 15 January 2015 at http://www.cbd.int/decision/cop/?id=7128, and amended at the COP-7 (Kuala Lumpur, Malaysia, 9–20 February 2004) accessed 15 January 2015 at http://www.cbd.int/decision/cop/default.shtml?id=7742.

[8] Given the driving force of the palm oil trade in inducing expansion of plantations into forests, the role of commercial trade law should become a high priority. Palm oil demand is growing and will continue to grow as the human population in Asia and worldwide grows. In the short term, traders should agree to support and adhere to the certification rules of the Roundtable on Sustainable Palm Oil (RSPO). Only 14 per cent of palm oil is certified. See Roundtable on Sustainable Palm Oil (RSPO), *RSPO Highlights Role of Indonesia in Increasing CSPO Demand* (RSPO, 2012) accessed 15 January 2015 at http://www.rspo.org/news-and-events/news/spo-highlights-role-of-indonesia-in-increasing-cspo-demand. In the longer term, agreements are needed to bar new overseas investment in converting wet peatland forest into palm oil or pulp plantations. The European Union (EU) examined the perverse impact of its EU Renewable Energy

II. TRANSBOUNDARY AIR POLLUTION: DRIVERS, IMPACTS AND THE NEED FOR BIOLOGICAL REMEDIES

Environmental law offers non-traditional remedies for abating air pollution while furthering sustainable development. Cooperation among States to restore and sustain biological resources can produce lasting 'remedies'. Assertions of State responsibility and political negotiations alone

---

Directive, which has been a driver of peatland destruction in Indonesia since the end of 2010. EC, *Commission Directive 2009/28/EC of 23 April 2009 on the promotion of the use of energy from renewable sources and amending and subsequently repealing Directives 2001/77/EC and 2003/30/EC*, [2009] OJ L 140/16 (EU Directive).

[9] The status of biofuels, including the full environmental costs of biofuels such as palm oil, can be raised in the deliberations of the still new International Renewable Energy Agency (IRENA). EU States participate in IRENA's work, and IRENA could assess the EU Directive. Standards for biofuels should require all externalities be accounted for, before certifying a biofuel as a renewable energy under IRENA's authority. This could help the EU and other countries, such as the United States of America (US), to re-examine the use of palm oil. One of the drivers of the forest fires could be reduced. A parallel approach could be used in the 1994 Energy Charter Treaty and its Protocol on Energy Efficiency and Related Environmental Aspects. Energy Charter Treaty and Energy Charter Protocol on Energy Efficiency and Related Environmental Aspects, adopted 17 December 1994, 2080 UNTS 95 (entered into force 16 April 1998).

[10] Forest livelihoods are significant. Some 40 million Indonesians are dependent on forest resources. Their food and jobs and homes and culture are forest related. Plantation conversions disrupt existing employment and land use patterns. Given the global stakes of GHG emissions, all international agencies and organisations, such as the UN's Children's Fund (UNICEF), need to be brought into this effort. The welfare of children and workers in areas affected by the fires could be examined by the UNICEF and the International Labor Organization. The palm oil industry has some 3.7 million workers currently, and the salaries are relatively low compared to plantation income; workers' salaries and living conditions are at poverty levels. Manginar Situmorang, 'Strengthening the Peasant and Plantation Worker's Movement in North Sumatra', *Asia Monitor Resource Centre* (16 November 2010) accessed 15 January 2015 at www.amrc.org.hk/node/1005. Half of rural residents who lost their lands to plantations report a decline in the standard of living. Krystof Obidzinski, Rubeta Andriani, Heru Komarudin and Angus Andrianto, 'Environmental and Social Impacts of Oil Palm Plantations and Their Implications for Biofuel Production in Indonesia' (2012) 17(1) *Ecology and Sociology* at 10. A UN inquiry into labour or human rights abuses in the palm oil plantations may be appropriate.

are not likely to end the air pollution because they address the haze only as a symptom and do not reach underlying causes.

A. Air Pollution as an International Common Concern

Southeast Asia's air pollution is matched elsewhere. Air pollution in northern China from time to time extends to pollute the air in Korea, and also reaches Japan. Urban air pollution episodes in Western Europe often affect more than one nation. Forest fires in Mexico have caused pollution in the United States of America (US) and pollution from the US has affected Canada and vice versa. Even when urban air pollution is not yet so vast as to cross borders, it has emerged as a common problem for developed and developing nations alike.[11] The rapid growth of megacities in many countries is accompanied by recurring periods of air pollution.

Since 1992, nations and intergovernmental organisations have been slow to act on the 'sustainable development' recommendations in Chapter 9 of Agenda 21 regarding reducing atmospheric pollution.[12] Acute urban air pollution had already been identified as a priority of 'The Urban Challenge' described by the UN World Commission on Environment and Development in 1987.[13] In principle, regional consensus favours abating air pollution. The ASEAN Charter provides that one of its founding purposes is '[t]o promote sustainable development so as to ensure the protection of the region's environment, the sustainability of its natural

---

[11] The first UN Environment Assembly (UNEA) of the UN Environment Programme (UNEP) explicitly called on governments to take action to improve air quality and abate air pollution. See UNEA, 'Ministerial Outcome Document of the first session of the United Nations Environment Assembly of the United Nations Environment Programme' (27 June 2014) at 3, accessed 15 January 2015 at http://www.unep.org/unea/docs/UNEA-1%20outcome%20document.pdf; 'Summary of the First UN Environment Assembly of the UN Environment Programme' (2014) 16 *Earth Negotiations Bulletin* at 8, accessed 15 January 2015 at http://www.iisd.ca/vol16/enb16122e.html.

[12] *Agenda 21: Programme of Action for Sustainable Development*, UNGAOR, 46th Sess, Agenda Item 21, UN Doc A/Conf.151/26 (1992) (*Agenda 21*), Ch 9, at paras 9.9–9.21.

[13] UN World Commission on Environment and Development, *Our Common Future* (Oxford: Oxford University Press, 1987) at 235–5, Chapter 9 (*Our Common Future*).

resources, the preservation of its cultural heritage and the high quality of life of its peoples'.[14] Air pollution retards or denies sustainable development.[15]

Air pollution is an avoidable problem. Its causes can be identified and the pollution eliminated. Air pollution injures human health and even causes death. It violates the fundamental human right to health. While polluting the air, forest fires also entail myriad additional losses. The years from 1997 to 1998 witnessed vast air pollution, much of it from fires. This period experienced the periodic El Niño Southern Oscillation event, when prolonged dry weather conditions fanned fires worldwide.[16] Because forest fires were seen as isolated domestic issues, or acts of god, the fires of 1997–1998 were largely ignored by international agencies. The impact within Indonesia was devastating. The Center for International Forestry Research (CIFOR) estimated that 11.7 million hectares were lost to the fires.[17]

Abatement of transboundary pollution necessitates examining its causes. It is simultaneously a local, national and international phenomenon. Emissions locally produce dreadful environmental conditions that also incrementally degrade air quality elsewhere. Air pollution controls are imperative at local, regional or global scales of human activity.[18] Environmental law's task is to apply controls in congruent and coherent ways across all levels and scales of governance.[19]

---

[14] Charter of the Association of Southeast Asian Nations, adopted 20 November 2007 (entered into force 15 December 2008) (ASEAN Charter), Art 1(9).

[15] See *Our Common Future*, *supra* note 13 at 330: 'Human laws must be reformulated to keep human activities in harmony with the unchanging and universal laws of nature.'

[16] See Nicholas A Robinson, 'Forest Fires as a Common International Concern: Precedents for the Progressive Development of International Environmental Law' (2001) 18 *Pace Environmental Law Review* 459.

[17] See Luca Tacconi, 'Fires in Indonesia: Causes, Costs and Policy Implications' Center for International Forestry Research (CIFOR) Occasional Paper No 38 (2003), at V, accessed 15 January 2015 at http://www.cifor.org/fire/pdf/pdf23.pdf.

[18] See e.g., World Charter for Nature, GA Res 37/7, UNGAOR, 37th Sess, Supp No 51, UN Doc A/37/51 (1982), 22 ILM 455 (1983).

[19] When environmental law was first emerging within and among nations following the 1972 UN Stockholm Conference on the Human Environment, René Dubos observed that societies and individuals should 'think globally, and act locally'. See Noel Gough, 'Thinking/acting locally/globally: Western science and

Economists characterise such air pollution emissions as 'externalities'. The waste is external to the economic enterprise, which enjoys 'free' discard. Once the drivers of such harmful activities are identified, laws can be adopted to abate harmful discharges at their sources. Regulations that directly address the pollution fix costs that effectively 'internalise' the externalities. Other regulations address activity that is, indirectly to the pollution, constraining the opportunities to engage in polluting acts. Since ecological relationships are intertwined in complex ways, it is instructive to assess all applicable regimes that aim to sustain environmental integrity. Where one remedy is unavailable, others often can be invoked. Since direct controls on air pollution in ASEAN are currently lacking, indirect remedies must be pursued.

International environmental law aims to enhance socio-economic and cultural well-being. Yet States are slow to build their capacity for environmental stewardship. Air pollution has increased since the 1992 'Earth Summit'. Why have sustainable development policies been ineffective? Why are legal tools available to achieve the objectives of the ASEAN Charter little used?

B. Reasons for Neglect of Environmental Law Remedies

Ambient air pollution arises from many sources. It accumulates, spreads through weather conditions, and being diffuse, is often tolerated. Health impacts worsen slowly. Even when sources of emissions can be identified, specific injury to a human's health, or concrete economic losses, can rarely be ascribed to a specific source. Liability by linking cause and effect cannot be established. In the transboundary context, even when all sources may be found in one nation and State responsibility is assigned, emissions will not end. Compensation might be paid for past damages, but it is impractical to repeatedly seek compensation for injuries resulting from ongoing patterns of pollution.

To cope with diffuse pollution, environmental regimes curb practices that pollute. The economic burden on those regulated is clear and can appear heavy. The benefit to victims appears only gradually. This asymmetry induces intense negative responses from those whose economic interests are affected; they wish to continue 'business as usual'. Victims rarely defend pollution control regimes. Moreover, many nations have scant experience with environmental remedies and relatively few

environmental education in a global knowledge economy' (2002) 24(11) *International Journal of Science Education (Special Edition: Environmental Education and Science Education)* 1217.

specialists exist to deploy them. Political leaders often regard these laws as mere amenities, welcomed when economic times are robust and shunned when weak.

It must be acknowledged also that environmental stewardship is difficult. Unlike engineering solutions, pollution legal remedies cannot be had with the 'turn' of a spigot. Effective 'Environmental Management Systems'[20] require sustained and often intergenerational efforts. Implementing environmental laws requires capacity building, education and reiterative learning about how best to apply methodologies, such as ecosystem management or environmental impact assessment. When pollution is entrenched, years are needed to root out harmful practices. Adopting environmental laws, or adhering to treaties, is only the beginning of a sustained marathon for achieving and then sustaining environmental quality.

Although environmental laws offer foundations for sustainable development,[21] expediency and a focus on short-term economic growth preclude investments in restoring or sustaining the natural systems. The UN SDGs[22] aim to shift governments toward investing for human health and well-being. In order to attain the SDGs, environmental law remedies will become increasingly important. Unhealthy air pollution is a symptom of weak, unsustainable development.

---

[20] For the United Kingdom (UK), see UK Environment Agency, accessed 15 January 2015 at https://www.gov.uk/government/organisations/environment-agency. For the US, see US, Environmental Protection Agency, 'Environmental Management Systems', accessed 15 January 2015 at http://www.epa.gov/ems/. For generic models, see International Standards Organization, 'ISO 14000 – Environmental management', accessed 15 January 2015 at http://www.iso.org/iso/home/standards/management-standards/iso14000.htm.

[21] A high environmental quality of life is the aim of the wealth of nations. Advanced socio-economic development emerges and is sustained through myriad factors. Historian Niall Ferguson identified six factors that produced the European legacy of development: (i) fostering economic production and trade, (ii) providing an educated populace with scientific competences, (iii) *ensuring the rule of law* and means for peaceful settlement of disputes about rights, (iv) *ensuring good health* and nutrition for all, (v) providing material goods for individuals, and (vi) sustaining a work ethic for honest productivity. Niall Ferguson, *Civilization: The West and the Rest* (London: Penguin Books, 2011) at 13 (emphasis added).

[22] See SDGs, *supra* note 5.

Environmental law compliance and enforcement contributes to peaceful domestic relations, and is a dimension of peace, of national security.[23] Conversely, as environmental degradation accelerates,[24] economies lose resilience[25] and development falters.[26] Yet while Earth's natural systems are all interlinked, the regimes of nations are varied and at odds with nature. As the 1987 UN World Commission on Environment and Development put it: '[T]he Earth is one but the world is not.'[27] States have made great strides in establishing a latticework of multilateral environmental agreements (MEAs) and other treaties to realise 'sustainable development'.[28] However, air pollution has not yet been made a major focus of these efforts.

Countervailing influences are also at work. Even where laws are in place, States ascribe a low priority to environmental law enforcement. General principles of international law, such as the duties of cooperation or of equity, and in particular environmental principles of law, such as those contained in the 1992 Rio Declaration,[29] should offer guidance for strengthening environmental stewardship. However, where short-term economic gain is threatened, such general principles of law are avoided or implemented superficially.[30] For example, governmental incentives for

---

23 See the International Network on Environmental Compliance and Enforcement, accessed 15 January 2015 at www.inece.org.

24 The 'great acceleration' is noted in Steffen, Crutzen and McNeill, *supra* note 1.

25 See Nicholas A Robinson, 'The Resilience Principle', (2014) 5 *International Union for Conservation of Nature (IUCN) Academy of Environment Law (AEL) eJournal* 19, accessed 15 January 2015 at http://digitalcommons.pace.edu/cgi/viewcontent.cgi?article=1951&context=lawfaculty.

26 See UNEP 'Global Environmental Outlook 5 (GEO-5)' (June 2012) accessed 15 January 2015 at www.unep.org/geo.

27 *Our Common Future*, *supra* note 13 at 27.

28 Contemporary international environmental law is presented in Lal Kurukulasuriya and Nicholas A Robinson (eds), *Training Manual on International Environmental Law* (UNEP, 2006).

29 See 1992 Rio Declaration, *supra* note 2.

30 From the perspective of the environmental sciences, whenever States attempt to resolve environmental disputes too narrowly or superficially, they invariably fail. Some symptoms can be abated for a time, but the root causes reappear or collateral impacts undermine the solutions tried. Ecosystems are interrelated and a holistic understanding is required before addressing a specific problem. See Christina Voigt, *Rule of Law for Nature* (Cambridge: Cambridge University Press, 2013).

promoting trade often seek short-term economic gain by avoiding environmental rules. Controversies result, frustrating both trade and environmental protection, as illustrated by negotiations for the Trans-Pacific Partnership (TPP).[31] For trade to be environmentally sustainable, trade should adhere to treaty obligations to protect the environment.[32] The Doha Round of trade negotiations was to have elaborated rules for safeguarding environmental protection laws from being undermined by trade agreements, but States have yet to undertake this task.[33] The rich palm oil trade obscures its contributions to environmental degradation.

---

[31] Coral Davenport, 'Administration is Seen as Retreating on Environment in Talks on Pacific Trade', *The New York Times* (15 January 2014) accessed 15 January 2015 at http://www.nytimes.com/2014/01/15/us/politics/administration-is-seen-as-retreating-on-environment-in-talks-on-pacific-trade.html?_r=0; Shawn Donnan, 'Greens Fear US U-turn on Trade Deal Vows', *The Financial Times* (16 January 2014) accessed 15 January 2015 at http://www.ft.com/cms/s/0/561715b0-7e14-11e3-b409-00144feabdc0.html#axzz3BuA2tmlo. Malaysia, Singapore, Brunei and Vietnam are negotiators with Japan, Australia, New Zealand, Chile, Mexico, Canada and the US. The draft treaty would allow private trade deals to trump environmental laws, where investor-state trade dispute panels could override environmental laws (even those required under a multilateral environmental agreement). This sort of trade sectoralism was renounced by the World Trade Organization (WTO), and its recurrence in the TPP negotiations is an indicator that trade negotiators are so narrowly focused that they ignore the environmental laws of nations. If the TPP is willing to sacrifice environmental laws in favour of trade, then there is little logic in suggesting that investment in palm oil plantations, e.g., in Indonesia, should be governed by environmental laws that might restrict the growing conversion of forests to plantations to produce and trade transnationally in palm oil products.

[32] The most recent MEA, the 'Minamata Convention on Mercury' provides that 'this Convention and other international agreements in the field of environment and trade are mutually supportive'. 2013 Minamata Convention on Mercury, adopted 13 October 2013 (not yet entered into force), Preamble, at para 10 (2013 Minamata Convention).

[33] The Doha Round of trade negotiations under the General Agreement on Tariffs and Trade (GATT) at the 4th Ministerial Conference in November 2001 (Paragraph 31 of the Doha Declaration) mandated negotiations on the elimination of barriers to environmental goods and services. See WTO, Ministerial Conference, *Doha Declaration on the TRIPS Agreement and Public Health* (declared 14 November 2001), WT/MIN(01)/DEC/W/2, Fourth Session (Doha Declaration). Agreement on the application of Sanitary and Phytosanitary Measures (GATT Article XX), now broadened through environmental laws nationally and multilaterally, remains to be negotiated. See Daniel C Esty, *Greening the GATT* (Washington, DC: Peterson Institute for International Economics, 1994) at 53.

Before cooperation begins to resolve an environmental problem, much less becomes effective, there is inevitably a period when the problem is ignored. Parties pretend that it is someone else's responsibility, sometimes trading accusations about accountability, engendering growing mistrust.[34] This occurred in Europe in 1972 when Sweden taught the world of nations about transboundary 'acid rain'. Citing international rules of State responsibility, as expressed in the *Trail Smelter* case[35] or Principle 21 in the 1972 Stockholm Declaration,[36] did not stop diffuse emissions of sulphur dioxide from the industrialised Ruhr basin or the British Isles. The emissions continued as long as the factories operated. The air contaminants were carried north with the hydrologic cycle, poisoning the lakes of Sweden and impairing forest ecosystems.

To abate 'acid rain', States negotiated the 1979 Convention on Long-Range Transboundary Air Pollution[37] setting the stage for a 50 per cent reduction in the emissions that cause 'acid rain' in Europe. Canada and the US agreed to a 1991 bilateral Air Quality Agreement,[38] which has been implemented to reduce emissions by 50 per cent in North America. Nonetheless, most States still ignore transboundary air pollution.[39] Societies and nations would rather quietly 'export' their air pollution than contain contaminating emissions.

---

[34] These social patterns are discussed by Lord Eric Ashby in *Reconciling Man with the Environment* (Redwood, CA: Stanford University Press, 1978), and discussed in Nicholas A Robinson, 'Legal Systems, Decision-making, and the Science of Earth's Systems: Procedural Missing Links' (2001) 27(4) *Ecology Law Quarterly* 1077 at 1097.

[35] *Trail Smelter Case (United States of America, Canada)* (1938 and 1941) 3 RIAA 1905.

[36] Principle 21 of the 1972 Stockholm Declaration is deemed binding as customary international law: 'States have … the sovereign right to exploit their own resources pursuant to their own environmental policies, and the responsibility to ensure that activities within their jurisdiction or control do not cause damage to the environment of other States or of areas beyond the limits of national jurisdiction.' 1972 Stockholm Declaration, *supra* note 2.

[37] Convention on Long-Range Transboundary Air Pollution, adopted 13 November 1979, 1302 UNTS 217, 18 ILM 1442 (entered into force 16 March 1983).

[38] Agreement between the Government of the United States of America and the Government of Canada on Air Quality (US-Canada) (adopted and entered into force on 13 March 1991), 30 ILM 676.

[39] States today are acting slowly to abate the other 50 per cent of the emissions that are precursors of acid rain and smog in the northern hemisphere. Acid rain persists. Elsewhere, States have not acted at all with regard to the wholly unabated sources of 'acid rain' arising from China and Southeast Asia, all

In Southeast Asia, even though the 2002 ASEAN Haze Agreement has come into force, it lacks provisions to abate all air pollution. Effective laws from other regions deserve study. Where air pollution has been abated effectively, States have designed proactive systems of 'action plans', including collaborative capacity-building measures across public and private sectors. Under the federal Clean Air Act in the US,[40] all of its States have adopted 'State Implementation Plans'.[41] The EU developed relatively strong regional environmental laws through a series of five-year action plans. ASEAN has adopted many such action plans, across a wide range of environmental stewardship topics,[42] but has not adopted clean air measures like those in use in other regions.

C. Interstate Cooperation for Sustainability

Abatement of air pollution should consider the full spectrum of applicable environmental agreements and deploy available remedies.[43] Because Earth's ecological systems are interrelated, every environmental MEA or other treaty offers tools that can contribute, however indirectly, to resolving problems like the haze. Where remedies are unavailable under one legal framework, there will be other agreements through which cooperation continues in ways helpful to resolving an environmental problem. There are synergies among the MEAs, which States can employ to resolve complex environmental problems. This is especially clear under the 'biological' agreements surveyed here.

---

of which produce the Asian Brown Cloud over the Indian Ocean, eventually polluting all of South Asia with 'acid rain' in the monsoons.

[40] *Clean Air Act*, USC § 7401 et seq (1963).

[41] *Ibid* at Part A, 196.

[42] See Koh Kheng-Lian (ed), *ASEAN Environmental Law, Policy and Governance: Selected Documents*, Vol I (Singapore: World Scientific Publishing Company, 2009), and Vol II (Singapore: World Scientific Publishing Company, 2013). See also Koh Kheng-Lian, *Selected ASEAN Documents on the Environment* (Singapore: Asia-Pacific Centre for Environmental Law, 1996).

[43] A good example of this successfully at work is the implementation of the Vienna Convention to Protect the Stratospheric Ozone Layer and the Montreal Protocol. Vienna Convention for the Protection of the Ozone Layer, adopted 22 March 1985, 1513 UNTS 293, 26 ILM 1529 (entered into force 22 September 1988); Montreal Protocol on Substances that Deplete the Ozone Layer, adopted 16 September 1987, 1522 UNTS 3, 26 ILM 1550 (entered into force 1 January 1989). See James Gustave Speth and Peter M Haas, *Global Environmental Governance: Foundations of Contemporary Environmental Studies* (Island Press, 2006).

The use of such indirect, alternative remedies is often neglected. Governmental ministries, NGOs and intergovernmental organisations focus narrowly on their assigned sectors of responsibility. Environmental programmes rarely receive sustained support over long periods of time.[44] Few nations perceive or cultivate the synergies among the MEAs. It is difficult to perceive how legal tools in one decision-making forum can contribute to attaining objectives that are a primary focus of another entity. For example, those concerned with energy rarely devote attention to biodiversity, and vice versa, although demand for biofuels drives impacts on biological resources.

In 1992, Agenda 21 undertook the extraordinary task of creating a holistic, inter-sectoral action plan for sustainable development.[45] Many national laws do not yet function sustainably in line with Agenda 21's recommendations. Environmental protection reforms are opposed by vested economic interests and resisted by entrenched government bureaucracies, both favouring 'business as usual'. Although air pollution is a clear health hazard, Agenda 21 assigns no priority to air pollution. States tolerate incremental air pollution, perhaps on the assumption the winds will dissipate it and that it could always be abated whenever a nation wished. Twenty years later, the poverty of this approach is evident. Avoidance is not '20/20 vision'.

---

[44] The ASEAN Strategic Plan of Action 1994–1998, and subsequent action plans, did not provide sustained attention to abating the regional driving forces of environmental degradation, and so those drivers continued to produce damage. ASEAN, *Strategic Plan of Action on the Environment 1994–1998*, adopted 26 April 1994, accessed 15 January 2015 at http://cil.nus.edu.sg/1994/1994-strategic-plan-of-action-on-the-environment-1994-1998-adopted-on-26-april-1994-in-bandar-seri-begawan-brunei-darussalam-by-the-environment-ministers. The resources needed to maximise environmental education, to be able to enlist the citizenry in supporting abatement of problems, has also been less than what would be needed. However, excellent as they are, the ASEAN Environmental Education Action Plan 2000–2005 and Action Plan II (2008–2012), need greater support from ASEAN Member States to realise their potential. Similar shortfalls appear in other regions. ASEAN, *Environmental Education Action Plan (AEEAP) 2008–2012: Environmental Education for Sustainable Development*, adopted 6 September 2007, accessed 15 January 2015 at http://environment.asean.org/wp-content/uploads/2012/12/AEEAP-2008-2012_-ENVIRONMENTAL-EDUCATION-FOR-SUSTAINABLE-DEVELOPMENT.pdf.

[45] See *Agenda 21*, *supra* note 12. The *traveaux préparatoires* of the 1992 UN Conference on Environment and Development (UNCED) are published in Nicholas A Robinson, *Agenda 21 and the UNCED Proceedings* (New York: Oceana Publications, 1993), in six volumes.

Holistic analytic tools that could be used to avert new sources of air pollution, such as environmental impact assessment (EIA), have similarly been avoided. EIA is required of States in their national decision-making under Principle 17 of the 1992 Rio Declaration.[46] Many States have enacted legislation for EIA. The International Court of Justice deems EIA to be required as a binding obligation of customary international law.[47] EIA is a logical consequence of Principle 21 of the 1972 Stockholm Declaration:[48] how can one know if conduct that a State allows on its territory has an effect abroad if the State fails to study the environmental impacts of the actions? EIA is a 'system' suited to demonstrating that a State has complied with Principle 21 and applied the Precautionary Principle[49] or other general principles of international environmental law as set forth in the 1992 Rio Declaration. Without robustly deploying tools such as EIA, sustainable development will be elusive.

Applying EIA is a useful tool for averting or resolving transboundary disputes, as evidenced by the Espoo Convention on Transboundary Environmental Impact Assessment (Espoo Convention)[50] in the European context, or by the bilateral EIA agreements between States such as Canada and the US. Where States ignore EIA, as the US improvidently has done for trade negotiations,[51] governments often fail to identify environmental problems, which in turn cause them problems.[52] EIA is legally mandated in Indonesia, yet is not applied to the conversion of

---

[46] 1992 Rio Declaration, *supra* note 2, Principle 17.

[47] *Case concerning Pulp Mills on the River Uruguay (Argentina v Uruguay),* [2010] ICJ Rep 14 (*Pulp Mills*).

[48] See Louis Sohn, 'The Stockholm Declaration on the Human Environment' (1973) 14 *Harvard Journal of International Law* 485–93.

[49] Principle 15: 'In order to protect the environment, the precautionary approach shall be widely applied by States according to their capabilities. Where there are threats of serious or irreversible damage, lack of full scientific certainty shall not be used as a reason for postponing cost-effective measures to prevent environmental degradation.' 1992 Rio Declaration, *supra* note 2.

[50] Convention on Environmental Impact Assessment in a Transboundary Context, adopted 25 February 1991, 30 ILM 1461 (1991) (entered into force 10 September 1997) (1991 Espoo Convention).

[51] See *Delegation of Certain Authorities and Assignment of Certain Functions under the Trade Act of 2002*, 67 Fed Reg 225 (2002) accessed 15 January 2015 at http://www.gpo.gov/fdsys/pkg/FR-2002-11-21/pdf/02-29832.pdf.

[52] For political opposition within the US to the avoidance of EIA in the trade context, see http://www.citizen.org/trade/article_redirect.cfm?ID=10448 (accessed 15 January 2015).

peatland forests and wetlands into palm oil plantations. ASEAN needs its own version of the Espoo Convention.

Simply enacting air pollution or EIA laws, however, is not sufficient. Civil servants, private businesses and citizenry all need to learn to observe environmental laws. States need to sustain a culture of respect for the law, and have both the capacity to implement their laws *and* the integrity to observe the required norms. Where the 'rule of law' is weak, employing available environmental tools may be ineffective, or too formalistic. The UN General Assembly has called on all States to bolster their rule of law, at national and international levels.[53] States acknowledge that without the rule of law, sustainable development cannot be.[54]

'Environmental rule of law' is specifically endorsed by UNEP.[55] The rule of law is being emphasised in 2015 because it is the 800th anniversary of the Magna Carta (1215).[56] States increasingly recognise the right to a healthy environment and clean air, with 147 nations recognising the right to the environment within their constitutions. There are now more than 500 environmental courts and tribunals existing in some 50 nations.[57] Access to justice is fundamental to realising the UN SDGs. Sustaining respect for the rule of law makes possible the use of environmental remedies to abate the scourge of air pollution.

Environmental rights and remedies are more than theory. They are applied in specific, often troublesome circumstances. A survey of the causes of the haze in Southeast Asia will make clear which biological remedies are most relevant.

---

[53] *The Rule of Law at the National and International Levels*, GA Res 64/116, UNGAOR, 64th Sess, Supp No 49, UN Doc A/RES/64/116 (2010).

[54] *The Rule of Law at the National and International Levels*, GA Res 66/102, UNGAOR, 66th Sess, Supp No 49, UN Doc A/RES/66/102 (2012).

[55] Environmental Justice and Sustainable Development: A Global Symposium on Environmental Rule of Law, Summary and Key Messages (Document produced at the First Session of the UN Environment Assembly of the UNEP in Nairobi, Kenya, 23–27 June 2014), UN Doc UNEP/EA1/CRP.1.

[56] Andrea Martinez, Daniel Barstow Magraw Jr and Roy Edmund Brownell II (eds), *Magna Carta and the Rule of Law* (Chicago: American Bar Association Publishing, 2014). *Magna Carta* is the legal instrument that originated the 'rule of law'.

[57] In 2012, there were 350 environmental courts and tribunals in 41 countries. See Nicholas A Robinson, 'Ensuring Access to Justice through Environmental Courts' (2012) 29 *Pace Environmental Law Review* 363.

D. Drivers of Transboundary Air Pollution in Southeast Asia

The immediate transfrontier problems between Indonesia and the Malay Peninsula are air contaminants in smoke from forest fires. International cooperation to control forest fires has been the subject of bilateral cooperation elsewhere for many years.[58] States have learned to share their capacities to extinguish forest fires and their national or provincial firefighting resources. In Southeast Asia, however, the issue is not simply one about wild forest fires, which might be perceived as a 'common enemy'. The haze comes from agricultural practices: burning forests to clear them in order to plant palm oil or pulp wood plantations. Some benefit economically from the fires.

The 2002 ASEAN Haze Agreement[59] and the earlier 1995 ASEAN Cooperation Plan on Transboundary Pollution (1995 Plan)[60] provide a legal framework to abate air pollution. In addition, in cooperation with the EU, the region has a Forest Law Enforcement and Governance (FLEG) cooperative programme to combat unlawful forest exploitation or destruction. FLEG builds capacity and strengthens compliance with laws through a detailed action plan for 'Strengthening FLEG in ASEAN 2008–2015'.[61]

Many aspects of the 1995 Plan and the 1997 ASEAN Regional Haze Action Plan (1997 Plan) remain to be fully implemented two decades later. The 1995 Plan (timely detection of forest fires, prohibiting burning, sharing information during burning episodes and promoting investments in alternative uses of biomass) is not controversial and the 1997 Plan strives to prevent fires through 'better management policies and enforcement' and improved fire-monitoring and firefighting capacity. Educational measures, such as promoting 'zero burning' techniques for agricultural land conversion, have been met with insufficient success. National law enforcement remains weak in Indonesia and voluntary measures by planters are not working well. Malaysia and Singapore have provided technical assistance to provincial authorities in Sumatra. Singapore has assisted with monitoring fires through the ASEAN Specialized Meteorological Centre. These and other measures have been insufficient to engage Indonesia in effectively implementing the 2002 ASEAN Haze Agreement.

---

58 Robinson, *supra* note 16; Kurukulasuriya and Robinson, *supra* note 28 at 93.

59 2002 ASEAN Haze Agreement, *supra* note 3.

60 See Koh (2013), *supra* note 42 at 321.

61 *Ibid* at 459.

Why are the agreements not yet fully implemented? The answer is found in the complexities of Indonesia's fires, including the underlying driving forces that both deter action and stimulate more burning. What are the drivers that perpetuate the burning in Indonesia that in turn pollutes Malaysia, Singapore, Brunei and even the Philippines? Why is global biodiversity loss and the release of vast GHGs not yet a cause for more effective international cooperation? Answers are complex.

The Indonesian Ministry of Forests has granted many permits for converting forests to agricultural uses (2 million hectares worth in 2011 alone).[62] After Indonesia's President Susilo Bambang Yudhoyono issued a two-year moratorium on forest conversions, many problems prevented its effective implementation.[63] In May 2011, 103 clearings occurred in forests protected by the moratorium. The President's commendable efforts, and the fate of his REDD+ Presidential Working Group, require renewed commitments as Indonesia's next presidential administration takes office. Institutional governmental commitment is needed to curb the forest fires, root out corruption and assert the rule of law.

External forces drive acute deforestation. Indonesia leads the world today in palm oil production. The demand for palm oil is growing. Consumption of palm oil for food processing increases as populations grow in China and India. Consumption of palm oil as a biofuel increased in the EU. These markets together account for 49 per cent of total palm oil imports.[64] The EU's Directive requires use of biological, renewable fuels to reduce GHG emissions.[65] This EU Directive does not address the sustainability of the production systems for the renewable biofuels and fails to distinguish between lawful and sustainable production and illegal unsustainable practices. In addition, although the US Environmental Protection Agency in 2012 declined to list palm oil as a renewable

---

62 References on the many contesting domestic forces that perpetuate Indonesia's forest fires are provided in Sarah Guzicki and Nicholas Robinson, 'The Implementation Limitations of and Alternative Solutions for Indonesia's REDD+ Programme Concerning Peatland Restoration' (2013) *International Journal of Rural Law and Policy* (Special Edition) accessed 15 January 2015 at http://epress.lib.uts.edu.au/journals/index.php/ijrlp/article/view/3351.

63 Nani Afrida, 'President Calls for Better Synergy to Meet Climate Change Targets' *The Jakarta Post* (2 October 2011) accessed 15 January 2015 at www.thejakartapost.com/news/2011/10/01/president-calls-better-synergy-meet-climate-change-targets.html.

64 'Top Ten Palm Oil Importers 2012' *Agriculture Corner* (22 November 2012) accessed 15 January 2015 at http://www.agricorner.com/top-ten-palm-oil-exporters-2012/.

65 See EU Directive, *supra* note 8.

biofuel, the US Congress in 2013 reset a tax credit to advantage alternative fuel imports, making palm oil more attractive to import than soybean oil for biodiesel. Indonesia enjoys increasing palm oil imports to the US.[66]

The voluntary, non-governmental RSPO has produced guidance for sustainable production, but only 14 per cent of Indonesian sites are certified by RSPO as sustainable, and meanwhile Indonesian palm oil production increased by 8.11 per cent and exports grew by 9.50 per cent, with EU palm oil imports increasing by 12.09 per cent.[67] Imports to China are growing, having increased by 16 per cent in 2013. Eighty per cent of India's large consumption of palm oil, largely for cooking, comes from Indonesia, and the same applies to Pakistan's large imports and uses.[68] Singapore and other ASEAN Member States import palm oil for domestic uses also. Trading relations for palm oil disregard the adverse environmental impact of palm oil production. Each trader and consumer has tolerated the haze as have their governments.

To meet market demands, agricultural conversion of forestlands for producing palm oil increases rapidly in Indonesia. It is estimated that 100,000 hectares of peatlands are converted to plantations for pulpwood, palm oil and rice production each year in Indonesia, and 40 per cent of Indonesian forests have been lost in the past 50 years.[69] Land clearing for new rice production is important for Indonesian food security, but the acidity of peat waters has made conversion to rice paddies problematic.[70] Indonesia hopes to increase rice production from 37 million metric tons to 47 million metric tons annually as its national population grows, but

---

[66] 'US Blending Credit for Alternative Fuels Encourages Palm Oil Imports' *World Growth* (22 March 2013) accessed 15 January 2015 at http://worldgrowth.org/2013/03/trade-and-environment-newsletter-issue-18-march-2013/.

[67] 'Indonesian Palm Oil Production by Year' *IndexMundi,* accessed 15 January 2015 at http://www.indexmundi.com/agriculture/?country=id&commodity=palm-oil&graph=production.

[68] Guzicki and Robinson, *supra* note 62 at 11.

[69] *Ibid.*

[70] Conversions have had some problems. In 1996, the 'Mega Rice Project' in Kalimantan Province began to convert one million hectares of peatlands into rice paddies. Excessive drainage was caused, when acid water drained off and the peat dried out, causing spikes in local air emissions of carbon and sulphur pollutants. 'Indonesia: Environmental Profile' *Mongabay,* accessed 15 January 2015 at http://rainforests.mongabay.com/20indonesia.htm.

some 13 per cent of the population currently lacks adequate nutrition.[71] Palm plantations compete with land for rice production.

The loss of biodiversity from conversions of forest to plantations in Indonesia is significant. Between 80 and 90 per cent of native flora and fauna is lost in these agricultural conversions.[72] Since sustaining biodiversity is a 'common concern of mankind', the loss of Indonesian habitat is of concern. Moreover, much of the plantation lands are situated on peatlands. Indonesia holds 50 per cent of the world's tropical peat, accumulated over the entire era of the Holocene (10,000 years or more).[73] The peat wetlands sequester ancient carbon (196 tons of carbon per hectare, or 30 times more than that of ground forests).[74] Indonesia releases some 2.5 billion metric tons of carbon dioxide from peat soils each year.[75] Fires that are used to clear forestlands often burn out of control (as was the case with the fires of 1997–98 following the El Niño event).

Indonesia's peat gas releases are becoming so voluminous that they place the nation on a scale comparable to the emissions of China or the US, making Indonesia one of the world's major emitters of GHGs.[76] Only by shifting new palm oil lands to non-peatlands and re-wetting the drained peatlands can these GHG emissions be curbed and stabilised. President Susilo Bambang Yudhoyono in 2009 announced plans to cut Indonesian GHG emissions by 26 per cent by 2020 and by 41 per cent if international aid was provided.[77] 'Business as usual' has delayed implementation of these goals.

In addition to GHG emissions, water pollution has grown locally. Environmental impacts include the release of the waste from palm oil

---

71 Rudy Ruitenberg, 'Indonesia Wants to Boost Rice Production by 10 Million Tons, Suswono Says' *Bloomberg* (21 January 2012) accessed 15 January 2015 at http://www.bloomberg.com/news/2012-01-21/indonesia-seeks-10-million-ton-rice-crop-increase-suswono-says.html.

72 Guzicki and Robinson, *supra* note 62 at 4.

73 *Ibid* at 1.

74 *Ibid.*

75 *Ibid.*

76 *Ibid.*

77 Indonesia hosted the Conference of the parties of the UN Framework Convention on Climate Change, set up the National Council on Climate Change, and adopted a National Action Plan. Jessica Brown and Leo Peskett, 'Climate Finance in Indonesia: Lesson for the Future of Public Finance for Climate Change Mitigation', Working Paper (Overseas Development Institute, February 2011) accessed 15 January 2015 at http://www.edc2020.eu/fileadmin/publications/EDC_2020_-_Working_Paper_No_11_-_Climate_Finance_in_Indonesia. pdf.

production. For every metric ton of palm oil, some 2.5 metric tons of waste result, most discharged untreated into open waters.[78] Locally, drinking water supplies are polluted and labour conditions for workers are also problematic.

This summary overview of the principal drivers and impacts of the forest fires demonstrates that several fields of law are implicated by the transboundary air pollution.[79] Lacking effective enforcement under Indonesian law, it is necessary to muster international cooperation under environmental agreements to build the capacity and action that can result in abating the air pollution. Biological agreements offer several potentially effective 'remedies'.

III. REMEDIES FOR SUSTAINING BIOLOGICAL DIVERSITY: COOPERATION FOR COMMON CONCERNS

One way to curb the burning of forests is to remove all peatlands from agricultural use, by invoking laws that conserve peat wetlands and forests as habitats for many species of plants and animals. Peatlands are so valuable as GHG sinks and habitats for biodiversity that none should be used for any other purpose. Biological preservation of such lands would prevent new conversions to agricultural uses and avert attendant fires. Given the increasing destruction of forests in Indonesia, proactive efforts to preserve forestlands require vigorous action. If Indonesia could take remaining forest peatlands out of exploitation, it could forestall future forest fires. Plantation developers drain peatlands before setting them afire. Peat destruction is purposeful.

Indonesia could set aside all peatlands, acquiring areas privately held through exercise of eminent domain. National revenues and international financial support could be provided to compensate the government for

78 'Palm Oil & Soil and Water Pollution' *World Wildlife Fund* (2004) accessed 15 January 2015 at http://wwf.panda.org/what_we_do/footprint/agriculture/palm_oil/environmental_impacts/soil_water_pollution/.

79 E.g., regional cooperation laws under ASEAN, international trade law, international laws on finance and direct private investment, international environmental law, biodiversity law, climate change law, public health law, international laws on food security, laws on food law, sustainable development laws and standards, laws for cooperation in criminal law enforcement, among other legal sectors. Cooperation to strengthen the rule of law and combat corruption is a high priority for all sectors.

setting aside these areas. The government, in due course, could compensate private parties with economic interests in peatlands. Such preservation measures would shift the demand for new commercial agricultural land conversion from these wetlands and forests to other places. That process would need to be managed so that collateral and consequential issues would be addressed. But, in the short term, the forest fires could be ended.

Such a peat preservation remedy would require a major international effort, supplementing the generous commitment Norway has made to provide financing to avert future peatland losses. Downwind States like Brunei, Malaysia and Singapore may need to persist in their proposals to the UN and other international organisations to ensure that they undertake cooperative efforts. In principle, all affected States agree to expand protected areas, as shown in the 2007 Declaration on the Heart of Borneo between Brunei, Indonesia, Malaysia and the Philippines[80] and the 2007 Regional Action Plan for ASEAN Heritage Parks and Protected Areas.[81]

Even if the peatlands were set aside by Indonesia's executive action, and even though the 2002 ASEAN Haze Agreement has been ratified by Indonesia and entered into force, there would still be a need to use additional, complementary legal tools to encourage and reinforce actions that avert new fires. The economic drivers for new palm plantations need to be met with countermeasures. International environmental laws providing biological 'remedies' include the following:

(i) the UN Educational, Scientific and Cultural Organization (UNESCO) World Heritage Convention,[82] and regionally through the ASEAN Heritage Parks Program for Protected Area Management (ASEAN Heritage);[83]

---

[80] World Wide Fund for Nature, Declaration on the Heart of Borneo, adopted 12 February 2007; Koh (2013), *supra* note 42 at 105.

[81] ASEAN Centre for Biodiversity, *Regional Action Plan for ASEAN Heritage Parks and Protected Areas* (Parañaque, Philippines: Dolmar Press, 2008); Koh (2013), *supra* note 42 at 43.

[82] Convention Concerning the Protection of the World Cultural and Natural Heritage, adopted 16 November 1972, 1037 UNTS 151 (entered into force 17 December 1975) (UNESCO Convention).

[83] ASEAN, *ASEAN Heritage Parks and Protected Area Management*, accessed 15 January 2015 at http://www.aseanbiodiversity.org/index.php?option=com_content&view=article&id=68:asean-heritage-parks-and-protected-area-management&catid=11:biodiversity-concerns&Itemid=87.

(ii) the Ramsar Convention on Wetlands of International Importance;[84]
(iii) the Convention on International Trade in Endangered Species (CITES);[85]
(iv) the Convention on Biological Diversity (CBD);[86]
(v) the REDD+ initiative (Reducing Emissions from Deforestation and Degradation, plus conservation), emerging under the UN Framework Convention on Climate Change (UNFCCC), the CBD and also voluntary REDD+ initiatives;[87]
(vi) initiatives through the International Union for Conservation of Nature (IUCN) Commission on Protected Areas and the IUCN World Parks Congress;
(vii) action under the United Nations Food and Agriculture Organization (FAO);
(viii) measures to protect public health under the World Health Organization (WHO);
(ix) EIA; and
(x) the UN Sustainable Development Goals (SDGs).[88]

Invoking cooperative biological conservation duties can enable States to fashion 'remedies' under public international environmental law to reduce or prevent air pollution from fires.

A. Designation and Conservation of World Heritage Protected Areas

Indonesia can designate more of Sumatra and Kalimantan as world heritage areas under the terms of the 1972 Convention on the Protection of the World Cultural and Natural Heritage.[89] ASEAN Member States

---

[84] Ramsar Convention, *supra* note 7.

[85] Convention on International Trade in Endangered Species of Wild Fauna and Flora, adopted 3 March 1973, 993 UNTS 243 (entered into force 1 July 1975) (CITES).

[86] Convention on Biological Diversity, *supra* note 7.

[87] *Reducing Emissions from Deforestation and Forest Degradation PLUS,* accessed 15 January 2015 at http://www.un-redd.org/aboutredd/tabid/102614/default.aspx (REDD+).

[88] International Union for Conservation of Nature (IUCN) World Commission on Protected Areas, accessed 15 January 2015 at http://www.iucn.org/about/work/programmes/gpap_home/gpap_wcpa.

[89] The Convention in Article 2 defines natural heritage as 'physical and biological formations or groups of such formations, which are of outstanding universal value, from the aesthetic or scientific point of view … (or) which constitute the habitat of threatened species of animals and plants of outstanding

have participated in designating ASEAN Heritage natural protected areas, which is a hallmark of ASEAN cooperation. Indonesia has actively participated in the UNESCO Convention. It is timely to survey all potential heritage sites in areas where peatlands or endangered species habitats are being converted into plantations, and encourage States to nominate more sites for the World Heritage List.

In the ASEAN region, world heritage resources benefit tourism. World heritage is being lost faster than it can be recorded and conserved. It would behove ASEAN Member States to nominate sites and work with the World Heritage Committee to establish a proactive programme, so that designated heritage areas would not need to be named to the 'List of World Heritage in Danger'. Support via contributions to the UNESCO World Heritage Fund, or via establishing a new, special, voluntary fund could be established to protect natural peat forests as biomes in need of protection. Restoration of adjacent, former peat forests could be sought. Peatlands need to be recognised as being a world heritage biome of ancient origin.

An overlay of all areas of outstanding universal world heritage value could be prepared to identify sites where loss to agricultural conversion is possible. These sites might be given priority attention. Saving such sites, and building a stewardship programme to conserve them, would have the collateral benefit of averting future forest fires.

B. Wetland Management

The protection of wetlands of international importance offers another avenue for international cooperation. UNESCO has a Memorandum of Understanding with the Bureau of the Ramsar Convention for designating wetlands that can be designated under both the UNESCO Convention and the Ramsar Convention.[90] The Ramsar Convention expressly includes areas of peatland and water in Article 1.[91] Climate change, including coastal wetland impacts associated with sea level rise, threatens wetlands.

---

universal value … (or) natural sites of outstanding universal value from the point of view of science, conservation or natural beauty'. Anthropogenic impacts increasingly threaten both natural and cultural resources of outstanding universal value. UNESCO Convention, *supra* note 82.

[90] Memorandum of Understanding between UNESCO and the Bureau of the Ramsar Convention, signed and entered into force on 14 May 1999, accessed 23 January 2015 at http://archive.ramsar.org/cda/en/ramsar-documents-mous-ramsar-mou-with-the/main/ramsar/1-31-115%5E21517_4000_0__.

[91] Ramsar Convention, *supra* note 7, Art 1.

A special programme to give priority to identifying and protecting the coastal mangroves of Indonesia and the inland peatland wetlands would find support in the Ramsar Convention. The Ramsar Convention respects man-made wetlands, so former peat wetlands could be restored and a swap of peatlands for non-peatlands could be explored so that palm plantations could be moved to sustainable sites. Some accommodation is needed to provide for palm oil production on peatland areas. Local employment could emerge to re-wet and conserve peat, and relocate plantations elsewhere.

Collaboration between the Ramsar Secretariat and the Food and Agricultural Organization of the UN (FAO) might be enlisted to help identify protocols for managing the agricultural domain in ways that safeguard peat wetlands and forests. This approach would result in averting future forest fires.

C. Endangered Species Protection

CITES[92] offers avenues for cooperation on habitat protection. ASEAN is a bio-rich region, with many rare species of animals and plants. Peatland forests are habitat for many endangered and threatened species, and habitat destruction is a form of 'taking' species as much as poaching is. The remnant habitat of the orang-utan and other species are already the object of many intergovernmental and non-governmental efforts at conservation. If these habitats and species are to be saved, much more action must be taken.

ASEAN promulgated a Regional Action Plan (2005–2010) and launched an ASEAN Wildlife Law Enforcement Network (2005).[93] The scale of activity is still insufficient to save species. Under CITES, proactive measures can be taken to restrict the trade in endangered species. The next CITES Conference of the Parties (COP) could consider scientific surveys of species endangered by the forest fires and by the conversion of peat forest lands to palm oil agriculture. Where warranted, species can then be listed in the CITES's Appendices, regulating any

---

[92] See CITES, *supra* note 85.

[93] ASEAN, *ASEAN Regional Action Plan on Trade in Wild Fauna and Flora (2005–2010)*, adopted 3 May 2005, accessed 15 January 2015 at http://www.asean-wen.org/index.php/regional-agreements-protocols/file/2-asean-regional-action-plan-on-trade-in-wild-flora-and-fauna-2005–2010. See ASEAN Wildlife Enforcement Network, accessed 15 January 2015 at http://www.asean-wen.org/; Koh (2013), *supra* note 42 at 89, 99.

trade and advancing the need for habitat protection. National regulation to protect the habitats of these species can be advanced also.

Sites of endangered species and their habitats might well coincide with listings under the Ramsar Convention or the UNESCO Convention. These overlays of protection enhance the likelihood of protection, with the collateral benefit of conserving lands subject to forest fires.

D. Sustaining Biological Diversity through the CBD

Article 3 of the CBD restates each State's duty to ensure that activities do not harm other States' territory or the commons.[94] Indonesia has pledged to set aside 200,000 square kilometres of coastal and marine habitats under the CBD.[95] If financing and capacity building emerged to facilitate action, Indonesia could set aside vastly more lands. Brazil has done so at a vast scale already under its domestic legal frameworks, so there is precedent and experience to draw upon in planning to do so.

The CBD is the first holistic, ecosystem-based regime for the sustainable use of biological resources. The CBD's COP has encouraged cooperation among all the MEAs that address components of biological diversity, such as the UNESCO, Ramsar and CITES regimes. Attention should be devoted to how the National Biodiversity Strategies and Action Plans might address the conservation of peatland forests and shift agricultural land conversions elsewhere.

The CBD's COP adopted the Aichi Biodiversity Targets (Aichi Targets).[96] Rather than introducing new topics to deal with deforestation and forest fires in Indonesia, there is ample scope within the Aichi Targets to do much to avert the forest fires. Moreover, there is a 2020 deadline in the Aichi Targets, so there is agreement to act in a short period of time. Aichi Target 5, for example, calls for reducing the rate of habitat loss, including forests, by 50 per cent by 2020.[97] Target 15 calls for restoration

---

[94] Convention on Biological Diversity, *supra* note 7, Art 3.

[95] Jatna Supriatna, 'International Year of Biodiversity and Indonesia' *The Jakarta Post* (23 January 2010) accessed 15 January 2015 at http://www.thejakartapost.com/news/2010/01/23/int039l-year-biodiversity-and-indonesia.html.

[96] Conference of the Parties to the Convention on Biological Diversity, Tenth Meeting, *Decision X/2: The Strategic Plan for Biodiversity 2011–2020* (18–29 October 2010) accessed 15 January 2015 at http://www.cbd.int/decision/cop/?id=12268; Conference of the Parties to the Convention on Biological Diversity, Tenth Meeting, *Revised and Updated Strategic Plan: Technical Rationale and Suggested Milestones and Indicators, Note of the Executive Secretary,* UNEP/CBD/COP/10/9 (10 July 2010) (*Aichi Biodiversity Targets*).

[97] *Aichi Biodiversity Targets*, *ibid* at Target 5.

of at least 15 per cent of degraded habitats also by 2020.[98] UNESCO,[99] Ramsar,[100] CITES,[101] the Parties to the International Treaty on Plant Genetic Resources for FAO[102] and the UN[103] have endorsed these measures. This constitutes an opportunity to organise cooperative interstate action and funding both to strengthen the protection of ecological habitats before they are lost to deforestation for palm oil production and also to restore peatland forests, in particular those areas converted unlawfully under Indonesian law.

The CBD's *in situ* protection provisions for peat habitats can also be aligned with the sites identified under the UNESCO and Ramsar regimes. One important aspect of the CBD is the access to genetic resources and benefit sharing. It is worth exploring how advanced scientific laboratory research internationally might be cooperatively linked to the efforts to establish the biotechnology potential of species found in Indonesia's peat wetland and forest areas. Indonesia has a national interest in biological research that may discover valuable biotechnological applications. Peatland losses endanger Indonesia's potential intellectual property interests, and that concern may stimulate collateral efforts to avert peatland conversions and curb forest fires.

---

98 *Ibid* at Target 15.

99 UNESCO, World Heritage Committee on the Convention Concerning the Protection of the World Cultural and Natural Heritage, *Decision 37 COM 5A, Report of the World Heritage Centre on its activities and the implementation of the World Heritage Committee's Decisions*, 37th Sess, WHC-13/37.COM/20 (5 July 2013).

100 Conference of the Parties to the Ramsar Convention on Wetlands, 11th Mtg, Resolution XI.6, Partnerships and synergies with Multilateral Environmental Agreements and other institutions (13 July 2012) accessed 23 January 2015 at http://archive.ramsar.org/pdf/cop11/res/cop11-res06-e.pdf.

101 Conference of the Parties to the CITES, 16th Mtg, *Cooperation of CITES with other biodiversity-related conventions,* Resolution Conf. 16.4 -1 (14 March 2013).

102 Governing Body of the International Treaty on Plant Genetic Resources for Food and Agriculture, Cooperation with Other Bodies and International Organisations Including with the International Agricultural Research Centres of the Consultative Group on International Agricultural Research and Other International Institutions that Signed Agreements Under Article 15 of the Treaty, Res 8/2011, FAO, 4th Sess (18 March 2011).

103 Convention on Biological Diversity, GA Res 65/161, UNGAOR, 65th Sess, Supp No 49, UN Doc A/RES/65/161 (2011).

E. Reducing Emissions from Deforestation and Degradation, plus Conservation

Forests and wetlands in Indonesia are potentially more valuable 'as is' than as converted to palm plantations, by attracting public and private investment in sustaining forest photosynthesis, carbon sequestration, ecosystem services and biodiversity. REDD+ is promoted both by the COPs of the UNFCCC and the CBD. Private sector enterprises are currently offsetting their emission of GHGs by investing in the preservation of forests with their capacity, through photosynthesis, to sequester an amount of gases comparable to their emissions. Many NGOs are sponsoring programmes to encourage such investments. A few governments, including Indonesia, have also engaged in such investments. Yet agreement on how to extend REDD+ from these early prototypes to a wider regional or international level has been slow.

It took decades for the COP to the UNFCCC to decide that issues of 'Land-Use and Land-Use Change and Forestry' are important for climate change mitigation. Forest and land degradation contribute significantly to climate change. REDD was proposed as an approach to focus on the carbon sequestered in trees. In order to keep trees healthy, an approach to conserve the entire forest ecosystem was deemed essential; hence, the 'plus' was added to the REDD. In parallel deliberations, COP-9 of the CBD recognised that REDD+ was a valuable approach to safeguarding biodiversity in forests.[104] Both the CBD's COP and the IUCN have advanced REDD+ as a strategy for biological conservation.[105]

Developing nations have envisioned REDD+ as a regime that can produce financial investments to pay them for conserving their forests. Industrial and other emitters have been slow to invest in 'offsets' for their polluting externalities. There is no legal obligation yet to do so. As yet, there are also no agreed means to measure volumes of emissions or how to match the carbon sequestration capacity in different forest habitats. Little agreement exists on the measurements for determining how long a forest can sequester a comparable amount of carbon dioxide or on how to monitor and verify that the forest protection is attained. Theoretical proposals have been (and are still being) richly debated about how to

---

[104] Conference of the Parties to the CBD, *Decisions Adopted by the Conference of the Parties to the CBD on Biological Diversity at its Ninth Meeting*, UNEP/CBD/COP/DEC/IX/16 (9 October 2008), para B.11.

[105] IUCN, Approach to REDD+, accessed 15 January 2015 at http://www.iucn.org/about/work/programmes/forest/fp_our_work/fp_our_work_thematic/redd/iucns_approach_to_redd_plus_/.

generate investment money for REDD+ investments from the sale of carbon credits in markets, by auctions or by a public fund from taxpayers. Debates persist over creating either a compliance market or a voluntary credit market. Systems for verified emission reductions remain to be prescribed.

While debates continue, intergovernmental agreements on instituting REDD+ proceed too slowly to provide much relief from forest fires. Each REDD+ deal is individually negotiated, designed on a voluntary basis. Such bespoke deals are hard to replicate or systematise. In order to advance measures to launch REDD+, the Oslo Climate and Forest Conference in 2010 advanced a REDD+ Partnership and the Letter of Intent with funding (US$30 million) to finance the preparation of the REDD+ system in Indonesia.[106] The fine-tuning of the design of a regulated market system has deferred States from enacting national laws or entering into international agreements on a REDD+ regime. REDD+ is a work in progress, still at the incubator stage.

Nonetheless, REDD+ offers promise. So much attention is devoted to REDD+ that other biological ‘remedies’ are neglected. ASEAN Member States have acknowledged the importance of peat wetlands and forests both on climate change and transboundary pollution issues, and have an agreed strategy and action plan for peatlands.[107] ASEAN also has a ‘Common Position Paper on Reducing Emissions from Deforestation and Forest Degradation in Developing Countries’, which emphasises the need for capacity building and flexible approaches to defining REDD+ in light of different socio-economic development levels of States.[108]

Indonesia has been exemplary in the development of its capacity for facilitating REDD+ investments. Kuntoro Makusubroto, the head of the Indonesian Presidential Unit for Development, Monitoring and Oversight (UKP4), chairs the National REDD+ Task Force.[109] The Task Force’s

---

[106] Joint press release from Norway, Guyana and the World Bank, *Guyana REDD+ Investment Fund (GRIF) established* (9 October 2010) accessed 15 January 2015 at http://www.regjeringen.no/en/archive/Stoltenbergs-2nd-Government/Ministry-of-the-Environment/Nyheter-og-pressemeldinger/nyheter/2010/guyana-redd-investment-fund-grif-establi.html?regj_oss=2&id=620610.

[107] Koh (2013), *supra* note 42 at 361–87.

[108] ASEAN, *ASEAN Common Position Paper On Reducing Emission from Deforestation and Forest Degradation (REDD) in Developing Countries* (REDD, 2008) accessed 15 January 2015 at http://unfccc.int/files/kyoto_protocol/application/pdf/indonesiaaseanredd060608.pdf; Koh (2013), *supra* note 42 at 869.

[109] Presidential Decree, No 19 of 2010 (20 September 2010) formed the REDD+ Task Force, and was extended by a Second Decree on 8 September

landmark report of 2013 maps out the extensive work that Indonesia has undertaken to design a system to protect forests and prevent forest fires.[110] Through 2010, Indonesia had launched 40 REDD+ projects funded by private donors, NGOs, bilateral governmental grants (for example, Australia, Japan, Germany, Norway and South Korea) and the World Bank. These voluntary efforts include some 19 projects in Sumatra, involving the World Wide Fund for Nature (WWF) and others.[111]

Much effort has gone into preparing 'one good map' for Indonesia of its forests and the uses allowed in their forests (parks, plantation concessions, and so on). Many local authorities and national agencies are involved in assembling the data for these maps; this is a complex and unavoidably slow process if all the details on the maps are to be verified (verification of maps has so far been done selectively). The UN Development Programme (UNDP) has been assisting Indonesia in assembling the capacity for gathering remote sensing data for mapping also.[112] While preparing the map, little has been done to implement REDD+ because the map is deemed essential for a national approach to implanting REDD+ on a national scale. Public access for maps, as the process continues, has been available through the government's website.[113]

Indonesia has taken measures to preserve primary forests and peatlands, even while the mapping and REDD+ preparation are underway. On 20 May 2011, the Indonesian President mandated a two-year suspension

---

2011, which also worked on the implementation process for the Indonesia-Norway Partnership. Republic of Indonesia, Presidential Decree, No 19 of 2010 (20 September 2010) accessed 15 January 2015 at http://www.kemendagri.go.id/produk-hukum/2011/01/07/satuan-tugas-persiapan-pembentukan-kelembagaan-redd; Republic of Indonesia, Presidential Decree, No 25 of 2011 (8 September 2011) accessed 15 January 2015 at http://www.kemendagri.go.id/produk-hukum/2011/12/05/satuan-tugas-persiapan-kelembagaan.

[110] Republic of Indonesia, National REDD+ Task Force, *REDD+ In Indonesia – A Catalyst for Change* (2013) accessed 15 January 2015 at http://www.unorcid.org/upload/doc_lib/20121206145331_REDD%20BOOK%202012.pdf.

[111] See e.g., SEKALA's well-maintained website, Forest Climate Centre, accessed 15 January 2015 at http://www.forestclimatecenter.org.

[112] See UN-REDD Programme Indonesia, *REDD+ in Indonesia: Information, Monitoring & Measurement, Reporting and Verification (MRV),* accessed 15 January 2015 at http://www.unredd.net/index.php?option=com_docman&task=doc_download&gid=8862&Itemid=53.

[113] See Republic of Indonesia, *Geospatial Information*, accessed 15 January 2015 at http://tanahair.indonesia.go.id/home/.

of concessions for new plantations.[114] During the moratorium, measures to enhance law enforcement were undertaken. However, the resources were not available to enforce the moratorium everywhere and forest conversions to plantations with burning continued. Indonesia is committed to attaining a major reduction in GHG emissions by 2020, and REDD+ is to be part of that.[115] The UNDP and Norway are ready to finance major investments in REDD+.[116] Indonesia has set up FREDDI, an umbrella trust fund, a 'fund of funds', to aggregate funds available for REDD+ investments.[117]

It is not likely, however, that the REDD+ regime will be ready in Indonesia in the near future in a way that can contribute to ending the forest fires in Sumatra or Kalimantan. A REDD+ agency, with authority at the presidential level, is yet to be established. Monitoring and enforcement measures need to be enhanced, and further measures to enhance the rule of law strengthened. The new government, elected in 2014, needs to clarify how it will support Indonesia's REDD+ programme.

Brunei, Malaysia and Singapore therefore cannot wait for REDD+ to become functional and effective. States, their businesses and NGOs could work with Norway and others, including the WWF or other NGOs, to set up voluntary REDD+ investments in Kalimantan and Sumatra to push back against the commercial investors who seek to destroy forests and peatlands for palm oil plantations, and who cause forest fires. Regionally, States such as Singapore can encourage its citizens and companies to financially support some REDD+ projects in Sumatra.

The loss of forestlands to palm oil plantations represents a serious form of deforestation and degradation. Efforts could be made to establish a fund to make payments to preserve the ecosystem services of undeveloped peat forest wetlands which can be valued in excess of their value as agricultural palm oil plantations. Discussions through the various

---

114 Republic of Indonesia, Presidential Instruction 10 of 2011, *Suspension of Granting of New Licenses and Improvement of Governance of Natural Primary Forest and Peat Land* (20 May 2011).

115 Guzicki and Robinson, *supra* note 62 at 5.

116 *Ibid.*

117 See UNORCID, *Press Release: REDD+ in Indonesia: Shifting Gears – From Preparedness to Implementation Dinner Gathering at Warsaw Climate Conference* (25 November 2013) accessed 15 January 2015 at http://www.unorcid.org/index.php/media-centre/pressreleases/247-press-release-redd-in-indonesia-shifting-gears-from-preparedness-to-implementation-dinner-gathering-at-warsaw-climate-conference.

REDD+ channels, both non-governmental and intergovernmental, could explore with the government of Indonesia the idea of acquiring title to these areas using its sovereign power of eminent domain. International funds, provided by Norway and other States, and perhaps also through the Global Environmental Facility, could pay for the acquisition of peatlands and setting them aside for REDD+ stewardship. This would have the benefit of ending the land conversions and use of fires.

Since palm oil production is very profitable, it may be that Indonesia (as an exporting State) and Malaysia and Singapore and others (as importing States) should be encouraged to tax the exporting palm oil plantations and market transactions to produce revenues that could be deployed to subsidise the preservation of undeveloped peat wetlands and forests. These taxes would serve to 'internalise' the externalities of the palm oil plantations and slow future conversions. Tax revenues could pay for Indonesia's domestic REDD+ programme. Currently, palm oil plantations taxes produce US$495,000 per year in revenues, with plantations underpaying their taxes significantly due to corruption.[118] With the Indonesian export of palm oil amounting to US$14.5 billion per year in 2008,[119] there is ample room to assess a more realistic tax on palm oil plantations and dedicate the revenues to FREDDI. There is also an urgency to purchase or set aside all peatland forest and prevent further conversions, because once the lucrative palm oil production begins, opposition to restoring the peatlands would be significant.

There is precedent for a national practice to re-wet and preserve wetlands as soon as possible. The UK has decided to re-wet its peatlands and conserve them in order to prevent the release of GHGs in the peat.[120] As host to the COP of the UNFCCC in Bali, Indonesia promised to mitigate GHG releases and indicated that it needs financial assistance to advance its REDD+ objectives.[121] International funding toward this objective would further implementation of its duties under the UNFCCC and have biodiversity benefits.

---

[118] Guzicki and Robinson, *supra* note 62 at 14.

[119] World Growth Palm Oil Green Development Campaign, *The Economic Benefits of Palm Oil to Indonesia* (World Growth, February 2011) at 11, accessed 15 January 2015 at http://worldgrowth.org/site/wp-content/uploads/2012/06/WG_Indonesian_Palm_Oil_Benefits_Report-2_11.pdf.

[120] See IUCN UK Peatland Programme, accessed 15 January 2015 at http://www.iucn-uk-peatlandprogramme.org/.

[121] Guzicki and Robinson, *supra* note 62 at 5.

One immediate way to preclude further forest fire pollution is to finance REDD+ projects to preserve peatlands and cooperate to accelerate its implementation in Indonesia. Until REDD+ is implemented robustly in a bio-rich nation, such as Indonesia or Brazil, its use elsewhere will be slow and halting. REDD+ needs to be applied on a scale that shows its effectiveness. Doing so in Sumatra, for example, would be a biological 'remedy' to avert recurrences of the haze.

F. Stewardship through the IUCN's World Commission on Protected Areas

The IUCN and its World Commission on Protected Areas provide scientific and professional stewardship for parks and other protected areas. Every ten years, the Commission organises the IUCN World Parks Congress as a forum to study best practices and explore such ecosystem approaches to international cooperation for protected areas, while the Commission meets every four years at the IUCN World Conservation Congress.[122] These meetings afford opportunities to frame proposals for biodiversity conservation through protected areas. Protection for peat wetlands and forests deserve the Commission's attention. The IUCN could explore the viability of using various environmental remedies and provide peer review by scientists, protected area managers and environmental lawyers. Review by the IUCN furthers international understanding about how protected area conservation can serve multiple objectives.

Recent Scottish experience in preserving peat offers lessons for Indonesia.[123] The IUCN Committee for the UK, and leading IUCN NGO members such as the Royal Society for the Protection of Birds, have led the UK's peatlands restoration and preservation programme.[124] The UK Heritage Lottery Fund allocated funds to purchase peatlands, as did the Scottish government.[125] The plan is well advanced to restore one million hectares of peatlands by 2020.[126] British peatlands are being restored and preserved both for conserving biodiversity and for mitigating GHG

---

122 See http://www.iucn.org/about/work/programmes/gpap_home/gpap_events/gpap_wpc (accessed 15 January 2015).

123 *Ibid*.

124 See IUCN–UK, *UK Peatland Restoration Demonstrating Success* at 22, accessed 15 January 2015 at http://www.iucn-uk-peatlandprogramme.org/sites/all/files/IUCN%20Demonstrating%20Success%20Booklet.pdf.

125 *Ibid* at 54.

126 *Ibid* at 7.

emissions; the latter is significant, for a 5 per cent reduction in the carbon stored in Scotland's peat would equal the UK's total GHG emissions.

Based on the Scottish approach, a programme could be launched in Indonesia to purchase peatlands, either through exercise of eminent domain or voluntary acquisition, for the purpose of restoring and preserving them. Since 91 per cent of all palm oil plantations are in Sumatra or Kalimantan, this programme could focus on these areas. To win public support for such measures nationally across Indonesia, it would be prudent to develop forest preservation programmes and REDD+ payments for other Indonesian regions as well.

G. Cooperation toward Food and Water Security under FAO

FAO could be engaged in proactive measures to apply FAO's Soils Charter, a soft law instrument, to end burning as a technique of forest clearing and build peatland stewardship. FAO could facilitate planning to move plantations from peat soils and re-wet the drained peat soils to restore them. States could provide supplemental funding for FAO to undertake this work, which also could produce land use practices that avert the use of forest fires. Indonesia has been an FAO member since 1949, while Singapore joined FAO in 2013.[127] Indonesia would need to request FAO's involvement.

FAO also assists States in addressing food shortages as their population grows. This is an important concern for Indonesia. ASEAN has an Agreement on Food Security Reserve and cooperative plans with FAO on food security in the region.[128] FAO would depend upon an infusion of international aid to assist Indonesian efforts to increase its food production. Conversion of forests for palm oil plantations does nothing to help Indonesia to build its capacity to feed its growing population. 'Free market' forces and direct overseas investment in palm oil plantations undermine the UN's SDGs for sustaining food production and ensuring potable water.

---

[127] FAO, Legal Office, *FAO Members*, accessed 15 January 2015 at http://www.fao.org/legal/home/fao-members/en/.

[128] Agreement on the ASEAN Food Security Reserve, adopted 4 October 1979, accessed 15 January 2015 at http://cil.nus.edu.sg/1979/1979-agreement-on-the-asean-food-security-reserve-signed-on-4-october-1979-in-new-york-united-states-of-america-by-the-foreign-ministers/; Koh (2013), *supra* note 42 at 477–540. The UN World Food Programme meets acute shortages, but is not equipped to help with the trends of national food production.

FAO could facilitate studies of policies assigning priority for forest conversions for rice and other agricultural production, and banning new palm oil until domestic food production is at levels adequate for the needs of the Indonesian people. These issues could also be raised in the context of the UN Food Programme or through other multinational or bilateral cooperation. International cooperation can explore how to address the market dysfunctionality of privileging commercial production of a non-food over basic food needs. These efforts could have the collateral benefit of curtailing or ending the unlawful use of forest fires to clear lands for palm oil plantations.

H. The 'Airpocalypse': Addressing Unhealthy Air Quality Levels under WHO

The air quality in Kuala Lumpur, Singapore, London, Paris, New Delhi, Sao Paulo, Beijing and other Chinese cities, as well as cities in Riau Province and other Indonesian locations, all coincidentally reached unhealthy air quality levels at different times in 2014. Pollution of cities from agricultural burning is apt to grow; the Intergovernmental Panel on Climate Change (IPCC) projects that forest fires and their pollution of air will increase as a consequence of climate change.[129] Air quality control has become a matter of common international concern, as well as being an international issue with transboundary impacts.

Contaminated urban air is so extreme as to be an '*airpocalypse*'. Public health officials measure impacts on human lungs and the WHO and national health officials report the dangers of air pollution.[130] Urgent measures to eliminate emissions of fine particulate matter are essential to restoring urban air quality, and end threats to public health hazards. Forest fires emit particulates, which are hazards to public health. WHO could facilitate sharing national expertise on air quality and public health.[131] The health of the lungs of those who breathe polluted air

---

129 Martin Lewis Parry, Osvaldo F Canziani, Jean P Palutikof, Paul J van der Linden and Clair E Hanson (eds), *Contribution of Working Group II to the Fourth Assessment Report of the Intergovernmental Panel on Climate Change* (Cambridge: Cambridge University Press, 2007) at 402.

130 See WHO, *Ambient (outdoor) Air Quality and Health*, accessed 15 January 2015 at http://www.who.int/mediacentre/factsheets/fs313/en/.

131 Cities such as Los Angeles have succeeded in cleaning their once routinely unhealthy air without diminishing sustainable economic growth and renewal. The South Coast Air Basin in California, deals with all sources of air pollution, including smoke from forest fires. Mexico City has worked diligently

worldwide make this topic of priority interest to many nations, China and India among them.

Besides WHO, other agencies have mandates relevant to air quality. Since 1974, the UN Habitat Programme facilitates exchanges of expertise on sustaining cities and their civilisations.[132] Air pollution undermines urban life. UNEP may also make a contribution. At its inaugural meeting in June 2014, the UN Environment Assembly called upon the UNEP to begin efforts to assist nations in abating air pollution.[133] Within their existing mandates, the UN Habitat, UNEP and the WHO, could enlist national expertise needed to abate all sources of air pollution.

WHO has adopted recommendations for air quality for particulate matter,[134] but WHO does not yet recommend public health methods for identifying and establishing practical local or regional air pollution abatement implementation plans. WHO needs to enlist the expertise available in States experienced in sustaining air quality. For example, extreme air pollution across the US was abated through the Clean Air Act (1970), using 'State Implementation Plans' to attain healthy air in defined 'air sheds' known as Air Quality Control Regions, which transcend city limits and cross state lines.[135]

WHO's Constitution authorises the World Health Assembly to adopt regulations to prevent the international spread of disease.[136] Asthma and other lung diseases are growing in urban areas internationally due to air pollution. WHO regulations adopted under its Constitution are binding on all member States, unless they opt out within a short period of time after

---

to abate its urban air pollution in the Federal District. Regions in North America and in Europe have learned to address transfrontier air pollution when it affects air quality.

132 See UN Habitat, 'Chapter 1: Improving Information and Expertise for Air Quality Management' in *The Handbook*, accessed 15 January 2015 at http://ww2.unhabitat.org/wuf/2006/aqm/hb_chap1.htm.

133 United Nations Environment Assembly of the United Nations Environment Programme – First Session (23–27 June 2014), *Proceedings of the United Nations Environment Assembly of the United Nations Environment Programme at its First Session* UNEP/EA.1/10 (2 September 2014) at 40.

134 Guidelines are for two size classes of particles 10 microns (PM10) and 2.5 microns in size (PM2.5). See WHO, *WHO Air quality guidelines for particulate matter, ozone, nitrogen dioxide and sulfur dioxide* (2005) at 9, accessed 15 January 2015 at http://whqlibdoc.who.int/hq/2006/WHO_SDE_PHE_OEH_06.02_eng.pdf?ua=1.

135 *Clean Air Act, supra* note 40.

136 WHO Constitution, Art 21, accessed 15 January 2015 at http://apps.who.int/gb/bd/PDF/bd47/EN/constitution-en.pdf?ua=1.

the regulations are adopted. International Health Regulations (IHRs) apply to only a few infectious diseases and do not address urban air pollutants as diseases. However, the IHRs also take an 'all hazards' approach and can examine urban air pollution as a cause of disease. The definitions in Article 1 of the IHRs support developing an application to urban air pollution.[137]

WHO could be invited to study how air quality is maintained to protect public health in selected cities and frame appropriate recommendations to protect human health from air pollutants. WHO has already identified in general the necessary national preparations appropriate for implementing the IHRs (for example, legislation, policy planning and risk preparedness, and public health infrastructure, including cross-sector coordination). WHO's authority to act already exists and only awaits national public health ministries to activate it.[138] WHO, perhaps in concert with UN Habitat and UNEP, can accelerate action to protect lungs and to prevent forest fires.

I. EIA: Addressing the Environmental Impact of Development

EIA is routinely used in North America and in Western Europe to anticipate and avoid transboundary air and water pollution or other environmental impacts. EIA can engage upwind and downwind States in accordance with international law.[139] Principle 17 of the 1992 Rio Declaration requires EIA in each State's national environmental decision-making.[140] In Southeast Asia, the Asian Development Bank (ADB) and the World Bank both apply robust EIA procedures.[141]

---

[137] WHO, *International Health Regulations* (WHO, 2005), Art 1, accessed 15 January 2015 at http://www.who.int/ihr/publications/9789241596664/en/. *Disease* is 'an illness or medical condition, irrespective of origin or source, that presents or could present significant harm to humans'; an *event* is 'a manifestation of disease or an occurrence that creates the potential for disease'; and *public health risk* is 'a likelihood of an event that may affect adversely the health of human populations, with emphasis on one which may spread internationally or may present a serious and direct danger'.

[138] See generally Lawrence O Gostin, *Global Health Law* (Cambridge, MA: Harvard University Press, 2014).

[139] *Pulp Mills*, *supra* note 47.

[140] 1992 Rio Declaration, *supra* note 2, Principle 17.

[141] See ADB, Environmental Impact Assessment for Developing Countries in Asia, accessed 15 January 2015 at http://www.adb.org/publications/environmental-impact-assessment-developing-countries-asia.

Notwithstanding the legal mandates, States in ASEAN often lack capacity to undertake EIA. For example, despite Article 206 in UNCLOS mandating the use of EIA, virtually no ASEAN State protects shared marine resources through the use of EIA. ADB could assist the region in designing and implementing capacity-building programmes to use EIA to assess the causes of transboundary pollution and abate them.

Brunei, Indonesia, Malaysia and Singapore can further sustainable development and abate the haze by more effectively observing their duty to undertake EIA. Effective EIA in Indonesia could identify alternatives to forest conversions and the use of fire in every forest conversion permit. EIA requires examination of alternatives to proposed actions, as well as ways to mitigate adverse impacts. ASEAN could ensure that transboundary EIA is readily available and can be enhanced by encouraging bilateral agreements among neighbouring States. Multilaterally enhancement opportunities also exist. Regionally, the UN Economic and Social Commission for Asia and the Pacific (ESCAP) could invite the UN Economic Commission for Europe (ECE), the entity which led in the creation of the Espoo Convention, to collaborate in establishing a comparable transboundary EIA agreement for Asia.[142] EIA is also part of the earlier 1985 ASEAN Agreement on Conservation of Nature and Natural Resources, which awaits ratification by several ASEAN Member States.[143]

Air pollution in the ASEAN region is a consequence, in part, of the failure to implement EIA obligations. Weak EIA practices are inconsistent with ASEAN policy. Environmental cooperation is one of the ten priority areas for the years 2009–15, agreed for the ASEAN Socio-Cultural Community under the ASEAN Charter.[144] The ASEAN 2020

---

[142] Espoo negotiations began during the Cold War, with an 'academic' seminar in Warsaw under the auspices of the UN ECE, at which I, for the US, and my Soviet colleagues each presented papers favouring the development of an EIA tool. The Union of Soviet Socialist Republics (USSR) already had '*ecological expertize*' as a technique, and the US had the National Environmental Policy Act. If the US and USSR and European States can agree on an Espoo transborder legal system for EIA, certainly ASEAN with its robust cooperation systems and ESCAP can do the same.

[143] Agreement on the Conservation of Nature and Natural Resources, adopted 9 July 1985 (not yet entered into force) accessed 15 January 2015 at http://cil.nus.edu.sg/1985/1985-agreement-on-the-conservation-of-nature-and-natural-resources-signed-on-9-july-1985-in-kuala-lumpur-malaysia-by-the-foreign-ministers/.

[144] 2009 ASEAN Socio-Cultural Community (ASCC) Blueprint 2009–2015, accessed 15 January 2015 at http://www.asean.org/archive/5187-19.pdf.

Vision calls for a green ASEAN.[145] ASEAN policy is aligned to protect biodiversity and end the fires. ASEAN's multilateral networks could help.[146] Yet, despite ASEANs' elaborate process for consultation and coordination on many environmental and other issues, implementation lags. The 'ASEAN Way' has been to let each ASEAN Member State develop at its own pace and in its own ways. Where rule of law is weak, the implementation of otherwise sound environmental legislation is commensurately weak. The 'ASEAN Way' in deference to each Member State's domestic political cultures has had an unintended effect of excusing State conduct that harms other States. Indonesia has apologised for the haze, but has been unable to prevent it. Indonesia's inaction is matched by inaction in other ASEAN Member States.

For the ASEAN Charter to be implemented, ASEAN needs to do more to match policy with action. EIA is a prime example. Economic forces, in the absence of strong rule of law, have stalled these cooperative measures. ASEAN cooperation is a necessary step, and that step has been taken in principle with the 2002 ASEAN Haze Agreement, but it is not a sufficient one. Perhaps the SDGs can motivate ASEAN to use the biological 'remedies' available to abate the haze.

J. SDGs: Addressing Environmental Protection through Sustainable Development

The adoption of SDGs by the UN General Assembly provides opportunities to emphasise the importance of the environmental MEAs and international environmental law.[147] The SDGs give substance to the

---

[145] *1997 ASEAN Vision 2020*, accessed 15 January 2015 at http://www.asean.org/news/item/asean-vision-2020.

[146] The ministerial working groups of ASEAN could discuss with ASEAN Dialogue Partners, such as the EU, to select the preservation of Indonesian peatlands as a first order global priority. The preservation of the peatlands would, *de facto*, address transboundary air pollution issues.

[147] SDGs, *supra* note 5. Some of the more relevant SDGs are described herein. SDG 3 is to 'ensure healthy lives' and SDG 3.9 seeks by 2030 to substantially reduce deaths and illnesses from air pollution. SDG 11 is to 'make cities and human settlements inclusive, safe, resilient and sustainable,' with measures to reform transport systems (SDG 11.2) and protect nature heritage (SDG 11.4) and pay special attention to air quality (SDG 11.6). SDG 15 is to 'protect, restore and promote sustainable use of terrestrial ecosystems, sustainably manage forests, combat desertification, and halt and reverse land degradation and halt biodiversity loss', reinforcing virtually all the biological remedies available through international environmental law. SDG 15.1 in particular calls

general policies endorsed at the UN Rio+20 Conference in its outcome document, 'The Future We Want'.[148] The UN's sustainable development agenda from 2015 forward offers support for the biological 'remedies' explored above. As the SDGs' UN Open Working Group noted in recommending the draft SDGs to the UN General Assembly, 'Rio+20 affirmed the conviction that in order to achieve a just balance among the economic, social and environmental needs of present and future generations, it is necessary to promote harmony with nature'.[149]

The SDGs can stimulate all UN Member States, and global or regional intergovernmental organisations, to cooperate in practical ways, such as pursuing the biological 'remedies' outlined here to abate air pollution, especially with regard to transboundary air pollution. The ecosystem management approach is one way to do so, and offers means for States to rigorously observe Principle 21 of the 1992 Rio Declaration. Furthering the SDGs supports abating transboundary pollution.

The UNDP understands that the environment and development are one. UNDP is becoming a major player in building the capacity to observe the rule of law. It is important to build on UNDP's leadership and apply these positions to the issues of transboundary air pollution. The SDGs need to do more than simply embrace the Principles of the 1992 Rio Declaration. The new UN SDGs reaffirm that environmental protection is the foundation for sustainability. States can apply the SDGs rigorously to abate pollution that harms the public health, or impairs the recharge of aquifers and other sources of fresh, potable water, and groundwater. Loss of peatland causes such harm. UN Member States can direct UNDP to build capacity for States to protect public health and root out the causes of air pollution. Such an emphasis would make it more likely that the sources of transboundary air pollution from forest fires would be abated.

---

for conservation, restoration and sustainable use of wetlands; SDG 15.2 and 15.5 mandate restoration of degraded forests and natural habitats and SDG 15.9 provides for integrating resource protection into all land use regimes. SDG 16 mandates strengthening access to justice and the rule of law (SDG 16.3).

[148] *The Future We Want*, GA Res 66/288, UNGAOR, 66th Sess, Supp No 49, UN Doc A/RES/66/288 (2012).

[149] United Nations Sustainable Development Goals, Open Working Group on Sustainable Development Goals, *Introduction to the Proposal of The Open Working Group for Sustainable Development Goals* (19 July 2014) at Preamble, para 9, accessed 15 January 2015 at http://sustainabledevelopment.un.org/content/documents/4518SDGs_FINAL_Proposal%20of%20OWG_19%20July%20at%201320hrsver3.pdf.

IV. CONCLUDING THOUGHTS

National and transboundary air pollution, especially in the context of the Indonesian forest fires, is a symptom of deeper environmental problems. To treat the symptom, one must know the causes and resolve the problem at the cause. The diverse drivers of the haze are not amenable to a single solution, since admittedly it is unlikely all peatlands will be set aside as protected areas at once. Several complementary remedies need to be pursued in tandem. Multiple biological 'remedies' to forest fires exist and could be pursued in mutually reinforcing ways.

Continuing business as usual is increasingly unsustainable as health emergencies stir up public unrest about air pollution. Since at least the 1990s, Indonesia as an upwind State has relied upon 'dilution as the solution to pollution'. It was expedient to neglect duties to cooperate to abate pollution. Within Indonesia, losses from the fires, local air pollution health hazards, and biodiversity impacts all militate against treating the plantation conversions any longer as tolerable practices. The conversion to palm oil plantations sacrifices Indonesia's long-term biological wealth for short-term profits. The loss of peat imperils the stability of the climate worldwide. These realities are manifestations of the unsustainability of allowing palm plantations conversions to escape environmental oversight and regulation.

States have the capacity and authority – and the duty – to cooperate more holistically, to make common cause to sustain Earth's ecosystems and protect the public's health. Observing this duty builds an epistemic community among their civil servants and peoples, who know how to cooperate together and willingly do so to further mutual environmental commitments. This accounts for the success of the Vienna Convention and its Montreal Protocol to protect the Stratospheric Ozone Layer.[150] It is the reason why IUCN's World Commission on Protected Areas is effective.

Robust implementation of available biological remedies can energise the existing community of scientists, environmentalists, civil servants and ecosystem managers that already is endeavouring to save Indonesia's biodiversity for present and future generations. The duty to cooperate under international law exists in all sectors. Biological 'remedies' exist to be advanced through multiple avenues of collaboration.

Cooperation itself enhances the confidence to augment collaborative efforts. Forest fires are not local, parochial events. They harm the health

---

150 Speth and Haas, *supra* note 43 at 91.

and economies of *all* nations. States will either incur costs to contribute financially to control fires, or they will incur the costs in the form of injuries. Implementing agreed biological remedies offers cost-effective ways to end the scourge of the haze. Safeguarding Sumatra's biological heritage offers benefits for Indonesians and for the world and may signal an end to the haze. Building sustainable practices is a slow but sure way to end violations of Principle 21 of the 1992 Rio Declaration. International environmental cooperation works where recourse to State responsibility wanes.

Index